Crusader!
Last of the Gunfighters

CRUSADER!
LAST OF THE GUNFIGHTERS

Paul T. Gillcrist, Rear Admiral (USN, Ret.)

Foreword by Senator John H. Glenn, (Lt.Col. USMC, Ret.)

Schiffer Military/Aviation History
Atglen, PA

Acknowledgements

That I survived all those years as a carrier pilot is not so much due to the fortunate accumulation of aviation experience, as to the almost grim determination of so many people to ensure that I didn't get myself killed along the way. Their numbers are legion. Naming any of them would be to risk offending so many others who do not deserve to be thus slighted.

But, they range from diligent flight instructors, to skillful landing signal officers, to sharp eyed wingmen, to section leaders who made my big sky a classroom, to understanding squadron Skippers, CAGs and carrier Commanding Officers, to alert plane captains, to thorough catapult officers and the list goes on and on. To all of them . . . a tip of the hat and my deepest gratitude. You are the greatest fraternity in the world. I will never forget any of you . . . never!

And, finally, to all the Crusader MiG killers! If this book becomes a success, it will be due to your individual inputs. But, far greater than that, your contribution to your country, and to the history of aerial combat will ultimately withstand the test of time!

Dedication
To my son, Tom . . . you make me proud!

Page 1: The author in a VF-53 F-8E over Mt. Fujiyama, Japan. U.S.S. Hancock, 1967.

Title page: Two VF-154 F-8s on catapult for a dawn launch, U.S.S. Hancock (CVA-35), Eastern Pacific, 1957.
(Photo by D. Winicker, courtesy of J. Miottel)

Book Design by Robert Biondi

Copyright © 1995 by Paul T. Gillcrist.
Library of Congress Catalog Number: 94-69728

All rights reserved. No part of this work may be reproduced or used in any forms or by any means – graphic, electronic or mechanical, including photocopying or information storage and retrieval systems – without written permission from the copyright holder.

Printed in China.
ISBN: 0-88740-766-8

We are interested in hearing from authors with book ideas on related topics.

Published by Schiffer Publishing Ltd.
77 Lower Valley Road
Atglen, PA 19310
Please write for a free catalog.
This book may be purchased from the publisher.
Please include $2.95 postage.
Try your bookstore first.

CONTENTS

    Dedication   4
    Acknowledgements   4
    Foreword   9
    Introduction   10

PART I: CARRIER AVIATION   12
Chapter 1: The Jet Age   16
Chapter 2: The Fighter Requirement   18
Chapter 3: The "Twenty-Seven Charlie"   19
Chapter 4: Air-to-Air Weaponry   27
Chapter 5: FAGU and Topgun   30

PART II: THE EARLY YEARS   36
Chapter 6: ". . . A Need For Speed"   39
Chapter 7: Patuxent River   52
Chapter 8: The Great Shoot-Out   56
Chapter 9: Near Miss   60
Chapter 10: A Record of Sorts   64

PART III: THE WORLD FAMOUS SEAGOING BOOMERANGS   70
Chapter 11: "Gator"   73
Chapter 12: "Blue Moon"   77
Chapter 13: "Beaver"   83
Chapter 14: "Gitmo"   89
Chapter 15: Shark Bait   96
Chapter 16: Scramble   99
Chapter 17: Lost Opportunity   102
Chapter 18: Gunsight   104
Chapter 19: Night Noises   106
Chapter 20: Dead Stick   108
Chapter 21: Crash Landings   113
Chapter 22: Morte Tu Capo   117
Chapter 23: Aerial Refueling   120
Chapter 24: Harmonica   124
Chapter 25: Barricade   126
Chapter 26: Keelhauled   134
Chapter 27: Cold Cat Shot   137
Chapter 28: Bingo   141

PART IV: THE IRON ANGELS   144
Chapter 29: "Buying the Farm"   147
Chapter 30: Night Vision   151
Chapter 31: Night, Over-water Ejection   153
Chapter 32: Folded Wings Flight   158
Chapter 33: South China Sea Gran Prix   163

U.S.S. Hancock (CVA-19) on Yankee Station, 1967. (U.S. Navy Photograph)

PART V: AERIAL COMBAT 166

Chapter 34: Commander Harold "Hal" L. Marr, VF-211 169
Chapter 35: Lieutenant Eugene "Gene" J. Chancy, VF-211 172
Chapter 36: Lieutenant (JG) Phillip "The Skinny Guinea" Vampatella, VF-211 174
Chapter 37: Commander Richard "Dick" M. Bellinger, VF-162 176
Chapter 38: Commander Marshall "Mo" O. Wright, VF-211 178
Chapter 39: Commander Paul Speer, VF-211 181
Chapter 40: Lieutenant (JG) Joseph "Joe" M. Shea, VF-211 185
Chapter 41: Commander Bobby C. Lee, VF-24 187
Chapter 42: Lieutenant Phillip R. Wood, VF-24 189
Chapter 43: Lieutenant Commander Marion "Red" H. Isaacks, VF-24 194
Chapter 44: Lieutenant Commander Robert L. Kirkwood, VF-24 197
Chapter 45: Lieutenant Commander Ray G. Hubbard, VF-211 198
Chapter 46: Lieutenant Commander R.W. Schaffert, VF-111 202
Chapter 47: Lieutenant Richard E. Wyman, VF-162 208
Chapter 48: Commander Lowell "Moose" R. Myers, VF-51 211
Chapter 49: Lieutenant Commander John B. Nichols, VF-191 215
Chapter 50: Commander Guy Cane, VF-53 218
Chapter 51: Lieutenant Norman K. McCoy, VF-51 220
Chapter 52: Lieutenant Anthony "Tony" J. Nargi, VF-111 225
Chapter 53: Lieutenant (JG) Phil Dempewolf, VF-24 (probable) 226
Chapter 54: Lieutenant (JG) Gerald Tucker, The Last Hurrah 228
Chapter 55: The Box Score 230

PART VI: THE SOUTHEAST ASIAN AIR WAR: A PERSPECTIVE 232

PART VII: THE RESERVES 248

PART VIII: NASA 250

Chapter 56: The Supercritical Wing 253
Chapter 57: Digital Fly-By-Wire Crusader 256
Chapter 58: The Oblique Wing 259
Chapter 59: Thunderbird Aviation 261

PART IX: THE FRENCH CONNECTION 262

PART X: THE RECORD 270

PART XI: "IT AIN'T OVER..." 275

EPILOGUE 277

Sources 279

Appendices 281
Appendix I: Vital Statistics 281
Appendix II: F-8 Squadron Combat Deployments 283

*Four VF-194 F-8Es from the U.S.S. Hancock, 1966.
(Photo courtesy of René Francillon)*

PART I
CARRIER AVIATION

". . . any man who may be asked in this century what he did to make his life worthwhile, I think can respond with a good deal of pride and satisfaction: 'I served in the United States Navy.'"

President John F. Kennedy
1917-1963

From the fall of 1952 until the fall of 1985 I was involved, either directly or tangentially, with carrier aviation. Within that thirty-three year period was a lesser time frame during which I was directly involved in that exciting profession.

My first carrier landing occurred in the fall of 1953. My last carrier landing occurred in the fall of 1980 – twenty-seven years later. In that twenty-seven year interval, I was privileged to fly from sixteen different aircraft carriers and in over seventy-one different types of tactical airplanes. My flying experiences filled five pilot's log books and over six thousand flight hours.

In one of those log books are one hundred sixty-seven entries made in green ink. Those are the combat missions in the Tonkin Gulf. I wouldn't have missed those for the world! In reviewing the lives of my peers in civilian life; I have to admit that I have had the best of it . . . by a country mile! It has been a wonderful way to live one's life!

There are those who will tell you, referring to the era when I entered naval aviation, that they were the "good old days." Maybe so. Our major accident rate was over twenty-five times higher than it is today! But, that never seemed to bother us too much. After all, you could fly under the Oakland Bay Bridge in broad daylight . . . in formation; and not get cashiered. And, you could set your hair on fire at the officer's club without being "interviewed" by a man from the Defense Department Inspector General's office.

Yes, I guess, in retrospect, those really were "the good old days!"

Maritime Air Superiority

As a flag officer in the Pentagon in the early 1980s I spent, it seems, an inordinate amount of time trying to convince people (principally Navy people) to stop using the expression "fleet air defense" when talking about Navy fighter airplanes. The expression, however accurate, invited the listener to draw the inference that the fleet needed protection before it could go about its business of sinking other ships, shooting down airplanes, wrecking things, killing people and, in general, punishing the enemy. The next logical step in calling the expensive F-14 weapons system a "fleet air defense" fighter, is to question the logic of buying expensive battle groups if, in fact, they needed so much expensive protection before they can be put to use in serving the na-

VF-13 F-8E in the Mediterranean Sea, 1964, U.S.S. Shangri La (CVA-38). (U.S. Navy photo)

PART I: CARRIER AVIATION

VF-194 F-8E over the Tonkin Gulf circa 1969, U.S.S. Hancock (CVA-19) (Photo courtesy of René Francillon)

VFP-63 RF-8G over the Sea of Japan, circa 1969. (Photo by Jan Jacobs)

tional interests of the United States. The U.S. Navy has had a running battle with certain elements in our government (DoD, the Congress and the other armed services) over the last twenty-five years or more over whether we need carrier battle forces in the first place. So, as OP-50, I spent a great deal of my time entreating other senior naval officers to use instead, the expression "maritime air superiority."

That was before I learned about the incident with the Shah of Iran. It seems that in the early 1970s the Shah of Iran was shopping for a great many weapons systems for his own defense establishment. He bought brand new, gold-plated DD-963 class (Ayatollah Class as they came to be called later) destroyers and all sorts of airplanes. Included in his shopping list was a new fighter to replace their aging force of F-4 Phantom IIs. Naturally, the Navy was interested in showing him our new F-14 Tomcat; and the Air Force was interested in showing him their new F-15 Eagle.

Since the Imperial Iranian Air Force flew the U.S. Air Force version of the Phantom II; and since they were organized after the fashion of the U.S. Air Force, it is understandable that the IIAF would favor the F-15, over the F-14 . . . which they did. Furthermore, they made no bones about it! However, the Navy's F-14 was just a little farther along in the development process than the F-15 which permitted the Navy to show their airplane off a bit more. But, there was a showdown at Andrews Air Force Base where the two airplanes put on back-to-back flight demonstrations, observed, of course, by the Shah, himself. In all honesty, the F-14 aircrew outdid the F-15 pilot more in showmanship than in sheer airplane performance.

Of course, the Shah was a very intelligent man and knew the benefits which an F-14/AWG-9/PHOENIX system would provide in defending the borders of his country from Soviet overflights by MiG-25s. He made his decision in a carefully crafted speech to the Navy and Air Force participants. He told them his country needed an air superiority fighter such as the F-15. (This, of course made the U.S. Air Force participants almost wet their pants). Then he deflated them by adding that his country also needed an air supremacy fighter such as the Navy's F-14!

Ever since then I have used the expression "maritime air supremacy" in describing the mission of a Navy fighter. So, whether we are talking about the airspace over a battlefield, or deep behind enemy lines or even over the open ocean in the vicinity of a carrier battle force, the Navy fighter needs to achieve and maintain air supremacy for as long as is necessary to get the job done!

The Air Force's mission is considerably different. It has the mechanism in place to replace its attrition fighter forces much more easily than the Navy can. The carrier battle force must be more autonomous and self-sustaining. Its attrition fighter forces may not be so easy to replace. Therefore, it needs to be so much better than any forecast enemy fighter force that the exchange ratios will be concomitantly high, negating the need to do any substantial turnover of replacement hardware during the brief prosecution of the mission. The Navy's fighter force must consist of maritime air supremacy platforms. That is the basic assumption upon which the F-8 requirements were formulated several years earlier.

2
THE FIGHTER REQUIREMENT

There is a basic principle of naval warfare, or of any aspect of warfare, which simply says: "The superior force always wins!" But, it is a mistake to assume that superior relates only to equipment. There are at least two other ingredients in the superiority equation: training (includes tactics) and leadership (includes strategy).

Two or perhaps three major engagements in the war in the Pacific had a great effect on what the fighter requirements were for a Navy fighter as the transition to jet powered airplanes progressed in the late 1940s and early 1950s. To achieve and maintain maritime air supremacy not only over one's own battle forces but, also over the enemy objective area became a major design factor.

This, of course means that the fighter capability must exceed (in an overall assessment) that of the enemy's fighters. The "overall" disclaimer is there because carrier-based fighters are generally bigger, heavier and longer range than their land-based counterparts. They are also smaller in numbers which is the reason for the "supremacy" versus "superiority" distinction.

If a Navy fighter force cannot achieve an exchange ratio, for example against the North Korean threat of the 1990s, of over 4:1, then simple mathematics almost guarantees losing a protracted conflict. That may sound like heresy, but I believe it to be true.

The single factor that drove the development of the F-14/Phoenix system was the Soviet-built, air-launched anti-ship missile. The single element that drives the development of the next Navy tactical strike fighter will include the quantum increase in fighter capability represented by the technology of the Su-35 and MiG-29 type fighters!

Of course, it would be foolish to expect a U.S. versus North Korean scenario to be limited to U.S. naval forces alone. However, an attrition rate of as little as 3% or 4% would run a battle force commander out of fighters as well as fighter aircrews in a matter of days. So, the U.S. Navy's dilemma, as always, is that it absolutely must develop "world beater" tactical aircraft to be considered a viable instrument of U.S. national policy! The number bandied about by the defense analysts in the 1990s is that a carrier battle group costs the tax payer about $1 billion per year to operate. If that number is even close to being correct, it points out the absolute necessity of "maritime supremacy" for its tactical aircraft!

3

THE TWENTY-SEVEN CHARLIE

Just as jet propulsion changed the character of naval warfare forever; so also did three innovations (all conceived by our British brothers) change the character of carrier operations forever.

After the Second World War began the U.S. Navy recognized a need for a class of attack carrier not as large as the *Saratoga* but with enough capability to defeat the known threat of the Imperial Japanese Navy. The Navy needed them in a hurry, and the miracle of the *Essex*-class story is that before the war ended, the Navy had ordered thirty two of them of which the keels for twenty five were laid down. The total commissioned was twenty four. Keels were laid for U.S.S. *Reprisal* (CV-48) and *Iwo Jima* (CV-49) on July 1, 1944 and January 29, 1945 respectively. Both were canceled on 11 August, 1945. Six more *Essex*-class carriers, the last six still in the planning stage were canceled on 27 March, 1945.

On 31 January 1942 the U.S.S. *Essex* (CV-9), the first of the class was commissioned, only twenty months after the keel had been laid. A total of twenty-four of these magnificent ships was built in a shorter span of time than it takes to build one of today's modern CVs. Of the twenty four, only six were fully modified into the SCB-27C configuration. Of those, only five were employed in the strike mission in combat operations in Southeast Asia. Those

Sixteen VF-13 and VF-62 F-8Es from the U.S.S. Shangri La in the Mediterranean Sea, 1964. (U.S. Navy photo)

19

Two views of the U.S.S. Intrepid (CVS-11 - Special Attack Carrier) South China Sea, 1968. (U.S. Navy Photo)

ideal for a carrier airplane never entered into the equation. As a result, the F-8 accident rate in 1965 was eight times greater than its counterpart, the heavier F-4 Phantom which was assigned to the large-deck carriers. But, despite the difficulty of executing consistently good carrier-arrested landings in the Crusader, once the wheels and flaps came up and the plane assumed the clean configuration, it was the most combat effective fighter in the Vietnam arena. F-8 pilots without exception fell in love with it after a few flights. The wise Crusader pilots, however, worked very hard at perfecting their landing skills and keeping them up, because the alternative was too hideous to contemplate.

The Twenty-seven Charlies, for all their limitations, did their share and more in carrying the air war to the enemy in Southeast Asia. There were many times when an *Essex*-class carrier shared duty on Yankee Station with a larger carrier in the late 1960s, that the smaller ship out-flew the bigger one.

On the afternoon of the second of August in 1964 the U.S. Navy destroyer, U.S.S. *Maddox* (DD-731) detected three unidentified contacts on her surface search radar. *Maddox*, in the vicinity of Hon Me Island, off the coast of North Vietnam, noted that the three unidentified surface contacts were approaching at high speed and went to "general quarters." The contacts were identified as North Vietnamese torpedo boats. *Maddox* fired a warning shot and began to take evasive action just before torpedoes were reportedly launched from two of the attackers. *Maddox*'s five-inch batteries took the torpedo boats under fire and estimated moderate damage to one of the retiring attackers. Fortunately, the U.S.S. *Ticonderoga* (CVA-14) a Twenty-seven Charlie class attack carrier, was in the vicinity and vectored a flight of four F-8E Crusaders to the scene. The four fighters were on a training mission, loaded with five-inch Zuni rockets and twenty millimeter gun ammunition. They rolled in on the retreating patrol boats. One boat was reported sunk. A Twenty-seven Charlie was the first U.S. aircraft carrier to allegedly draw blood in Southeast Asia.

The U.S.S. *Constellation*, a big carrier, was ordered to sortie from Hong Kong in support of *Ticonderoga*. The destroyer U.S.S. *Turner Joy* (DD-951) accompanied "Connie." Two days later, in the early evening, *Maddox* and *Turner Joy* were reportedly attacked by an estimated five torpedo boats. Torpedoes were reported but neither destroyer was hit. The battle lasted several hours with two torpedo boats reported sunk. As a result of the two unprovoked attacks, President Johnson, directed a response. The response was air attacks by strike aircraft from *Constellation* and *Ticonderoga* against four torpedo boat bases and their fuel depot at the port of Vinh. The southernmost installation, Quang Khe, just north of the DMZ was also struck. The other installations ranged all the way to Hon Gai which was a few miles north of the major port of Haiphong. That same day the now-famous Tonkin Gulf Resolution was passed by a nearly unanimous vote from both Houses of the Congress. When the sun finally set that day, A-1H Skyraiders, A-4 Skyhawks and F-8 Crusaders reported sinking or seriously damaging twenty-five PT boats and destroyed a major portion of their refueling and port facilities.

Later that fall, U.S.S. *Hancock* (CVA-19), another Twenty-seven Charlie, U.S.S. *Ranger* (CVA-61) and U.S.S. *Coral Sea* (CVA-43), in response to a North Vietnamese attack on a U.S. Military Advisor's compound at Pleiku, mounted the first large scale joint U.S. Navy, South Vietnamese Air Force air attack on the military

CHAPTER 3: THE TWENTY-SEVEN CHARLIE

barracks and staging facility at Dong Hoi, just north of the DMZ.

By December 1965 the U.S. Navy's first nuclear-powered aircraft carrier, U.S.S. *Enterprise* (CVN-65) arrived in the Tonkin Gulf and the combined air units of *Enterprise*, U.S.S. *Kittyhawk* (CVA-63) and *Ticonderoga* struck, for the first time, a thermal power plant at Vong Bi. Two A-4 Skyhawks were lost on that raid and the U.S. Navy carrier tacticians were forced to rethink the merits of daylight low level bombing tactics against highly defended targets. On 24 December, just two days after that strike, a bombing pause was initiated unilaterally by the U.S. President Johnson's national security advisor's played their trump card and waited for the North Vietnamese to "reciprocate." The North Vietnamese response was to tell the U.S. to "stuff it." What was "accomplished" during the bombing pause, was a massive thirty-seven day build-up of North Vietnamese air defenses which undid whatever good may have been accomplished by South Vietnamese and U.S. Navy tactical aviators in the previous four months.

In aerial combat, strike aircraft from the *Essex*-class carriers held their own against the MiG threat in Southeast Asia. During the Vietnam War there were over 1,300 air-to-air incidents involving U.S. tactical aircraft and North Vietnamese MiGs. The greatest portion of these involved U.S. Air Force tactical aircraft. The bulk of the aerial engagements involved MiGs and F-4 Phantoms, either Navy or Air Force. A smaller but significant percentage of the Navy MiG encounters involved Crusaders, operating from the *Essex*-class carriers. The first MiG kill by an F-8 occurred 12 June 1966 when Commander Hal Marr, Commanding Officer of Fighter Squadron Two Hundred Eleven, flying from U.S.S. *Hancock*, shot down a MiG-17 north of Haiphong. Nine days later, his wingman Lieutenant (junior grade) Philip Vampatella, nicknamed "The Skinny Guinea," bagged another MiG-17. Later that year, on 9 October, Commander Dick Bellinger, CO of Fighter Squadron One Hundred Sixty-Two, flying an F-8 from U.S.S. *Oriskany* (CVA-34), another Twenty-seven Charlie, scored the first Navy victory over the newer, hotter MiG-21. *Oriskany* had been operating in support of yet another Twenty-seven Charlie, U.S.S. *Intrepid* (CVS-11).

Intrepid, an anti-submarine carrier, had been sent into combat with A-4 Skyhawks serving the roles of both strike and fighter missions. The *Intrepid* experiment as a limited attack carrier merely proved that she needed support in all but the most benign strike environments. But the word "limited" didn't seem to keep *Intrepid*'s strike fighters out of "harms way." The newly-formed Air Wing Ten was comprised of two light attack squadrons VA-34 and VA-18 equipped with A-4s and one strike fighter squadron VSF-3, also equipped with A-4s operating in the fighter role. These truly "intrepid" aviators participated in numerous strikes on targets in "Indian Country" along with the larger, more capable attack carriers. VA-34 lost four airplanes to enemy action during its six month deployment.

The U.S.S. Bon Homme Richard (CVA-31) 13 June 1969, Gulf of Tonkin. (U.S. Navy Photo)

Above: The U.S.S. Oriskany (CVA-34) South China Sea 1969. (U.S. Navy Photo)

Partially as a result of those A-4 losses, the *Intrepid* embarked on its 1966-1967 WestPac deployment with a detachment of F-8C Crusaders from Fighter Squadron One Hundred Eleven (VF-111). Called "Detachment Charlie", the three-airplane detachment was led by its colorful Officer-in-Charge, Lieutenant Commander Foster "Tooter" Teague with three other pilots. The intensified tempo of air strikes in the Red River Valley kept the four pilots busy, often flying three flights a day into Indian Country. During this deployment "Tooter" Teague was shot down by ground fire and recovered in the ensuing Search and Rescue (SAR) operation. The next deployment of the U.S.S. *Intrepid* saw the size of "Detachment Charlie", under its new O-in-C, Lieutenant Commander "Dusty" Rhodes, increase to seven pilots and six F-8Cs. In this, her last combat deployment, the venerable Twenty-Seven Charlie joined the ranks of the MiG killers when F-8C driver Lieutenant Tony Nargi shot down a MiG-21 in September 1968.

During the Vietnam War the United States Navy experienced three disasters on three of its aircraft carriers. Two of these were in the combat zone and the third occurred in work-up exercises while preparing for combat, The first of these tragic accidents occurred on a Twenty-seven Charlie class attack carrier and, as the causes were being analyzed in the subsequent investigation, there was much speculation in the news media regarding the vulnerability of this

The U.S.S. Oriskany (CVA-34) Tonkin Gulf, 1971. (U.S. Navy Photo)

CHAPTER 3: THE TWENTY-SEVEN CHARLIE

class of carrier relative to the newer, post-World War II classes. Subsequent events were to suggest otherwise.

On 26 October, 1966 a fire broke out in a parachute flare storage locker on board U.S.S. *Oriskany* (CVA-34). The storage space was on the hangar deck, and the intense heat of the flames set off ordnance stored on the hangar deck in preparation for strike operations. Fifty-four people lost their lives in this tragic accident and the ship was out of commission for thirty days. By the time repairs had been completed and she returned to combat operations the cost of repair had grown to seven million six hundred thousand dollars.

A second fiery tragedy occurred on 29 July 1967 aboard U.S.S. *Forrestal* (CVA-59) on Yankee Station, in her first week of combat operations. The fire started when some source of stray voltage was introduced to a ZUNI rocket pod suspended on an airplane on the flight deck. The rocket motor ignited and started a fire which engulfed the surrounding aircraft. Before the holocaust was finally extinguished one hundred thirty-four men lost their lives and six aircraft were destroyed. It was sixty days before *Forrestal* returned to combat operations and the repairs cost seventy-two million dollars.

The third of these tragedies occurred in the Navy's first nuclear-powered carrier U.S.S. *Enterprise* (CVN-65) off Hawaii on 14 January 1968. *Enterprise* was starting up its airplanes preparatory for a launch and training exercise. Hot gases from a jet-powered air starting unit ignited the rocket motor of a five-inch Zuni rocket on a VF-96 F-4B Phantom. Fire spread quickly through the group of aircraft loaded with ordnance and parked on the fantail (aft of the flight deck). After three hours the fire had been put out and *Enter-*

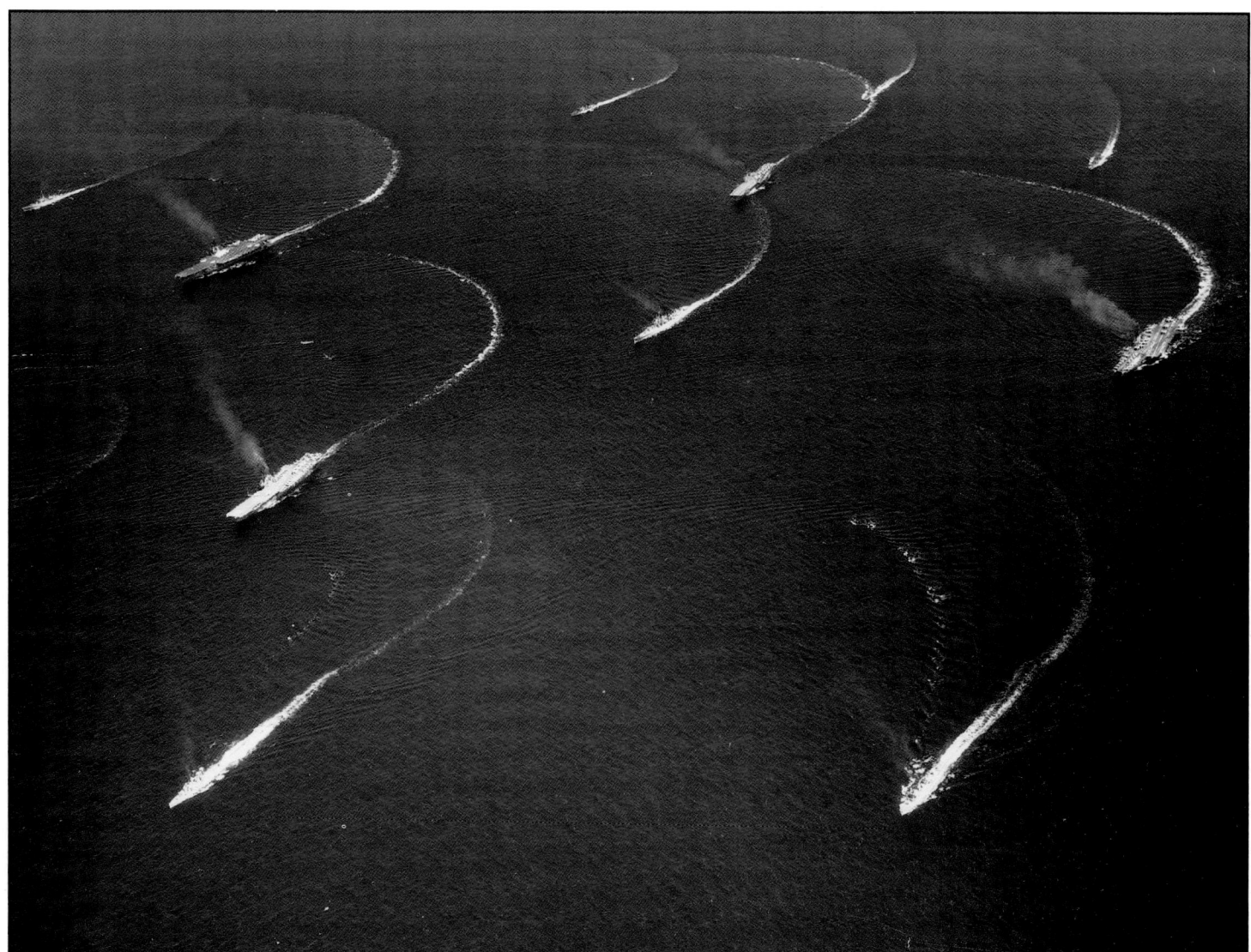

Rare four carrier formation photograph of Task Force 77 in the South China Sea, 1966. At the top left of the photo is the U.S.S. Constellation (CVA-64). At left center is the U.S.S. Ticonderoga (CVA-14). At the top right is the U.S.S. Hancock (CVA-19). At bottom right is the U.S.S. Coral Sea (CVA-43). "Connie" is a Forrestal class carrier, and Coral Sea is a Midway class carrier. Hancock and Ticonderoga are Twenty-Seven Charlies of the Essex class. (Photo courtesy of D. Winiker)

prise prudently put into Pearl Harbor for repairs. Forty-nine days later she returned to the combat zone. The cost was heavy; fifteen men were lost, six airplanes were destroyed and the repairs cost fifty-six million dollars.

The U.S. media and opponents of aircraft carriers had a heyday with this tragedy. They laid heavy emphasis on the volatility and vulnerability of a carrier flight deck loaded with jet engines, jet fuel and explosives. The real message was never told. *Enterprise* had taken a terrible beating. Nine 1,000 pound bombs had exploded and had dealt a mortal blow to the steel flight deck and surrounding aircraft. In true combat, *Enterprise* could and would have slapped steel plates over the bomb holes, bulldozed damaged aircraft over the side and continued flight operations. The real lesson of how much punishment an aircraft carrier could absorb was somehow lost in this torrent of self-serving rhetoric.

In all three of these tragic accidents, procedural errors were identified and steps taken to preclude recurrences. In addition a flag-level review board was established in the Pentagon to periodically review the progress made in several research and development programs which were initiated to correct deficiencies in the munitions carried on board U.S. naval vessels. One of these programs was aimed at rendering the explosive material used in the warheads of a number of weapons carried on U.S. naval ships less sensitive to the high temperatures associated with aircraft fires.

Another important shipboard fire safety program is associated with rendering all weapons less sensitive to ignition from radiated electromagnetic energy from shipboard electromagnetic emitters, such as radar gear.

A third, and on-going research and development program involves the introduction of jet engine powered sea water cannons mounted on the superstructure which can produce an enormous amount of sea water under tremendous pressure to extinguish flight deck fires, lower the temperature of over-heated ordnance and, if necessary, in extremis, literally wash burning aircraft and ordnance over the side. Development of this important program is nearing completion.

Much more stringent procedural ordnance-handling changes have rendered carrier flight decks much safer since the last fire of that type occurred in 1968.

I recall sitting up late one night in the intense humidity of my state room in U.S.S. *Ticonderoga* wondering what it would be like to sleep in air conditioned comfort. I noticed when I first arrived on board "Tico" how gaunt and tired all the aviators looked. A little thing like air conditioning in living spaces was terribly important for combat operations. I couldn't sleep because of the temperature and humidity so I picked up one of my favorites; an anthology of poems by Rudyard Kipling. I reread one of my favorites, "Tommy Atkins", a poem written about the sad treatment of the British soldier by an ungrateful British populace after World War I. One of the stanzas said:

"Oh, it's 'Tommy this' and 'Tommy that',
And, 'chuck him out, the brute',
But it's 'saviour of his country'
When the guns begin to shoot."

I was still chafing over the two carrier strikes in recent weeks when the Task Force planners used strike groups from *Essex*-class carriers to troll for MiGs and SAMs. Little or no credit was extended in the post strike message report to fleet headquarters. So I picked up my pen and began scribbling.

"The Twenty-Seven Charlie"

"It all began with Maddox in the fall of '64,
And before we really knew it, we'd bought ourselves a war.
The bigger decks could 'hack it', they were pretty fancy stuff,
'Til Lyndon raised the ante, then there simply weren't enough.
And aircrews from the CVNs observed with great dismay
The tired, ancient Twenty-seven Charlie saved the day.

They only had two catapults, and no conditioned air,
And one approached the ramp at night with precision and with care.
With hook to ramp displacement of half a dozen feet,
The 'one wire' cost you fifty cents and 'number four' was sweet.
Oh, it's 'Connie this' and 'Nimitz that' and 'aren't those big boats
*   beauts?*
But it's 'Twenty-seven Charlie' when the guns begin to shoot.

Her deck was stacked with Tinker Toys, F-8s and even Spads,
With tired, worn-out E-1Bs the only eyes we had.
She had no multi-engined jets to clutter up the scene
And tanking in an F-8 from a Spad was just obscene.
But when the Haiphong Highway Bridge was targeted one day
A Twenty-seven Charlie was sent in to the fray.

When Lyndon upped the ante next and airfields were 'legit',
And Kien An was picked out for the Hancock Group to hit,
They bagged our CAG, Dutch Netherland, with SAMs as thick as
*   flies*
While the sister strike from Enterprise encountered clear blue skies.
Altho the post-strike messages applauded 'Big E's' feat,
It failed to note a Twenty-seven Charlie took the heat.

With the final chapter starting about this dismal war
They're toting up the MiG kills to finalize the score.
With all their speed and power, the big boats didn't shine
For the Twenty-seven Charlies won it seventeen to nine
Oh, it's 'Sara, this' and 'Midway, that' and 'aren't those big boats
*   cute?'*
But, it's 'Twenty-seven Charlie' when the guns begin to shoot."

4
AIR-TO-AIR WEAPONRY

Jet propulsion and the carrier revolution changed the character of naval air warfare forever, but we continued to inflict damage to the enemy (from airplanes at least) by pointing a gun barrel at him. There was, of course, nothing wrong with that approach as airplanes first developed into weapons deliverers. The biplanes of the first world war effectively attrited each other with guns ... all sorts of then, ending finally with the machine gun. As the development of bigger, faster airplanes continued through the period between the two world wars, it was kept pace by the development of bigger caliber and faster firing guns. The thirty caliber machine guns firing at a rate of a few hundred rounds per minute gave way to the fifty caliber guns firing at six and seven hundred rounds per minute. Then the number of guns increased from one in the early biplanes to as many as twelve in the World War II Hawker Hurricanes. The difficult achievement of rates of fire in excess of say eight hundred rounds per minute was subverted by simply adding guns.

In the Korean War, the increase of the killing power of guns was achieved by increasing the damage done per round. Bigger calibers with higher muzzle energy began with the twenty millimeter cannons of U.S. fighters like the F9F Panthers. Then the deadly twenty-three millimeter gun appeared in Soviet-built MiG-15s and 17s and proved to be a very intimidating weapon. The biggest caliber of all, the deadly thirty-seven millimeter gun of the MiG-17 was a marvel of its time. The author has a chromium-plated 37 mm round from a MiG-17 sitting on his desk as a paper weight. I still shudder whenever I pick it up. I shudder because I can only imagine what damage it would have done to my F-8 Crusader when I was flying combat missions in Southeast Asia.

Then, rate of fire was determined to be the next, and probably the last, great technical breakthrough for guns. It made its debut in the 100 series U.S. Air Force fighters like the F-104. The mechanism for delivering this great breakthrough was not an new concept ... but rather an old one ... the Gatling gun.

When fired by one airplane at another it represented a very high energy stream of steel which, quite literally, could cause a

F-8 gunfighter poster which appeared at all of the Navy fighter bases in the early 1960s.

lightly constructed jet fighter to disintegrate. From then on all U.S. tactical fighters carried a single, deadly Gatling gun ... the M-61. The only exception to this rule was the F-5 Freedom Fighter which carried two conventional 20 mm guns.

However, it was becoming clear to aerial tacticians that sheer muzzle energy increases would not compensate for fundamental changes which were occurring to aerial warfare as a result purely of aircraft speed.

How distinctly I remember attending a lecture given by a Lieutenant Colonel Bolt, a U.S. Marine Corps fighter pilot who, during an exchange tour of duty with the U.S. Air Force in Korea, killed a few MiGs. At the time I was a weapons delivery instructor at what was then the predecessor to Topgun, an organization known as the Fleet Air Gunnery Unit, Pacific (FAGU). Colonel Bolt showed some air-to-air combat footage of one of his MiG engagements in an F-86. All of us were astonished when he opened fire at a range of easily three thousand feet. We were even more astonished when his target, a MiG-15 began smoking then burst into flames.

PART I: CARRIER AVIATION

AIM-9 Sidewinder missile being fired from an F-20. (Northrop photo0

At the time we were teaching open fire ranges with 20mm cannons of about 1200 feet! That was because it was about the average range of most kills by gunfire. Colonel Bolt proved that he could be very effective a far greater ranges. Nonetheless, he was in the minority as compared to firing ranges for other aces of the time.

It became apparent in the skies over North Korea, that as altitude and airspeed increased, the effectiveness of aerial gunfire decreased rapidly. Aerial weapons design engineers proved mathematically that it was virtually impossible for a fighter to shoot down a bomber flying at 40,000 feet at 450 knots true air speed. They proved the point by describing concentric circles around the target airplane known as isogees. These lines represented the "g" necessary for the fighter to pull long enough to generate a fire control solution for a gyro operated, lead pursuit gun weapon system. As the circles drew closer to the target the "g"s they represented increased. For example, to hit a Russian Bear bomber flying at 40,000 feet and 450 knots true airspeed (non-maneuvering) at an open fire range of say 1,500 feet, a fighter would have to sustain about five "g"s for the duration of the firing run. Current fighters couldn't do that very well. And, even if they could, they would end up nearly dead astern of the bomber well within easy gun solution for the bomber's tail gunner.

The U.S. Air Force came up with a solution to the problem; unguided rockets using a lead collision fire control solution. Now, the weapon could get to the target without the shooter pulling any "g"s at all. All he had to do was fly toward a point ahead of the target and fire. The problem then, became one of lethality. The unguided rocket had a low probability of hitting the target. So, the Air Force solution was to increase the lethality of the rocket in one of two ways; either large kill radius or lots of rockets.

Their F-106 interceptor used an unguided rocket with a nuclear warhead. The warhead gave the one shot system plenty of kill radius but the problem of the use of nuclear weapons became a political one. The rocket was called GENIE. Defending the homeland of the United States was the only scenario wherein the GENIE could be employed. An earlier solution to the problem was lots of rockets. The F-86D, a version of the Saberjet equipped with a relatively long range radar (in those days) and an afterburner, carried a pod with a cluster of 2.75 folding fin aircraft rockets (FFAR). This became the weapons system of choice for the North American Air Defense Command until it began to be replaced slowly by the F-106.

Other airplanes equipped with clusters of rockets were the F-94 and the ill-fated F-89. The problem was the rockets and their lead collision fire control system were highly ineffective. Although the Air Force did not publicize the incident, a pair of Air Defense Command F-89s was sent up to shoot down a U.S. Navy F6F Hellcat drone airplane which had gone out of control. When the Pacific Missile Range commander finally admitted to authorities that his target drone aircraft had wandered, uncontrolled off the range and was headed south toward San Diego the U.S. Air Force scrambled two of their highly touted F-89s to shoot it down. It was a simple enough tactical problem because the propeller-driven drone was putting along at a medium altitude at a sedate 175 knots. "No problem. The Air Defense Command said, "We'll handle it!"

After sitting dead astern of the hardy Hellcat and expending all of their rockets, all the Air Defense Command accomplished was the setting of several forest fires in the Laguna Mountains while the Hellcat putted off and, after running out of fuel, crashed, thank God, in an unpopulated area. CINCNORAD was a laughingstock for some time after that and Americans went to sleep at night feeling not quite as secure as they had felt before the incident. This occurred sometime in the 1956-1957 time frame.

Meanwhile, the U.S. Naval Weapons Center at China Lake, California was working hard on the development of a super secret air-to-air guided missile which actually homed on the infra-red

CHAPTER 4: AIR-TO-AIR WEAPONRY

emanations of the hot exhaust pipe of jet airplanes. It was called SIDEWINDER! This high tech marvel not only homed on the enemy airplane's tailpipe, it also contained an influence fuse and an expanding rod warhead which made it lethal if it only passed within, say, twenty-five feet of the target. Sidewinder revolutionized aerial warfare. A China Lake pilot demonstrated the missile to a group of us at FAGU with astounding results against an old Navy F9 fighter configured as a drone. This was in 1956.

Unfortunately, the Sidewinder came along too late for the Korean War. By the time of the Southeast Asian air war, a fourth generation Sidewinder called AIM-9D was in production. It proved to be between three and four times more effective than its competitor, a newly-developed radar-guided air-to-air missile called AIM-7 SPARROW. Since those days, China Lake has regularly improved Sidewinder at intervals of every three or four years. They are now at the AIM-9M version, each successive version representing a dramatic technological improvement. The same has occurred with Sparrow.

The author feels compelled to mention another operational air-to-air missile developed by the U.S. Air Force called FALCON. Unfortunately for Falcon and all those poor Air Force pilots who had to carry it into combat, the system required a contact hit to fuse and detonate. To my knowledge Falcon has never killed anything in combat. When I flew from Yankee Station to the Air Force Base at Ubon, Thailand on a liaison visit in 1967, the wing commander Colonel Robin Olds and his vice commander Chappie James would have gladly swapped every Falcon they had for just one of my AIM-9D Sidewinders. (They only had a limited quantity of older AIM-9Bs).

Notwithstanding Sparrow's improved performance, Sidewinder has continued to be the premiere air-to-air weapon of not only the free world air forces but also the former Warsaw Pact air forces who developed a crude imitation of Sidewinder. Thirty-eight years after the demonstration which I observed, Sidewinder continues to be the most combat effective air-to-air missile in the world.

In the late 1950s a system known as PHOENIX began development and appeared finally in the late 1960s as the weapon end of the F-14's AWG-9 weapons control system. This 1,000 pound monster was capable of killing at a range of 100 miles and had a launch-and-leave capability existent in no other radar-guided system in the world. The basic F-14 weapons system received its acid test in 1971

AIM-54 Phoenix missile being fired from an F-14A. (Hughes photo)

when it fired six Phoenix missiles almost simultaneously at six different targets out on the Pacific Missile Test scoring five kills.

In the 1970s through today there have been several semi-abortive air-to-air missile development programs whose goals included shorter ranges, higher agility, off-boresight firing capability, higher velocity etc. Of them only advanced short range air-to-air missile (ASRAAM) and advanced medium range air-to-air missile (AMRAAM) seem to be showing promise. The highly touted advanced air-to-air missile (AAAM) was recently canceled as a follow-on to Phoenix when the F-14D program was canceled. That will probably prove to have been a terrible mistake when historians (students of aerial combat) review the bidding in years to come.

The United States' European allies have come up with some air-to-air weaponry of their own and in joint development programs with each other and American engineers. But, as of the writing of this book, Sidewinder, Phoenix and Sparrow appear to be the premiere air-to-air weapons in existence. This is one of the reasons why the F-14 is still considered the premiere air supremacy fighter in the world today. It is the only tactical airplane in the world which carries all three of those weapons PLUS A GATLING GUN!

Despite enormous amounts of research, development, test and evaluation (RDT&E) money spent on an Air Force program called airborne laser (ABL), it appears that the day when a fighter pilot can down another airplane with a laser beam is still a long way off!

5
FAGU AND TOPGUN

It was a beautiful spring day in May 1950 when Lieutenant Commander Ralph L. "Larry" Walker reported to the Officer of the day (OOD) at Commander, Naval Air Forces, U.S. Pacific Fleet (COMNAVAIRPAC) headquarters at Naval Air Station, North Island, California. The beautiful city of Coronado never looked prettier. Larry was sitting on top of the world as he walked into the rotunda in the AIRPAC headquarters building. He had in his hands a set of orders directing him to detach from his tour of duty as a test pilot at the Flight Test Division of the Naval Air Test Center at Patuxent River, Maryland. The orders further directed him to report for duty to Fighter Squadron Fifty-One at Naval Air Station, North Island, California. Larry thought he had died and gone to heaven.

Certainly, his three years at Patuxent River had been fun and rewarding. He had spent the last two years test flying such interesting airplanes as the McDonnell F2H Banshee, the Vought F7U Cutlass and the early Grumman F9 Panther series of fighters.

But, now he had orders to Fighter Squadron Fifty-One, the world famous Screaming Eagles. It was a choice assignment. The squadron traced its history all the way back to the very beginnings of carrier aviation. It was a fleet squadron; a plus. Furthermore, it was homeported at North Island. What better, nicer place to fly Navy fighter airplanes? Larry laid the originals of his orders on the duty officer's desk to be stamped.

The duty officer studied them and said, smiling, "Commander, while you were enroute from your last duty station, a modification to your orders came in from the Bureau of Naval Personnel. Here they are, Sir, already endorsed by the Admiral, Sir." Larry frowned as he read the single piece of paper, a sinking feeling beginning to grow in the pit of his stomach.

Officer staff photo of FAGU photographed on the occasion of the first fleet air gunnery meet in 1957. Commander John Butts, front row center, was the commanding officer. Author is second from the left, rear row. FJ-3 in the background. (U.S. Navy photo)

CHAPTER 5: FAGU AND TOPGUN

Two fleet air gunnery unit F9F-8Ts over Imperial Valley, 1959. (U.S. Navy photo)

F9F-8Ts from FAGU carrying 1,000 lb. G.P. bombs over Imperial Valley, 1956. (U.S. Navy photo)

"God damn it", he swore under his breath, "What the hell are those clowns trying to do to me? I had the best set of orders in the world; a fleet fighter squadron based at North island. Now, they change them to a lousy FASRON (fleet air support squadron) and make me officer-in-charge of a detachment at Naval Auxiliary Air Station, El Centro, California . . . in the middle of the God damned desert. Jesus, what could be worse?" He was furious until the duty officer explained what the ORDMOD (orders modification) really meant.

Things were heating up in Korea. There was a strong need for a training school to build aerial gunnery training expertise in the fleet squadrons being equipped with brand new jet fighters. The U.S. Air Force, having already recognized the need, had formed up an aerial gunnery training squadron at Nellis Air Force Base near Las Vegas, Nevada. Nellis was close to many of the ordnance training ranges in the desert. They correctly viewed the new stable of Russian MiG jet fighters as the threat against which they needed to train.

Fleet air gunnery unit FJ-4B at NAAS El Centro, 1959. (FAGU photo)

The Navy chose El Centro because of its proximity to the Chocolate Mountain Gunnery Range currently being operated by Vincent Air Force Base at Yuma, Arizona, just fifty miles east. Larry's new orders were very important, said the duty officer. His detachment would consist initially of himself, two other pilots, a handful of FASRON SEVEN enlisted mechanics and one F9F-2 Panther jet fighter plane.

He was to form the nucleus of a composite squadron of airplanes representing all of the types existing in fleet squadrons. The detachment's mission was to train young pilots from each fleet squadron in the fine art of aerial gunnery and combat tactics.

Once he and his two officers had completed writing the training syllabus, it would be submitted to COMNAVAIRPAC for approval. As soon as approval was granted, additional enlisted men, pilots and airplanes would be assigned. The school would then open its doors and begin training pilots from both Atlantic and Pacific fleet squadrons.

It was a very important assignment, Larry admitted, but still, he was deeply disappointed at not going to the Screaming Eagles. Commander Lou Bower, AIRPAC Training officer, assured Larry that as soon as the school had begun cranking out crack gunnery training experts he would be sent on to a fleet squadron of his choosing. It was an offer he couldn't refuse, Commander Bower concluded. Larry grudgingly agreed.

While they were still working late into the night drafting the training syllabus, two more airplanes arrived along with some more enlisted personnel. The additional airplanes were an F4U Corsair (propeller-driven fighter built by the Chance-Vought Aircraft Company in Dallas); and a Douglas built AD-1 Skyraider (propeller-driven attack airplane).

In record time a syllabus was produced. The Air Force gunnery school at Nellis was very helpful. They sent pilots down with their brand new F-86 Saber jet fighters. Their reward for helping their Navy counterparts was a few rides in the Grumman built Panther jet.

By May of 1951 the Fleet Aerial Gunnery Unit, Pacific owned five brand new F9F-2 Panthers, five F4U Corsairs and five AD-6 Skyraiders. Along with the new airplanes came more pilots, ninety more enlisted personnel and the refurbishment of the old World War II hangars, barracks and training buildings. Things were moving fast and the first full class of students was graduated when Commander "Boogie" Hoffman arrived with orders in hand to relieve Larry Walker.

It was with mixed emotions that Larry pointed his old heap west from El Centro's main gate. As his car climbed the long grade into the Laguna Mountains, Larry was grateful to leave the intense heat of the Imperial valley and head for the cool tropical climate of San Diego.

Flight of two A-4s, and two F-16Ns from Topgun San Diego, 1990. (Photo by Shireman)

CHAPTER 5: FAGU AND TOPGUN

Topgun F-16N carrying ACMR pod and dummy Sidewinder, 1989. (Photo by Shireman)

Topgun F-16N on night adversary mission over air combat maneuvering range, Yuma, Arizona 1991. (Photo by Shireman)

It had been a hard year but a very memorable one for him. He had made a tremendous contribution to the combat readiness of U.S. naval carrier aviation. Little did he realize then, that just five years later, he would return with his own fleet squadron gunnery team and take first honors in the first fleet aerial gunnery meet. The author, a FAGU PAC instructor at the time, was the scoring official for the meet.

Topgun F-16N with maritime paint scheme, 1990. (Photo by Shireman)

Larry's team won over teams from all the fighter squadrons in both Atlantic and Pacific fleets. What an honor it was to be the best aerial gunners in the fleet . . . to be the best of the best! The title, appropriately bestowed on Larry Walker and each of his three teammates was "Topgun!"

In 1957 someone decided that this ought to be an annual event, a competition, between all fleet squadrons to determine which was the best in weapons delivery. They called the first event the Fleet Air Gunnery Meet. As an instructor at FAGU and a scoring official I was witness to some of the finest shooting I had ever seen. It was a roaring success; and all participants agreed to do it again next year. The following year the name of the competition was properly changed to the Fleet Air Weapons Meet. It was a marvelous opportunity for aviators from all over the country to meet, exchange views, match their skills and network with their peers. It was a morale booster and a great recruiting device for Naval Aviation. There were aerial demonstrations of new airplanes, and of course, the Blue Angels came to perform. The annual meet had become, or so we thought, a regular fixture of naval aviation.

In the spring of 1958, with a major reorganization of fleet training, it was determined that the newly formed replacement air group (RAG) would render the Fleet Air Gunnery Unit, Pacific superfluous. The reason was simple. The product of the RAG, the newly trained Category I pilot would be 100% combat ready and, therefore, in no need of further training. The thought process whereby that decision was made was foolish, stupid and, what's worse, demonstrated a woeful ignorance of both the needs of the fleet and the product of FAGU! With the bureaucratic stroke of a pen FAGU was abolished. Along with it went, of course, the annual weapons meet.

Now, let us advance our clocks seven years. The air war in Southeast Asia is in full swing and the U.S. Navy and Air Force come to the startling realization that they are achieving barely a 1:1

Two Topgun F-16Ns and one A-4 in desert camouflage paint schemes, 1989. (Photo by Shireman)

exchange ratio between highly trained F-4 aircrews and a rag-tag, third world air force flying outmoded Soviet MiG-17 fighters. What to do? Captain Frank Ault, author of the now famous Ault Report on weapons systems effectiveness, with all the zeal of a 16th century apostate, announced that they needed something like the old Fleet Air Gunnery training Unit, Pacific. Heavens!

Since the F-8 side of the fighter community seemed to be training more for the kind of air war which was evolving than the F-4 community, a decision was made to form a cadre of tactics instructors at the F-8 fleet replacement squadron, Fighter Squadron One Hundred Twenty-Four. The cadre was to be called, of all things, TOPGUN!

A few words need to be said about this decision . . . because it has become a little controversial as recent pundits have attempted to revise Naval Aviation history. The Crusader carried two air-to-air weapons; the gun and the Sidewinder. Both of these weapons were rear quarter systems. This means that Crusader tactics had to be employed which optimized its weapons. This meant maneuvering to bring one's own weapons to bear while, at the same time, nullifying the enemy's weapons systems . . . which, at that time were guns only.

The kinds of tactics which evolved from this fact were exactly the ones which proved to be effective in the unusual combat environment over North Vietnam!

The Phantom II, on the other hand, carried two weapons also; both air-to-air missiles, Sidewinder and Sparrow . . . and no gun. The Sparrow was its primary weapon, and the forward quarter (down the throat) was its most effective and primary shooting parameter. So, the F-4 community trained for its primary weapon and maneuvered to bring its primary weapon to bear . . . a forward quarter radar-guided missile. Furthermore, the missile's semi-active guidance system required that the target be illuminated by the airplane's

Two Topgun A-4s over Yuma, Arizona, 1988. (Photo by Shireman)

CHAPTER 5: FAGU AND TOPGUN

U.S. Marine Corps F-5E adversary aircraft over Yuma, Arizona, 1988. (Photo by Shireman)

New Topgun academic building – dedicated to Admiral Michaelis in 1992. (Photo by Shireman)

radar during the flight of the missile. In retrospect, most Phantom pilots admit that they failed to put enough training emphasis on the tactical employment of their secondary weapon . . . a rear quarter weapon. Therein lies the seed of the decision to put Topgun where it ended up, in VF-124. It is pointless, and incorrect to draw any other (pejorative) inference from that decision.

Headed up by a Lieutenant Commander, this VF-124 cadre began teaching fleet pilots how to win in combat in Southeast Asia. The program grew. The staff grew. Its reputation developed and; best of all, its graduates began appearing in the skies over North Vietnam winners far more often than losers. Professionals like Lieutenant Commanders John "Pirate" Nichols and Jerry "Devil" Houston sat down and wrote the tactics manual on how Navy fighter pilots should employ the "loose deuce" formation to win in the skies over North Vietnam. By the end of the war in 1972 the exchange ratio between F-4s and MiGs was a favorable 4:1!

In the ensuing years Topgun became an autonomous organization, headed up by next a Commander, then finally by a Captain who, in 1984 was assigned Echelon Two rank (equal to a three star Admiral) and reporting directly to the deputy Chief of Naval Operations (Air Warfare). Topgun had arrived. The concept of FAGU had come full circle!

PART II
THE EARLY YEARS

Success comes not from doing what you want to do, or what you like to do. Success comes from doing what you have to do."

Frank Tyger

Fleet pilot in full pressure suit, manning up at NAS Moffet Field, 1956. (Photo courtesy of John Miottel)

burner. "Duke" went supersonic and accelerated as he went to the turn in point 90 miles south of Mojave where he turned north for the first of two passes over the range.

The range controller radars gave him vectors to keep him right on track as he accelerated heading north in full afterburner. He crossed over the range course exactly on the money and recorded a speed of 1,018.553 miles per hour. After passing over the northern gate of the course, he began a 1.5 "g" turn which described a semi-circle 85 miles in diameter and flashed back over the course on his southerly run; again right on the money. The speed for this run was 1,012.303 miles per hour. This established his average speed for the two runs as 1,015.428 miles per hour, a new national record!

According to the FAI rules, a pilot is only allowed to vary his altitude during the pass by 100 meters or 328 feet. Not having a calibrated airspeed probe on this airplane "Duke" had a kneeboard card with the schedule for the altimeter versus indicated mach number that he would have to maintain on his cockpit altimeter in order to stay within the altitude restrictions for a record attempt.

Fortunately, "Duke" flew the first pass so precisely that he had a deviation of only 48 feet on the first pass. He assumes that he must have gotten a little nervous on the second pass because he had a deviation of 75 feet during the second pass. Both were well within the 328 foot limits set by FAI. Considering that he was traveling over one thousand miles per hour, "Duke" Windsor had literally flown through the eye of a needle. The airplane was a production model Crusader with a full load of ammunition and the four guns. No attempt was made to eclipse the world record set by the British Fairie Delta research airplane the previous fall after he had canceled his first record attempt. To do so would have required a speed of 1,143 miles per hour. FAI rules require that to be certified as a new record, an average speed of over one percent over the previous record must be established. DoD had instructed the Navy to get as close to 1,000 miles per hour as possible and still exceed it. The record setting team had done so handsomely.

When "Duke" Windsor finished his second pass he still had over 1,500 pounds of fuel remaining, a very comfortable margin. Had they wanted to, they could have cut it down to a lower total fuel upon completion and probably gotten fairly close to a world's record. However, the team felt satisfied that they had done a good job ... and there were few regrets, if any.

Three observer airplanes were airborne along the periphery of the range with FAI observers aboard. Two of the airplanes were T2Vs and the third was an F3D. Only one of the observers even saw the Crusader as it went by. The upper winds were 40 knots almost ninety degrees to the course heading. For that reason, there was very little wind effect on either of the two runs. By virtue of having made so many practice runs over the range "Duke" was very familiar with the many landmarks in the area's terrain features. So he was able to use them to keep himself fairly well lined up with the course ... with the radar assist from ground controllers.

Upon returning to the field, "Duke" Windsor made a pass down the runway and executed a victory roll. He hoped, after doing so, that he had not ruined all the barographs loaded into the airplane. But, he was feeling so exhilarated by the fact that his ground crew had already advised him by radio that he had indeed broken the national speed record and, in all likelihood, won the Thompson Trophy that he couldn't resist the temptation. He landed at 7:22 just 32 minutes after take-off.

Paul Thayer, Chance Vought's President, was on hand to congratulate him upon his return. Jack Ludwig and Curly Stomper, his plane captain and a crowd of others were to congratulate him on his flight and the success of PROJECT ONE GRAND! One interesting sidelight was that the British airplane which had set the mark the previous winter took five attempts to set the world's record. On its very last attempt it had just barely managed to exceed the existing mark by the prescribed margin. Had "Duke" been allowed to make his earlier attempt with the YF-8 the previous world record would probably have been exceeded and the British Fairey Delta would have probably been unable to exceed the Crusader's mark by one percent. But, life is full of "what ifs?"

Captain Bob Dose, U.S.N.
(his own words)

This is perhaps just another flying tale, but I think it has some interesting and even humorous aspects. It concerns a fast ride across the United States by two men in F-8U-1 "Crusaders." It was accomplished in a bit under three and one-half hours, a carrier-to-carrier cross country record which still stands. It's amazing how short 3 and 1/2 hours can be when you're having fun.

On June 6, 1957, the thirteenth anniversary of "D" day in Europe, Paul Miller and I were launched almost simultaneously in our F-8U-1 Crusaders from the *Bon Homme Richard*, fifty miles west of San Diego, California. We were catapulted toward the west into the wind and quickly turned eastward for our trek across the country to land on the *Saratoga* fifty miles east of Jacksonville, Florida. President Eisenhower was aboard the *Saratoga*.

This was fairly early in the days of supersonics. The planes, F-8U-1 Crusaders, were new and sleek and fast. Most important, they could travel unusually long distances with a minimum expenditure of fuel. In the vernacular, they had "long legs." They were a flashy plane – from release of brakes on takeoff to 40,000 feet took two and a half minutes. I loved the airplane.

We headed east climbing in burner. We had compared the performance of our two aircraft and it turned out that my old and trusty airplane was faster. It was the seventeenth Crusader ever built but it was faster than Paul Miller's new aircraft which was later off the production line. So we traded planes for the hop. I could then set full throttle and know that Paul could keep up with me.

CHAPTER 6: ". . . A NEED FOR SPEED"

Cdr. Bob Dose, C.O. of VX-3. (Chance-Vought photo)

The plan was to climb quickly to 43,000 feet, which was predicted by the weather guessers to be the best altitude for maximum cruising, and to hold mach 1.7 which was max for the plane, We would then climb, if necessary, to keep from exceeding that speed. Those early Crusaders, the F-8U-1s, had no dorsal fins and became directionally unstable above mach 1.7. The engineers had told us that if we exceeded mach 1.7, the plane would turn sideways and come unglued. We listened. We could measure the degree of instability as we approached max mach by putting on a little rudder and waiting for the plane to straighten itself out. Knowing that the automatic controls were already furtively applying a large amount of opposite rudder to bring the plane back to equilibrium, the time it took to stabilize was a good index of the nearness of too much speed.

We were scheduled to meet two AJ tankers over Dallas which was coincidentally just six miles off of the great circle route from San Diego to Jacksonville.

The flight was planned in detail so that we would make it in two legs with air-to-air refueling over Dallas. The first leg was from the carrier to Dallas, and the second from over Dallas to the *Saratoga*. On each leg we were programmed to go into burner for a calculated time toward the end of the leg, so that we arrived with a reasonable margin of fuel. Since Dallas is close to half way across the U.S. on the route we flew, that time was thirteen minutes on the first leg and seventeen minutes on the second. Afterburning time was longer on the second leg because that leg started at 25,000 feet instead of sea level. For the air-to-air refueling we had to dive from our ambient altitude of 43,000 feet or above to 25,000 feet and slow to sub-sonic speed for our rendezvous and plug-in with the slow tankers.

To regress a bit, toward the end of May, about a week before we were to make the cross-country, I received a phone call from a friend in the Operations Office of Op 05, DCNO Air. He asked me, the Commanding Officer of VX-3, if we could make a historic carrier-to-carrier cross-country flight about a week later, on June 6. I gulped, breathed deeply and replied, "Sure." We were committed.

In preparation for the flight, we left our home base at N.A.S. Atlantic City a couple of days later in order to go to the factory in Dallas to pick up Paul's new airplane and to modify my good old faithful steed so that it had a refueling probe.

Flying from NAS Dallas, we made one practice sortie simulating the conditions we expected on the actual cross country. We headed west until we were geographically where we expected to go into burner on the actual flight. We turned to the east and went into burner. I had been supersonic many times before, but not for as long nor on a flight plan. We picked up speed to mach 1.7, and headed toward Dallas. I found myself going through maps as if I were reading a magazine. All of a sudden there was Dallas below us. In those days we were believed safe from dropping shock waves if we were above 40,000 feet when supersonic. I throttled back and dove to rendezvous with the tankers at 25,000 feet. We went subsonic in the dive, but I made a mistake of ten thousand feet, pulling out at 15,000 feet instead of 25,000. We lowered a sonic boom on Dallas. As a matter of fact, John Conrad, the outstanding test pilot for Chance Vought, looked us up in the evening at BOQ and reported we had almost knocked him out of his bathtub.

The scheduled tankers were AJ-1s, built by North American. They had a reciprocating engine on each wing and a jet stinger in the tail. Our refueling speed was 250 knots indicated, which was absolute max speed for the tanker at 25 grand and almost down to stalling speed for the Crusaders. They made a strange couple when connected: the AJ was nose down and making its most possible speed and the Crusader was nose high, almost stalling and sloppy on the controls. What's more, the hose from the tanker was so short that it was necessary to close in on the tanker so that the tail of the tanker was about four feet directly above the cockpit of the Crusader. No big deal, but interesting. If the Crusader pilot had been forced to eject, he would have been shot directly into the tail of the AJ above him.

We took on a token load of fuel, broke off, and landed at NAS Dallas. The next day, we took off and flew non-stop to NAS Alameda where we landed with 1200 lbs. of fuel each. Our planes were hoisted aboard the carrier and the "Bonny Dick" departed the next morning for the San Diego area.

The following day, June 6, 1957, we launched from the *Bon Homme Richard* and headed east for the USS *Saratoga*, CVA-60. It was a beautiful sunny day. Paul quickly slid into loose formation and we climbed in burner to 43,000 feet, leveled off and came out of burner. Our turbine outlet temperatures were trimmed to 670 degrees instead of the normal 630. The engines delivered more power at the higher setting. We flew on a standard FAA flight plan. The way had been cleared by COMAIRPAC all the way across the country except for one hitch. Clearance had not been granted by the Air Force to cross White Sands Proving Ground. I contacted ground control about seven minutes short of our crossing the zone and warned them we were coming. We did not receive clearance until 35 seconds short of crossing. Not that it would have made any difference anyway. We were coming through regardless of Air Force stubbornness.

Just short of Dallas, we came out of burner, throttled back to idle and dove to 25,000 feet, pulling out at the correct altitude this time. I homed in on the tankers, keeping the homing needle of the ARA25 a bit to the right of dead center so that it would give me a swinging indication when we neared the tankers. We picked them up visually and plugged in. Taking on a full load of fuel I finished a few seconds before Paul and, knowing he could catch me, climbed in burner to 43,000 feet. We leveled off and picked up speed for the second leg of the journey. Somehow we got no help from the wind. It was later estimated that we had a three knot head wind on the first leg and a three knot tail wind for the second.

It was a beautiful cloudless day. Bright sky and a smooth ride. We went into burner over about the middle of Alabama and picked up our 1.7 mach. I was having difficulty holding it down to 1.7. The planes simply wanted to go faster. Not wishing to modulate, we started a slow cruise-climb holding 1.7. Somewhere over eastern Alabama or western Florida, I called in the routine report to FAA ground control. He asked me to verify the time over the next check point. When I duly verified the short ETA, he stepped completely out of character. He almost shouted, "What in the hell are you flying?"

I smugly replied, "The F-8U-1 Crusader." He didn't reply. As I said before, it was early in the days of supersonics, and I was enjoying the hell out of it.

To keep the speed down to 1.7, we climbed gradually to 47,500 feet. Suddenly, there was Jacksonville, but it was under the biggest, blackest, thunderbumper I ever saw. It towered way above us at our 47,500 feet. I would estimate the top of the build-up was more than 70,000 feet.

Throttling back to idle we commenced our dive to the carrier, rounding the huge cumulus to the north of it. There, laid out before us, was the formation of three carriers, with many accompanying ships, all heading south. The "Sara" was to the outside to the right. Remembering the flag bridge level of the carrier, I pulled out of the dive making about 650 indicated and passed the ship at that level and about one hundred feet out. There it was. Three hours and twenty-eight minutes. A record. I was pleased.

Flashy, eh? After all, the President, Ike Eisenhower, was aboard and alerted and we had to look good.

Then came grim reality. I looked ahead and realized I had an awful lot of speed to eat up somehow. I thought of doing a loop and putting the gear down and the wing up at the top of it, but that would have easily been 11,000 feet at the top and was hardly standard carrier procedure so that was ruled out. I had to make it look like good carrier discipline. How to do it? I could have shut down

CHAPTER 6: "... A NEED FOR SPEED"

Cdr. Bob Dose, Lieutenant Paul Miller, and President Eisenhower on U.S.S. Saratoga (CVA-60) off Jacksonville, Florida, after setting a new transcontinental, carrier-to-carrier speed record 6 June 1957. (Chance-Vought photo)

the engine but that risked the spectacle of not getting a relight and ejecting in front of the President. I decided there was a chance of getting down to speed if I made very hard turns and trusted in my guardian angel. It seemed to be the only sane solution.

So I went just a bit farther ahead of the carrier than normal and honked the plane into a seven-G turn. When I finished the 90 degree turn, I was down to something over 500 knots for the first cross-wind leg. I kept shouting to the poor blameless airplane, "Slow down, you SOB, the President's down there." Entering the downwind leg and again honking it into the turn, I came to the final down-wind leg at about 450 knots. Honking it again in the turn to the final cross-wind leg, the plane came out of the turn at about 300 knots, and the speed brake was finally becoming effective. I remember thinking, "Where were they when I needed them?" The things simply trailed at high speeds.

As I turned final, I was down to 220 knots and could raise the wing, drop the gear, and put flaps down. Half way down the glide slope I was down to the normal 142 knots approach speed. I added throttle for the first time since 47,500 feet, made one violent correction, and hit the deck. You can imagine my big sigh of relief when the hook caught the #3 wire. After all, the President was aboard.

Paul followed me closely, bless his heart, and landed aboard a few seconds later. The whole episode of my hairy carrier landing was reflected some time later when things quieted down a bit. The capable VX-3 Landing Signal Officer, Lt. Ken Sharp, remarked quietly, "Skipper, that was a dilly."

We folded our wings, and parked our aircraft and we were immediately surrounded by loud and rude press photographers and reporters. Wading through the mess, we were ushered to the base of the island, where we were greeted by the President with his famous big grin. He had come all the way down from the penthouse to the flight deck to greet us. He was surprisingly short, about 5 feet, 8 inches or so. He held his hand out to shake mine. In my exuberance of the moment, I squeezed his hand hard and heard it go crunch. He winced.

However, he recovered or perhaps covered quickly and said graciously, "You fellows got here almost before you took off, didn't you?" His question threw me, but I think he was confusing east-to-west time-zone travel with our west-to-east trek. He seemed, however, to be genuinely pleased and asked a few quite pertinent questions concerning speed, altitude, and performance, all the while smiling that big smile. He was a truly impressive gentleman. I was very proud to be serving under that Commander-in-Chief.

The carrier was a mish-mash of reporters and photographers, They pulled and pushed and tried to order us around. I got tired of it. I called pri fly and asked if they could fire us off so we could return to our home at NAS, Atlantic City. We didn't even check our engines. They had been running well for three and a half hours and it was only an hour or so to home. We launched on the angle cats and went proudly home.

Major John Glenn's Record

The idea for a cross-country speed record attempt came to Major Glenn while he was stationed at Naval Air Station, Patuxent River, Maryland as a project test pilot at the Armament Test Division of the Naval Air Test Center. He had been working on weapons test problems with the new Crusader at Pax River, as it was called, and, after being transferred to a staff assignment at the Navy's Bureau of Aeronautics in Washington, D.C. began giving the project serious thought.

Naturally, during this golden era of tactical aviation there was great competition between the U.S. Air Force and the U.S. Navy in matters aeronautical. Each service was working hard at developing new fighters and attack airplanes. What greater one-upsmanship could there be then, than to set some kind of world record with one's latest developments? The existing record for the flight from L.A. to N.Y was established by a U.S. Air Force F-84F, piloted by Lieutenant Colonel Robert Scott, which made the crossing in 3 hours, 44 minutes and 53.88 seconds. Glenn knew that his idea

46 PART II: THE EARLY YEARS

CHAPTER 6: "... A NEED FOR SPEED"

F-8U-1 at left and Glenn's F-8U-1P flown during Project Bullet. (U.S. Navy photo)

could only be sold on the basis of thorough homework and careful engineering analysis, most of which was performed by the contractor in Dallas, Chance-Vought Aircraft Company. In addition, by way of even more justification, the Bureau of Aeronautics generated a requirement for considerable data on a subject for which there was precious little empirical knowledge . . . sustained high speed jet engine performance.

Jet engines were new. Little was known about how those machines would perform at the extreme temperatures, pressures and loads imposed by high speed flight. Up until that point, fighter pilots used their afterburners very sparingly. Short bursts of afterburner provided the needed thrust to gain the upper hand over an opponent in aerial combat maneuvering. The fuel consumption of those engines in afterburner precluded their sustained use in normal operational flying.

The combined effect of high temperatures and enormous centrifugal loads on engine components such as compressor and turbine blades needed to be studied, tested and evaluated. That was the justification Major Glenn put forward when he tried out his proposal on his superiors. Finally, his presentation was given to the commander of the test center, Rear Admiral Thurston B. Clark. To John's utter delight it was accepted and approved. There was still a great deal of engineering homework to be done . . . but the project was off and running. I suspect the admiral had the vision to recognize the boost that an F-8 speed record would give to the Navy's efforts to develop its own tactical aircraft in the early stages of the jet age.

The Federal Aviation Institute was the organization which certified record attempts, and the National Aero Club in Washington,

OPPOSITE: *Major John Glenn in the F-8U-1P flown during Project Bullet. (U.S. Navy photo)*

D.C. was the donor of the award . . . the coveted Collier Trophy an annual award for aviation achievement. The specification of the record required that the air platform take-off from anyplace within 25 nautical miles of the city hall in Los Angeles and land at any place within 25 nautical miles of the city hall in New York. The Federation Aeronautique Internationale (FAI) then took the elapsed time achieved between those two places and calculated an average speed achieved. Then, it applied that average speed to the exact distance between the two city halls thus coming up with the exact course time. This adjusted time was then certified as the record time . . . if indeed it exceeded the existing record.

Of course, the key to success was in the aerial tankers. Those were the airplanes whose support would make or break the record attempt. Obviously, the tankers which were needed were the new KC-135 Air Force tankers (Boeing 707s modified for the tanking mission). Unfortunately, the Air Force explained to the Navy, they could not be made available to the Navy for the record attempt because of operational commitments. Correctly interpreted, the Air Force's refusal to help meant that there was no way they would help the U.S. Navy set any record which they themselves coveted. After all, it was rationalized, the Air Force's business was airplanes . . . the Navy's business was ships.

The only tanker available to the Navy was what was called the "Widow Maker", the North American AJ2. It was a venerable machine and an oddity of the times. It was a propeller-driven airplane with an extremely limited capability in altitude, speed and giveaway fuel capacity. It was a carrier-based tanker and apparently was going to be the tanker of necessity for the Navy's attempt at the record . . . if indeed an attempt were to be tried at all. The "Widow Maker" had gotten its name for the great number of widows it made from the Navy pilots who were killed trying to operate it from aircraft carriers. It was a three place heavy attack bomber which had been converted to a tanker as an ancillary mission. In addition to its two R-2800 radial reciprocating engines, The "Widow Maker" also had a small jet engine mounted in the fuselage with a tailpipe extending beyond the empennage. The jet engine was supposed to give the bomber a speed and altitude capability during the weapons delivery phase of its bombing profile which would improve its vulnerability to enemy fighters. In actual fact the puny engine added more weight and complexity to the airplane than the performance enhancement was worth.

One could say that the AJ2 was an engineering marvel in its time. However, by 1957 it was ancient and certainly not an adequate substitute for the marvelous Boeing KC-135 which could climb to 30,0000 feet and cruise at a mach number which was comparable to the cruise regimes of the new breed of supersonic fighters like the F-8 Crusader and the F-100 Supersabre. The Navy tankers came from two squadrons, Heavy Attack Squadron 6 (VAH-6) at NAS North Island, San Diego and Heavy Attack Squadron 11 at NAS Sanford, Florida.

Fortunately, the BUAER and Chance-Vought engineers knew that the enormous internal fuel load in the Crusader and its impressive maximum speed were more than a match for the F-100, even given the advantages of the Air Force tanker over the "Widow Maker." They knew that any record set by a Crusader was unlikely to be beaten by a Supersabre. So the team of Chance-Vought and the U.S. Navy went to work in earnest.

The track of direct route of the 2,456 mile flight and the estimated range of a Crusader in full afterburner called for three in-flight refuelings; one over Albuquerque, New Mexico, another over Olathe, Kansas and the last over Indianapolis, Indiana. It was obvious to John Glenn and the planners that extremely close coordination between the ground control team, the Crusaders and the refueling groups would be the key to success or the cause of failure of the entire mission.

In the late 1950s there was a network of what were called ground control intercept (GCI) radar sites all over the country. Although they were established for the North American Air Defense Command (NORAD), they were to become the basis for what we know today as the air route traffic control system used by the Federal Aviation Agency for commercial traffic control in the continental U.S. These radar sites offered only rudimentary radar coverage and control for what had now been officially named "Project Bullet." Some news media person calculated that John Glenn's record speed would amount to an average velocity greater than that of a .45 caliber bullet fired from a handgun at a velocity of 586 miles per hour. The GCI sites would not be able to provide the fine positioning control for the precision which everyone on the team knew would be essential if "Project Bullet" were to be carried off.

Major Glenn correctly concluded that what was absolutely necessary was good weather and clear visibility. He practiced rendezvousing with the AJ2 tankers using the radar sites to give him

Glenn in F-8U-1P during Project Bullet. (U.S. Navy photo)

rough positioning, then the VHF direction finding equipment on the tankers to give him direction information to the tanker. Visual acquisition was essential as soon as possible for the descent and approach during which deceleration to the tanker's speed had to be accomplished neither too early nor too late. He estimates that he made approximately fifty practice tanking runs. The Crusader was not the easiest airplane to aerial refuel. The in-flight refueling probe was positioned about thirty-six inches to the left of the pilot's left eye. It was also slightly above and behind the normal pilot field of view. So, it was necessary to fly an approach on the tanker's refueling basket based upon a learned visual picture of the tanker. It took practice, and John Glenn knew that better than anybody. He would have to perform three tankings flawlessly. Most fleet Crusader pilots experience a little fencing with the tanker's refueling basket every once in a while. If there is a little turbulence, the basket tends to dance around a bit making it substantially more difficult. Glenn knew that the fuel remaining at the end of each supersonic leg would be so critical that it would permit no delays. Everything had to go perfectly.

As I review the mechanics of "Project Bullet" with the hindsight of 1994 radar technology, fly-by-wire flight control systems, automatic flight stabilization technology and everything that has been accomplished in the field of electronics in the intervening thirty-seven years, I am absolutely amazed that the team pulled it off so smoothly. The operative word, I suspect, is "team." That had to be the key to success.

An incident occurred during one of John Glenn's practice tanking sessions which points out, better than words I could choose, the inherent hazards of jet aviation in those rough-and-tumble days of iron men and fickle airplanes.

John and the tanker crews had developed a technique of referencing a prominent geographical feature on the ground, like a bend in a river, and describing each other's position in reference to it for

CHAPTER 6: "...A NEED FOR SPEED"

final stages of the rendezvous. They had accomplished this part of it and Glenn visually acquired the tankers. His deceleration profile was letter perfect and the sleek Crusader coasted smartly up to the extended refueling basket. The transition from 1.35 times the speed of sound at 50,000 feet to 190 knots at 26,000 feet went perfectly.

Without a lost second, Glenn made a perfect refueling attempt and the probe hit the basket dead center. Glenn added power and settled into position to take a token amount of fuel before breaking off for another practice attempt. At precisely this moment Major Glenn noticed a puff of black smoke come from the starboard engine of the AJ2. There followed a series of puffs. Then the same things happened to the left engine. By this time Glenn had backed out from the basket and moved to a comfortable distance so he could watch this strange phenomenon.

The pilot announced to Glenn that he had some kind of a problem. He wasn't getting full power from either engine and had begun a gradual descent. With a full tanker store full of jet fuel (and no way to dump it) the tanker pilot had no recourse but to descend, evidently hoping to find a lower altitude where he could maintain level flight.

Major Glenn, flying wing on the troubled airplane, kept a running commentary on the smoke trail now being left by the engines. The airplane kept descending. The pilot seemed unable to determine the cause of the loss of power nor solve it. Finally, in exasperation the pilot warned Glenn that they might have to bail out if they could not level out. Passing three thousand feet, terribly low for a manual parachute bailout from the "Widow Maker", the pilot made the tough decision to leave the airplane. The two crewmen stepped over the side. Then the pilot trimmed the airplane nose-down to cause it to crash in an uninhabited area, and left the stricken airplane. In John Glenn's words, "... the airplane exploded like an atomic bomb." All of the jet fuel in the tanker store went up like a fourth of July fireworks display.

When Major Glenn and his flying partner Lieutenant Commander Charlie Demmler were finally satisfied with their rendezvous and refueling proficiency they began making final preparations for the flight which was set for 6 July 1957. As mentioned previously, weather was going to be the determining go-no-go criterion. The 6th of July came and went. The weather wasn't good enough. Several more times the launch was canceled for weather. Finally, on 14 July an enormous high pressure air mass crossed the west coast headed east. Gradually the entire United States opened up. From coast to coast Americans were looking up at clear blue skies when John Glenn and Charlie Demmler launched from Los Alamitos Naval Air Station near El Segundo, California. "Project Bullet" was finally under way!

The first leg from Los Alamitos to the tanker rendezvous over Albuquerque, New Mexico went like clockwork. The weather was perfect. There was not a cloud in the sky over the great American desert as the two Crusaders took off in full afterburner and climbed at a constant indicated mach number of 0.9 through 38,000 feet. They eased the nose over at that altitude and in just moments went supersonic. Now, with sound left behind them they began to accelerate to an indicated mach number of 1.35 times the speed of sound.

At this point the airplane wanted to accelerate past mach 1.35 so the pair of Crusaders began what is called a cruise climb. They left the power at full afterburner and, holding a constant 1.35 on the mach meter, began a steady climb. The airplanes had already passed 50,000 feet when the computer-generated flight profile called for the commencement of a descent to refueling altitude in just a few minutes. At 50,000 feet the air is so thin that the observed airspeed indicator only registers one half of the true airspeed of an airplane. The air temperature is over 55 degrees below zero. A human being exposed to the lowered pressure without some sort of protection would die instantly as the air captured in his blood turned instantly to vapor causing the blood to boil and the brain to turn to jelly. The sky is also an unearthly blue; deep, cobalt, exquisite blue. The Crusaders were clicking off a nautical mile across the face of the planet every three and one half seconds. It was like some giant travelogue rolling under the airplanes as the vast California desert changed to the Arizona desert then changed rapidly to the New Mexico desert. Then it was suddenly time to descend.

This was the first anxious moment for the two aviators. There could be no time wasted in finding, rendezvousing, plugging in to the tankers, transferring fuel, disconnecting and, lighting afterburners, climbing back to 38,000 feet to accelerate to supersonic speed again. It was a now or never deal . . . they both knew. The exquisitely intricate business of putting the in-flight refueling probe in the dead center of the refueling basket on the very first try required the fine hands of a brain surgeon. Nothing could be messed up . . . not even slightly. But, this is where Charlie Demmler came to grief.

While John Glenn executed the first of three perfect refueling attempts, Charlie Demmler engaged the basket of the second tanker. The in-flight refueling probe hit the basket a little off center and with a little bit of side load induced by a random bit of clear air turbulence. The refueling probe was bent and wouldn't retract back into the recess in the port fuselage. The record attempt was over for Charlie Demmler. With the refueling probe stuck out in the airstream the airplane was limited to speeds far below the speed of sound. A dejected Charlie Demmler bingoed into Kirtland Air Force Base in Albuquerque, New Mexico and climbed out of his airplane.

Meanwhile, far above him and several hundred miles to the east, a single, lone Crusader plowed along through soundless air toward the second tanker rendezvous over Olathe, Kansas. That tanking evolution was as flawless as the first and, after taking on a full load of fuel, the sleek Crusader backed away from the tanker. Once clear of the basket a long sheet of flame shot out of the tailpipe as the afterburner lit and the fighter accelerated rapidly away from the tanker. They wished the young man in the cockpit good luck and God speed as his airplane began its long, graceful upward arc,

Glenn in F-8U-1P during Project Bullet. (U.S. Navy photo)

decreasing to nothing but a tiny black speck in a few seconds . . . then nothing but a long, thin, white vapor trail far ahead and above them. The third refueling didn't go nearly as well as the first two for John Glenn and he began to wonder whether his streak of luck was running out. The human body is an amazing creation. It can perform at super levels aided by adrenalin but, after a few hours fatigue begins to set in and only enormous self-discipline and personal concentration can keep the performance levels at high output for extended periods. But, luck is also important.

The visibility which he had enjoyed for the rendezvous' over Albuquerque and Olathe was degraded a little for his third and final refueling over Indianapolis, Indiana. The earth background and physical features which he had used so effectively in effecting early visual acquisition were not nearly so helpful over Indiana. Fuel was also a problem. The numbers given him by the Dallas engineers had been very accurate for the first two legs. But, as he flashed across the Indiana border he was several hundred pounds of fuel below that which had been predicted.

This time it was not just time and speed which were in jeopardy. This time he was running low on fuel. A foolhardy decision in the interest of pressing on with the record attempt could very easily cost him his life. This was added stress when he least needed any more. He had descended well below that altitude where he had spotted the tanker in the earlier two attempts . . . and still there was no tanker in sight. Finally, with an enormous sigh of relief, Major Glenn spotted the pair of airplanes and slid neatly into position for an approach to the tanker. He was perilously low on fuel and knew that if he messed up the tanking evolution, there probably would not be enough fuel remaining to make it even to an ignominious landing at the nearest emergency airfield.

Concentration was again the deciding factor and, with another enormous sigh of relief, he noted with pleasure, the refueling probe slip into the dead center of the refueling basket. Home safe again, he thought as he saw the green refueling light illuminate and felt the surge of fuel pouring into his near empty tanks. The fuel gauge read 400 pounds when the needle stopped its descent across the dial and began an inexorable climb back upwards. But, two thirds of the way through the refueling process the tanker began to act up. Fuel pressure and flow dropped way below its design levels and the rate of climb of the fuel needle slowed noticeably. The tanker pilot was all apologies. Would the Major like to back out from the probe, re-engage and try again? Or would the major like to back out and

CHAPTER 6: "... A NEED FOR SPEED"

try the other tanker? A quick decision was needed. Time was running out. The sweep second hand was marching inexorably around the dial of the eight day clock on his instrument panel. A decision was needed now!

John Glenn made the decision. He was short almost 1,500 pounds of fuel and it looked like the tanker could not transfer the remainder fast enough to him to make it worth while. There wasn't time to try the other tanker. His record clock was ticking. Checking one more time the numbers on his kneeboard, the estimated tailwind for this leg of the journey and the fuel reserves predicted by the engineers for his arrival at destination, he made his decision . . . press on! He would go with what he had and take his chances! The Crusader backed away from the refueling basket, the sheet of flame shot out from the tailpipe and John Glenn was off and on his way into the history books . . . one way or another!

Mrs. Glenn had gotten a call from her son, telling her that he would be coming by very shortly, but nine miles above the family home in New Concord, Ohio. He also told her he would be coming by in excess of the speed of sound and maybe, just maybe, if she listened she might hear the sound of her son's passage. Predicting sonic booms in those days was a black art. Mrs. Glenn just happened to mention the fact to a few of her friends and neighbors in and around New Concord, Ohio.

The generation of a supersonic shock wave is dependent upon many factors; like temperature, density, pressure of the air; and the altitude, speed and the general aspect of the velocity vector of the airplane. In general, it was thought that if the airplane stayed above thirty thousand feet the sonic boom would have a negligible effect on the populace on the earth's surface.

But, the telephone rang in Mrs. Glenn's home at the appointed hour. It was from Mrs. Gilooley, a neighbor. She informed Mrs. Glenn that her son Johnny had just dropped a bomb somewhere nearby!

There was a small post office in a small town in Pennsylvania whose windows were shattered by John Glenn's sonic boom!

The rest of his flight is now history. But, the anxiety over the fuel problem bugged him to the bitter, cliff-hanging end of the flight. Officials from the Federation Aeronautique Internacionale (FAI) were to certify compliance with the rules for the setting of such records. In order to ensure that he had complied fully, John Glenn had to literally fly though the eye of a needle. He had to pass through a gate which extended up from the runway at Naval Air Station, Floyd Bennett, New York to a height of no more than 200 feet. At a speed of seven hundred miles per hour a height of only 200 feet can seem perilously close to the ground. John literally held his breath as he flew his Crusader successfully through the needle's eye and then looked with horror on a fuel gauge that read almost empty.

There was time only to throttle back to idle power, execute an enormous chandelle turn to the downwind leg, hold his breath until the airspeed decreased enough for him to lower the wheels and raise the wing. Not daring to touch the throttle, the intrepid Major planted the Crusader right on the numbers. The new world record had just been set with a time of three hours, twenty-three minutes and eight point four seconds. When the verification officials calculated the final numbers, John Glenn had averaged a speed of 725.55 miles an hour over the 2,456 mile course and had beaten the existing record set by the U.S. Air Force by almost twenty-one minutes. This, of course was just prologue for the records he would set in space. But, the greatest accolade that could be passed on for his performance that day came from a veteran Naval Aviator who reviewed the record of the flight.

"He flew a perfect mission!"

Author's Note: There are numerous interesting sidelights of a story which are often unearthed inadvertently in the course of doing normal research. In the case of John Glenn's famous flight I learned that the F-8U-1P (RF-8A) in which he made history was returned years later to NAS Key West during the Cuban Missile crisis in 1962 from NADC Johnsville and was actually flown by Tad Riley, the hero of the chapter called "Blue Moon." Ten years later it returned to the fleet a second time as a refurbished RF-8G. The only distinguishing feature of the airplane was a small plaque affixed to it which identified the airplane as the original record-breaking airplane. From an interview with "Turtle" Redditt, one of Naval Aviation's more distinguished landing signal officers, the airplane in question showed up for recovery one warm afternoon aboard the U.S.S. *Oriskany* in the Tonkin Gulf in 1972. The pilot was Tom Scott who began what had become known as a classic F-8 carrier arrested landing pass. The LSO's notebook showed a high fast start with an over correction on the glide slope along with a deceleration at midpoint in the approach. Then the two critical remarks appeared in LSO's shorthand: NEPIG, EGAR (not enough power in the groove, eased gun at ramp). The final, remark was added after the smoke, flame and flying debris had abated. It said, simply: RAMP STRIKE! John Glenn's famous airplane which probably ought to be hanging in a museum, sits instead in a bed of prehistoric silt under a few hundred fathoms of Tonkin Gulf water. Not the end one would reserve for a historic, record-setting airplane . . . but a monument, nonetheless, to the intrepid young man who took it on its last flight!

7

Patuxent River

Patuxent River is to Naval Aviation as Edwards is to the U.S. Air Force. Two of my first projects, after graduating from the test pilot school and reporting in to the Flight Test Division, were to certify contractor compliance with the spin requirements and the structural demonstration points of the F-8U-2.

The PLAA (positive low angle of attack) and PHAA (positive high angle of attack) demonstrations were simply structural data points which the contractor was obliged to demonstrate prior to continuing with full rate production of the airplane. These two points were very clearly specified in the contract. The company pilot, a man named Jack Walton had a reasonably simple task . . . to fly the demonstration point. For this he was paid a handsome bonus.

Jack had only one problem. He must not exceed the demonstration point . . . and for a very good reason; it might cost him his life! That, of course, is why he received the bonus in the first place. One of the demonstration points for the F-8U-2 was 825 knots equivalent airspeed and 6.75 "g"s! The design engineers at Dallas would not guarantee that if the demonstration point were exceeded by so much as a tenth of a "g", the airplane would not disintegrate into a puff of smoke and a thousand tiny fragments.

As in all structural demonstrations, the test pilot went through a series of flights during which he worked up to the demonstration point in increments. The work-up flights were conducted at the contractor facility at Dallas, Texas. When the pilot declared himself ready to go for the big point, the airplane was transported to the Naval Air Test Center at Patuxent River, Maryland, where he was required to demonstrate the data point under the watchful eye of a Navy test pilot who would be the certification officer. That's where I came in. I was the certification officer.

The demonstration airplane had an instrumentation package which was supposed to verify the data point. It was carefully examined prior to flight, sealed and, after the flight landed, the package was unsealed and examined for proof of compliance.

The data point for PLAA, naturally, was out at the very edge of the flight envelope . . . in fact, slightly beyond that permitted by the Navy pilot's handbook – NATOPS.

As certifying officer. I had to be airborne and in the reasonable vicinity of the demonstration airplane when it made the demonstration point. To do this I was given an F-11F-1 airplane. My airplane would not go that fast. So, Jack and I practiced a series of build-up flights in which we both arrived over the same spot on the eastern Maryland shore at the point where he made his 6.75 "g" pull up. He started over the Virginia Capes at 40,000 feet and I was fifty miles ahead of him. We both went into full afterburner and flew separate profiles which put us over the same place at the same time.

I was always able to pick him up visually a few miles away and we would gradually come together at a geographical point, at an altitude of 1,500 feet over the ground and 825 knots (KEAS). Of course, I was 100 knots slower than he was. But, I was close enough to observe the maneuver and then escort him back to Patuxent River where an inspecting officer opened up the data package and extracted the necessary proof of compliance.

I always wondered why there were not more noise complaints called in from the poor inhabitants of the eastern shore. The supersonic booms from such a high speed, low altitude pass must have been spectacular!

There was always something curious about the assiduous way Jack checked the weather prior to each flight. In particular, he checked on the surface winds in the area of the eastern shore of Maryland where he would make his pull-up maneuver. Anytime the meteorologist estimated that the surface winds would be in excess of fifteen knots, he canceled the mission. Finally, I asked him why. He answered that company policy required it. When I persisted and asked why, he told me that a parachute landing in wind conditions exceeding fifteen knots would be hazardous . . . sprained ankles and broken legs and such. After several cancellations for these reasons I became frustrated and told him there would be very little left of him hanging in a parachute if something went wrong at 1,500 feet over the ground and 825 knots!

Jack finally finished off the PLAA demonstration and went on to the simpler, and less hazardous PHAA maneuver. When he was finished he returned to Dallas . . . and an even more colorful company pilot, Don Schultz arrived.

52

CHAPTER 7: PATUXENT RIVER

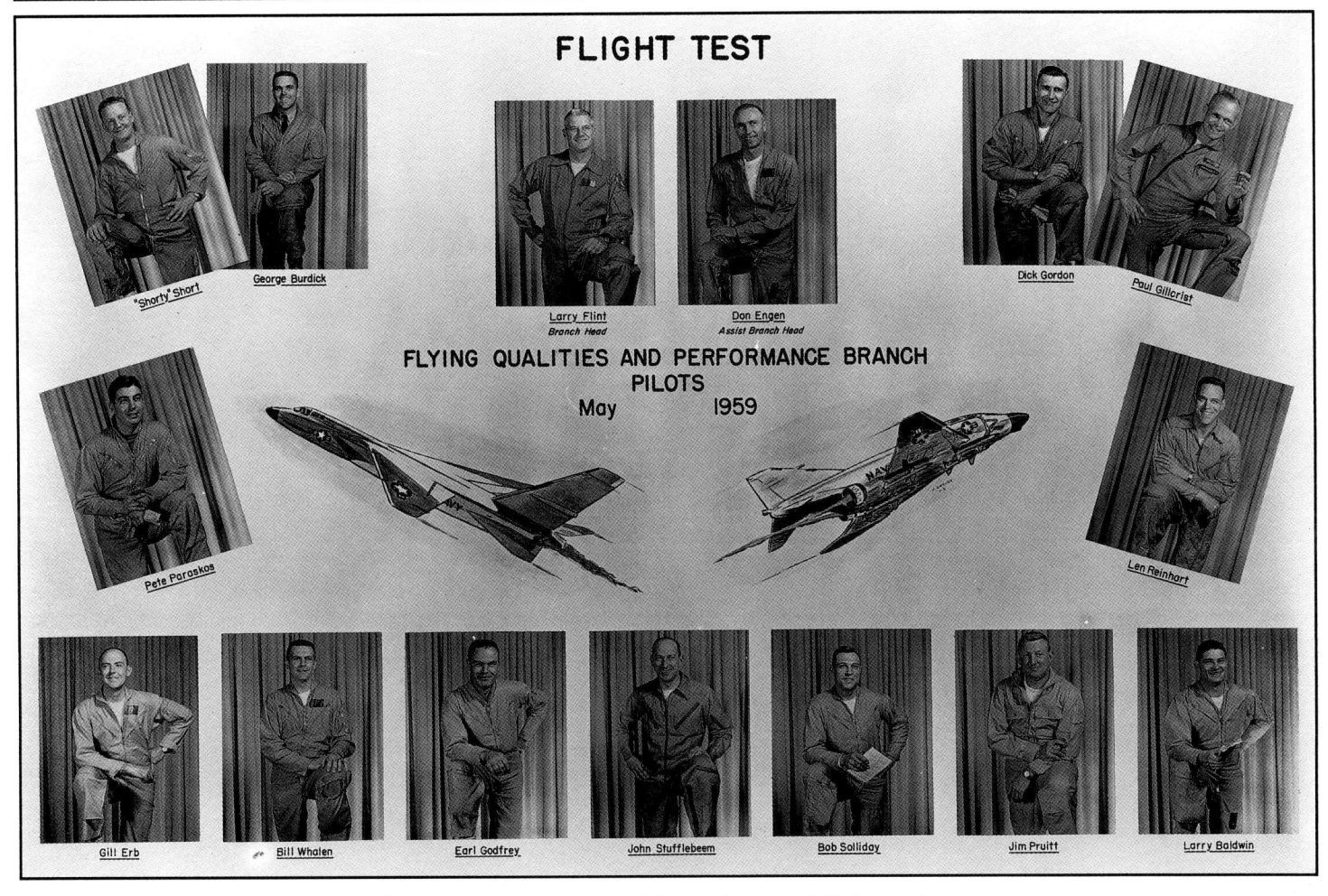

Carrier branch (newly formed F, Q & P) roster. (U.S. Navy photo)

Schultz was the F-8U-2 spin demonstration pilot . . . and he was a piece of work! I will never forget our first meeting. He showed up in the conference room at the Flight Test Division and met the division head, Captain Bob Elder as well as the branch head of the Flying Qualities and Performance branch, Larry Flint.

As a Lieutenant in those days, I thought Captain Elder was next to God in the pecking order. Therefore, I was duly horrified when Don Schultz, having looked at the Navy's list of spin entry requirements, announced in a loud voice:

"Anyone who deliberately hammerheads an F-8 is a stupid Shit!" Bob Elder never batted an eye as he responded that if a maneuver could be performed, inevitably a young fleet pilot would perform it, albeit inadvertently. (It was a sort of a variation of "Murphy's Law"). Therefore, Bob concluded, we need to know if that young pilot will be able to recover from the ensuing maneuver, and, if so, how?

Needless to say, Don Schultz demonstrated a hammerhead maneuver and successfully recovered from it. I certified that maneuver also. In fact, I certified all ten of the spin demonstration flights in the F-8U-2 which occurred between 2 February and 10 March 1959. In the course of those ten flights we became close acquaintances.

Don Schultz was an unusual guy. Unbeknownst to me, when he showed up for the spin demonstration program, he asked for a spin familiarization flight in the T-28 which the test pilot school owned. This was granted on a weekend (a Saturday) so that he could be ready to go on Monday morning. The poor bastard from the test pilot school staff who flew in the back seat spent several hours in the air with Don Schultz who, they said, was queer for inverted spins. How many inverted spins he did that Saturday may never be known. I'm glad it wasn't me who flew with him.

The F-8 spin airplane was also something to behold. Externally it had what can only be described as a steel rail mounted to either side of the fuselage from a position below the wing root middle point to a position just forward of the leading edge of the UHT where it meets the fuselage. The rail was intended to keep the spin parachute from pulling the tail off the airplane when it deployed. The spin chute was packed and stowed in a housing in the vertical stabilizer just above the tailpipe. Otherwise, there were no external features which would distinguish the airplane from other F-8s.

But, the cockpit was quite another story. XF-8U-1 Bureau Number 138900 was like no other F-8 in the world. The cockpit had obviously been configured by Don Schultz at a point in time when little was known about how much yaw rate would be generated by a Crusader in a flat spin; and how much that rate would translate to what is called "eyeballs out" "g" ("g" forces that throw the pilot forward against his shoulder and lap restraint). The company spin pilot and test engineers must have been expecting the worst (not a bad philosophy) when they configured the cockpit because it resembled something out of Star Wars.

The first thing I noticed when Don Schultz showed it to me rather proudly was the strange seat restraint arrangement. Instead of the standard four point Crusader system there was an old World War II seat belt/shoulder harness system with another three inch wide nylon strap coming up between the legs. All five points connected up together at the lap belt connection. The strap between the legs, Don explained, was to keep the pilot from sliding out under the lap belt under "eyeballs out" "g" forces. Then there was still another three inch wide strap which came across the chest at the breast level. I forget what Don told me that was for. Then, there was another restraint arrangement at the sides of his helmet holding it against the back of the seat. Don explained that this was to keep the head from being slammed about during a post stall gyration. So much for pilot restraint.

Arrayed across the instrument panel glare shield were four large metal loop handles, each three inches wide. Don explained that the one on the left armed the spin chute. The second one deployed the chute. The third one released it. And, the fourth one was a back-up to the third. Each handle had to be turned ninety degrees before it could be pulled. When I asked why the back-up release handle, Don explained that the airplane would not fly with the chute deployed. It would just go straight down. Not a good way to end one's career, Don commented.

The instrumentation was equally elaborate. On either side of the instrument panel (about halfway up) was a large white light (maybe two inches in diameter). Don told me they were yaw lights. If the one on the right illuminated, the airplane was yawing right. It was easy, Don explained, to get confused as to which way an airplane was yawing especially if it was in an inverted spin.

There were two cockpit altimeters and a green light associated with one of them. The green light was supposed to go out when the airplane descended through 20,000 feet.

But, the "piece de resistance" by far was the "dead man's switch." This was a paddle switch device on the forward side of the control stick grip. This paddle switch was electrically connected to a covered, guarded, two-position toggle switch on the left console which electrically armed the ejection seat! The way it worked, Don explained, was that, prior to entering a spin, the pilot grasped and squeezed the paddle switch holding it closed. Then, with his left hand he armed the toggle switch. Thereafter, if the pilot, for any reason, released his grip on the paddle switch, the ejection seat automatically fired. This feature, intended to save a disabled pilot, horrified me with its implications.

On completion of the demonstration flights, the plan was for the airplane to be turned over to the Navy and a select few pilots in the carrier branch (FQ&P), including me, were going to do an evaluation of the spin characteristics of the F-8U-2 for ninety days.

During the ten flight spin demonstration program, Don Schultz did probably forty spins. My job was to chase him, keep him in sight, keep him clear of other airplanes, count the turns and, above all, to tell him when he had descended below twenty thousand feet. This last thing I failed to do on the one occasion when Don found it necessary to deploy the spin chute. My chase airplane was a Grumman F11F-1 and my technique was to enter a tight descending turn around the spinning airplane, staying level with him, counting turns over the radio and watching his altitude. I even had devised a helmet-mounted camera which I turned on before each spin demo. All I had to do was keep my head pointed at the airplane and the movie camera recorded the maneuver. This turned out to be not nearly as easy as it sounds.

Don was furious at me the one time I failed him. It was a vertical entry with a set of control inputs which I have forgotten. The plan was to hold the pro spin controls for five turns then attempt a recovery. Spin entry was 40,000 feet. An inverted flat spin was generated. At the end of the fifth turn Don applied spin recovery controls and, after two more turns, nothing happened. Meanwhile I was dutifully calling out turns with my thumb holding down the radio button on my throttle. At turn number nine, Don later told me, he passed through twenty thousand feet and armed the spin chute, still holding spin recovery controls.

There was no way he could call me because I had the radio button depressed. At turn number eleven, he deployed the chute and the airplane recovered immediately into a "hanging in the chute" nose-down attitude. Then, using the third handle, he attempted to release the spin chute. No luck! Using the back-up release handle, Don again attempted to release the chute. Still no luck!!! Now, he was getting low for any kind of a recovery and was still hanging in the chute! At that point I realized we were in serious trouble and saw a long sheet of flame come out of the tailpipe. Don had lit the afterburner, and the nylon chute melted like a wisp of spider web. The airplane fell vertically a few more thousand feet and finally pulled out. We went home and he made his usual simulated flame-out approach and landing. He always did this for fear that he may have done some damage to the engine during the spin. His airplane touched down still trailing a length of steel cable. It took him some time before he would speak to me.

"Why didn't you call passing twenty thousand feet?", he demanded. All I could say, rather sheepishly, was that I was so fasci-

CHAPTER 7: PATUXENT RIVER

nated by his maneuver that I simply forgot. Don finished up on 6 March and went home, leaving his airplane behind him as agreed upon.

Dick Gordon was given the spin evaluation project and I was one of his spin pilots. The first thing we did was disarm the stupid "dead man's" switch. During our spin test program we learned a great deal about the spin characteristics of the F-8U-2 (with ventrals). I like to believe that our report which went out to the fleet did some good. Certainly, the number of departure/stall/spin/ejection incidents decreased.

The F-8 had a wild spin characteristic. Depending on the mode of entry (usually it was during high angle of attack flight in air combat maneuvering) the maneuver could vary from relatively benign to horrible. The first thing the F-8 did was enter a post stall gyration until it decelerated to autorotative spin speed of about 170-180 knots. The gyration could be an end over end tumbling motion and was wild to watch . . . even wilder to experience. One of the things we found out was that if the pilot let go the stick during the post stall gyration the airplane would eventually recover (most of the time). So, the method of handling what came to be called a "departure from controlled flight" was taught in both fleet replacement squadrons and was unerringly successful. The method was to reach up to the instrument panel with your right (stick) hand and wind the eight day clock. One had to let go the stick to do that and usually that was all it took.

But, if recovery didn't occur right away, the Crusader pilot was in for the ride of his life. There has never been another Navy airplane that I know of that was quite so wild!

There were other flights in the Crusader during my tour of duty at Patuxent River. In all I accumulated over 100 flight hours in six different models of the Crusader. My log book also shows that, in my three years there I flew 37 different models of aircraft.

In many respects it was what I call the golden age of carrier aviation. During that brief period the U.S. Navy introduced a number of spectacular airplanes. First, there were the Phantom II, called then the F-4H-1; the Super Crusader, called the F-8U-3; the Vigilante, called then the A-5J; the Intruder, called the A-2F-1; the Tiger, called the F11F-1; the Crusader, called the F-8U-1/2; the Skyhawk, called the A-4D-1/2; the Corsair II, called the A-7; the Demon, called then the F3H and the Fury, called then the FJ-4B.

It was a good time to be a Navy test pilot. There was a constant sense of excitement in the air. We were doing important things for naval aviation . . . and we knew it!

8
THE GREAT SHOOT-OUT

In the spring of 1958 the brightest moment of U.S. Naval Aviation occurred like a lightning flash . . . and then died somewhat ignominiously. There is no doubt that more new tactical airplanes were introduced into the fleet in the 1950s that at any other time in the 84 years of carrier aviation. The spate of new airplanes was indicative of the yearning by this country to become preeminent in the field of international aerospace.

Historians are quick to point out that in 1962 the new American president John F. Kennedy threw down the challenge to the American people to lead the world in space by being first to go to the moon. To this country's credit it responded to that challenge and achieved world dominance in space exploration, seizing it from the grasp of the Soviet Union at the very last minute.

But, what historians do not point out is that the decade of the 1950s saw equally difficult, if less dramatic, achievements in the general field of aerospace, and more specifically, the field of tactical aviation.

When I arrived at Patuxent River, the U.S. Navy was in the final stages of this revolution in carrier aviation. The piece de resistance of this revolution was what I choose to call the great shoot-out between the Chance-Vought F-8U-3 and the McDonnell F-4H-1. Indeed, while I was completing the course of instruction at what was then called Test Pilot Training, startling events were occurring without me! I believe all of us in Class 20 felt the same. We all felt we were missing out on all of these watershed events.

Test pilots at the Carrier Branch of the Flight Test Division were busy conducting the NPEs (Navy Preliminary Evaluations) of the F-8U-3 and F-4H-1 almost simultaneously. In fact, follow-on NPEs of these two airplanes placed enormous pressures on the two prime contractors by pitting the two in a final side-by-side shoot-out at Edwards Air Force Base in the spring of 1958.

The Congress directed that only one of the two airplanes could be placed into production in the fleet. The Chief of the Bureau of Aeronautics, Rear Admiral Robert E. Dixon and Rear Admiral Thurston B. Clark, Commander Naval Air Test Center decided to form a team of Patuxent River test pilots who would conduct a joint competitive evaluation of the two airplanes. The idea was that a more valid and credible selection decision could be made if the

F-8U-3 at Edwards Air Force Base, 1957. (U.S. Navy photo)

CHAPTER 8: THE GREAT SHOOT-OUT

F-8U-3 Crusader III, 1957. (U.S. Navy photo)

The 7th of April 1959 persists in my memory with startling clarity. it was one of those glorious spring days that make one feel glad to be alive. The weather was perfect for flying and I was on the flight schedule for a good project flight . . . albeit a short one.

The project pilot for the stability and control testing of the latest model of the Chance Vought Crusader, the F-8U-2NE, had offered me a flight or two in his airplane. He picked a relatively simple flight in recognition of the fact that I had precious few flight hours in the Crusader. It was always a hard thing to break into a new airplane. Most project pilots were selfish about their project airplanes; and also did not want to risk them in the hands of some "ham-fisted" new guy looking for flight time. Billy Lawrence wa

The 1,000th Crusader (F-8U-2NE) rolls off th

same group of pilots evaluated both airplanes. This joint venture came to be known as "the great shoot-out."

The team was headed by Captain Bob Elder, Director, Flight Test Division and included Captain Larry Flint, Head, Carrier Branch (at Flight Test) and project pilots Commander Don Engen, Lieutenant Commanders Bill Nichols and "Ace" O'Neal and Lieutenants Billy Lawrence and Dick Gordon. Nichols and O'Neal had the responsibility for the land-based carrier suitability part of the evaluation. The remainder of the team evaluated the flying qualities and performance portion of the tests.

The team received a preliminary indoctrination at the contractor facilities at St. Louis and Dallas, then proceeded to Edwards where, from September through November 1958, the great shoot-out was conducted. I have referred to this competition as the brightest moment of naval aviation. That statement may be a little bit of hyperbole for a young test pilot born in 1929.

But, if one reflects upon the path which carrier aviation has taken in the 1970s, 1980s and 1990s the conclusion is obvious. The great shoot-out was really the way to develop carrier airplanes. Indeed, the performance numbers generated by the Navy evaluators for the two airplanes showed a remarkable improvement in the brief time between the Congressional direction for a shoot-out and the completion of the competition.

The Navy decision to choose the F-4H-1 Phantom II as the winner over the F-8U-3 Crusader has been second-guessed many times since by knowledgeable critics. Indeed, the chief engineer of the Bureau of Aeronautics, Mr. George Spangenberg has said more than once that the Navy picked the wrong airplane. I remember, as a student at the test pilot school, going on a class tour of contractor facilities. When we got to Dallas, we were briefed by a brilliant young project engineer, named Conrad Lau, on the Chance-Vought entry. Then were regaled at the Phantom plant in St. Louis on the sterling qualities of the McDonnell entry. Our class was as divided perhaps as the Navy was over the relative merits of the two airplanes. I still believe it is fair to say that the decision was based more on the relative merits of the "two engines and two aircrewmen versus single seat and single engine" philosophy than on the specific merits of the airplanes themselves. This is not to say that, in the long run, the Navy made the wrong decision. I am not that prescient!

However, subsequent to my tour of duty at Patuxent River I proceeded on a path in the fleet which made me a Crusader pilot for twelve years, culminating in flying 167 combat missions in southeast Asia. As I go back and read accounts by the NPE joint team of their flights out to mach 2.1 in the F-8U-3 I have to wonder at the big question that it begs.

The F-8U-3 design corrected so many of the more horrible features of the F-8U-1 and 2 that any Crusader pilot's mouth would have to water at the thought of taking the big airplane into combat over North Vietnam.

engine which had been installed in an FJ4-F test bed at Patu
River a year earlier. It was intended for inclusion in the A-3J-1
lante program to give the heavy attack airplane a high speed,
altitude deep strike sprint capability.

In the spring of 1958 the Navy prepared for the upcoming
by designating Lieutenant Billy Lawrence as the F-8U-3 pr
officer; and Lieutenant Dick Gordon as the F-4H-1 project of
The general plan of attack was to have a team of pilots, head
by the project officer fly their airplane exclusively. Then there v
be three pilots, Elder, Flint and Engen, who would fly bot
planes. The individual airplane teams were expected to be a
cates of their own airplanes. The overall trio, Elder, Flint and E
would give the relative comparison evaluation which would,
fully, help in the down-selection process of choosing a winne

The team left Patuxent River for Naval Air Station, No
Virginia in early September for the fitting of full pressure
suits and some high altitude decompression training. The air
were expected to fly to altitudes where such equipment precau
were deemed necessary.

After completion of training at Norfolk, the team went
Louis for training on the airplane systems in the F-4H-1. T
the team went to Dallas for similar indoctrination training c
plane systems in the F-8U-3.

The project pilots, Gordon and Lawrence were each g
"courtesy" flight in the "other airplane" to better help them i
own testing deliberations. After that they were excluded fro
ing anything but their own airplanes.

The flying began in earnest on 15 September 1958 with
scheduled for Bob Elder in the F-4H-1 and Larry Flint in the
3.

Because of a conservative approach by the McDonnell
neers, it was decided to limit the speed to something less than
2.0. The Vought engineers were not that conservative. Th
how the F-8U-3 was first to get to Mach 2.0! It was no big

However, what was perhaps a minor consideration was t
that Larry Flint felt a little under the weather that particular
ing and asked his project pilot, Lieutenant Billy Lawrence
the flight. Billy ran his airplane out to the predetermined Mac
of 2.0 and became the first Navy pilot to fly a Navy airp
Mach 2.0! It had clearly not been planned that way!

Don Engen got his first flights on 16 September by flyi
flight each in both airplanes. Between 16 September and 1(
ber, when the testing ended, he managed to get ten flights i
8U-3 and roughly the same number of flights in the F-4H-1.
one of Don Engen's flights in October he ran into a problem
probably had far-reaching implications for the ultimate outc
the competition

During supersonic accelerations, the F-8U-3 went throug
cal airflow in the engine intake duct between Mach 1.25 an
On this particular flight, he felt a rumbling sensation in the

10

A RECORD OF SORTS

I was busily working my way through some paperwork which I could no longer, in good faith, put off. Suddenly, I became aware of someone standing at my desk in the austere office in the Carrier Branch of the Flight Test Division. It was Lieutenant Commander Jake Ward and, as usual, he was grinning. The question which he softly drawled would turn a lackluster day into one that could have put my name in the aviation history record books, but for lack of an appropriate category.

"Want to go do some flying?," he asked. The look on my face was answer enough.

Jake had many of the really good projects in the carrier branch. Its name notwithstanding, the Carrier Branch did all of the flying qualities and performance testing of the Navy's stable of fixed wing carrier airplanes. Another branch in the Flight Test Division called Carrier Suitability, did all of the catapult and arresting gear work. Today, Jake wanted to finish up a project which entailed testing the stability and control of the photo-reconnaissance version of the Crusader, the F-8U-1P. The fighter version of the airplane, the F-8U-1, was already in the fleet. And, it had already been determined that the directional stability of the fighter began to degrade as mach number increased through 1.5; or one and one half times the speed of sound. Furthermore, the directional stability degraded even more drastically, at these speeds, if the angle of attack of the airplane were increased even slightly; as would occur in maneuvering flight.

Although the stability of the airplane was augmented, about the pitch and yaw axes, by an electronically controlled automatic flight stabilization system, a certain degree of unaugmented (inherent) stability was a government design specification. As such, it had to be demonstrated. Plans had already begun by the airplane's designers in Dallas to improve the inherent stability of the airplane by the addition of ventral fins. These were being completed for the fighter version. However, the photo-reconnaissance version of the airplane had some unique features which aggravated the directional stability of the basic airplane.

The photo-reconnaissance modification involved removing the four twenty millimeter cannons in the forward fuselage section of the fighter and replacing them with a bay full of cameras and photographic equipment. This meant radically changing the cross sec-

F-8U-1P chase aircraft, NATC Patuxent River, Maryland 1958. (U.S. Navy photo)

9
NEAR MISS

The 7th of April 1959 persists in my memory with startling clarity. it was one of those glorious spring days that make one feel glad to be alive. The weather was perfect for flying and I was on the flight schedule for a good project flight . . . albeit a short one.

The project pilot for the stability and control testing of the latest model of the Chance Vought Crusader, the F-8U-2NE, had offered me a flight or two in his airplane. He picked a relatively simple flight in recognition of the fact that I had precious few flight hours in the Crusader. It was always a hard thing to break into a new airplane. Most project pilots were selfish about their project airplanes; and also did not want to risk them in the hands of some "ham-fisted" new guy looking for flight time. Billy Lawrence was different, in that regard, and saw the look of yearning in my eyes whenever he caught me crawling around his airplane.

So, today was the day to gather a time history of an acceleration run from two hundred knots to Mach 1.8, almost twice the speed of sound. All I had to do, Billy told me, was to stay within fifty feet of the designated test altitude. Easy enough, I thought, as I left the line shack and walked across the warm concrete parking ramp toward that beautiful looking airplane. I recall thinking happily that this was the best of all possible worlds . . . and it was!

The climb to forty thousand feet was accomplished in basic engine (no afterburner) in order to conserve fuel for the acceleration run. The Patuxent River course rules, at that time, called for all supersonic speed runs to utilize a specific radial of the Patuxent

The 1,000th Crusader (F-8U-2NE) rolls off the assembly line in Dallas, Texas. (Chance-Vought photo)

F-8U-1 loaded with 5" ZUNI rockets, NATC Patuxent River, Maryland 1956. (Art Schoen, Chance-Vought)

River TACAN navigational facility. The radial ran southeast from the facility, down the Chesapeake bay, crossing the tip of the southern Maryland peninsula just north of the Hampton Roads tunnel. This procedure supposedly kept the sonic boom away from populated areas as much as possible; and gave the test airplanes plenty of room for speed runs.

Sonic booms were a relatively new phenomenon; and weren't yet being taken too seriously. Just after I passed over the Patuxent TACAN station headed outbound, I turned on the oscilloscope panel that would record the time history with wiggly lines on a piece of paper in the telemetering station back at the hangar.

I put the airplane precisely at forty thousand feet, exactly on the specified radial and exactly at 200 knots as I ran the throttle to full afterburner. As the airplane accelerated I knew I would have to stay ahead of keeping it trimmed for level flight and not stray more than fifty feet from the test altitude. If I messed this flight up I knew that Billy would never offer me another opportunity to fly any of his airplanes. This is going to be a flawless flight I vowed.

CHAPTER 9: NEAR MISS

Billy is going to be so impressed that next time he will offer me something more challenging. Little did I suspect that, in a few minutes I would come perilously close to never having "a next time"!

The acceleration run was progressing well. The altimeter needle bobbled a little as the plane went supersonic but, that was to be expected. Anyway, the altimeter needle stayed within limits. Just as the mach meter needle was passing 1.6, well down the Chesapeake Bay by now, something caught my attention, a tiny black speck directly in the center of the windscreen!

Time froze! Before I could translate perception to action, the black speck grew to monster size, rapidly filling the windscreen. It was the head on silhouette of a Crusader! It grew beyond the windscreen and passed directly overhead my airplane... so close that I knew we would touch as we passed. Still, my brain had not had time to send a message through my central nervous system to make muscles move and cause the control stick to take evasive action. Just as suddenly, the airplane was gone! A split second later there was a loud explosion as my airplane passed through the other plane's shock wave. The jolt caused the "g" meter to peg in both directions... positive and negative. I think my heart stopped beating for several seconds.

The Crusader must have passed no more than ten feet above me. He was obviously staying on altitude as I was doing. Only, I wondered what the hell he was thinking of making his speed run in the wrong direction. I checked afterwards, and learned that the pilot, a new guy also, from another test directorate had misread the course rules!

Ever since that day, I have wondered, what it would have felt like if we had collided! Would there have been a microsecond of pain, before my airplane and I were turned into tiny fragments of metal, fuel and human tissue? Who knows? In the life of a test pilot, there isn't any place for such thoughts... I knew that. So, I put it out of my mind.

But, every once in a long while, I remember that day, that split second of horror. There is one thing for certain... it would have been a spectacular ball of fire!

10

A Record of Sorts

I was busily working my way through some paperwork which I could no longer, in good faith, put off. Suddenly, I became aware of someone standing at my desk in the austere office in the Carrier Branch of the Flight Test Division. It was Lieutenant Commander Jake Ward and, as usual, he was grinning. The question which he softly drawled would turn a lackluster day into one that could have put my name in the aviation history record books, but for lack of an appropriate category.

"Want to go do some flying?," he asked. The look on my face was answer enough.

Jake had many of the really good projects in the carrier branch. Its name notwithstanding, the Carrier Branch did all of the flying qualities and performance testing of the Navy's stable of fixed wing carrier airplanes. Another branch in the Flight Test Division called Carrier Suitability, did all of the catapult and arresting gear work. Today, Jake wanted to finish up a project which entailed testing the stability and control of the photo-reconnaissance version of the Crusader, the F-8U-1P. The fighter version of the airplane, the F-8U-1, was already in the fleet. And, it had already been determined that the directional stability of the fighter began to degrade as mach number increased through 1.5; or one and one half times the speed of sound. Furthermore, the directional stability degraded even more drastically, at these speeds, if the angle of attack of the airplane were increased even slightly; as would occur in maneuvering flight.

Although the stability of the airplane was augmented, about the pitch and yaw axes, by an electronically controlled automatic flight stabilization system, a certain degree of unaugmented (inherent) stability was a government design specification. As such, it had to be demonstrated. Plans had already begun by the airplane's designers in Dallas to improve the inherent stability of the airplane by the addition of ventral fins. These were being completed for the fighter version. However, the photo-reconnaissance version of the airplane had some unique features which aggravated the directional stability of the basic airplane.

The photo-reconnaissance modification involved removing the four twenty millimeter cannons in the forward fuselage section of the fighter and replacing them with a bay full of cameras and photographic equipment. This meant radically changing the cross sec-

F-8U-1P chase aircraft, NATC Patuxent River, Maryland 1958. (U.S. Navy photo)

CHAPTER 10: A RECORD OF SORTS

F-8U-2N flight test project aircraft, NATC Patuxent River, Maryland, 1957. Note infra-red search and track dome in front of cockpit, and Bullpup air-to-surface missile. (Courtesy Art Schoen, Chance-Vought)

tional shape of the forward fuselage to a flat sided, flat bottomed configuration. The flat surfaces contained the windows through which the cameras looked.

Wind tunnel tests showed that this configuration made the directional stability problem of the basic airframe even greater. As mentioned earlier, the ultimate solution was ventral fins. These were two fixed planes extending downward and outward from the bottom of the rear fuselage section; and looked not unlike two arrow feathers. But, design changes took time and the operational capability provided by the F-8U-1P was badly needed in the fleet. So, Jake's project included the examination of the stability of the F-8U-1P in the supersonic region.

A subset of that project was to examine the effect on the airplane's stability of a supersonic shock wave generated by another airplane in formation. Today's flight was the final flight of this subset. The collection of data, some of it qualitative, would be done in an acceleration run out to a mach number of 1.7!

Our flight of two brand new production F-8U-1Ps took off at 10:30 AM on a nice, clear, sunny day in May 1959. The climb to the test altitude of forty thousand feet was uneventful. In those days, supersonic flight over land was permitted mainly because so little was known about the potential hazards of the phenomenon known as the sonic boom.

A supersonic corridor had been established on a radial of the Patuxent River TACAN station. We planned to conduct our test outbound on that corridor, heading southeast down the Chesapeake Bay, crossing the beach north of Norfolk thence out into the Atlantic Ocean naval operating area. Our flight conducted the climb to test altitude in a tear drop pattern crossing the TACAN station headed outbound at forty thousand feet. We lit our afterburners on his call and the flight went supersonic about twenty miles down the Chesapeake Bay.

The first thing I noticed was that the supersonic shock wave emanated from his airplane in a cone originating at the nose of his airplane. The angle of the cone varied with the mach number. The higher the mach number the narrower the cone. But, regardless of the mach number, the shock wave had a pronounced effect on an airplane flying in close proximity. It literally pushed my airplane in towards his. If I didn't make positive corrections, the shock wave would have caused our two planes to collide. At a mach number of 1.6 I had to maintain a ten degree bank angle away from Jake's airplane to keep from colliding with him.

The higher the mach number, the greater the correction was necessary (aileron and rudder) to keep a constant distance (10 to 100 feet) apart. There wasn't time to examine how far away from his airplane this shock wave effect was apparent. A second phenomenon became apparent and that was the power differential between the airplane creating the shock wave and the airplane flying behind it. I found myself gradually reducing power to keep from overrunning Jake's airplane. The higher the mach number, the lesser the thrust necessary to stay with him. As we passed mach 1.65 my throttle was at the very aft position of the afterburner modulation range. To reduce thrust any farther, I would have to come out of afterburner and, of course, that would have ended the test.

It was not unlike a phenomenon which race car drivers experience, in which they "draft" behind another car until the straightaway then, using the substantial power differential involved, accelerate past their opponent.

When I informed Jake of the power problem, he replied, "Okay, move out a little, go to full thrust and try to fly through my shock wave." In response, I moved my Crusader out another fifty feet and went to full afterburner. The airplane moved quickly ahead and, as it encountered Jake's shock wave, it yawed to the right so violently that my helmet was slammed against the left side of the canopy. The airplane's yaw stabilization system tried to correct for the yaw and caused a yawing excursion in the other direction of equal violence, slamming my helmet against the opposite side of the canopy. The excursions got increasingly violent and were rapidly going divergent when, in stark terror, I came out of afterburner and dropped a good two hundred feet behind Jake. The yawing motion stopped and I quickly engaged afterburner again and began moving back into position. All of this occurred in less than four or five seconds from touching the shock wave . . . but it scared the hell out of me. The yaw excursion rate was about three times per second. I knew instinctively that my airplane had been at the very ragged end of directional stability. Had the excursions been allowed to continue, there is no doubt in my mind that the stabilization system would have been overpowered and the airplane would simply have "swapped ends." At over one thousand knots, there would have been nothing but small pieces of the airplane and me in a few microseconds.

Jake must have known about what would happen because I noticed his shoulders shaking as though he were laughing. At the moment, I didn't think it was so funny!

Just as I was sliding back into close formation, Jake announced that we were at "bingo" fuel, the test was terminated and we were returning to base. He informed me he was coming out of afterburner and rolled into a steep bank away from me for a turn back to Patuxent. As I began to follow, I looked down at my airspeed indicator and saw 625 knots indicated. Jesus, I said to myself. I wonder if I could make it over the top! Without a moment's hesitation I hauled back on the control stick and began what I hoped would be a successful half Cuban eight. Jake had his belly to me and couldn't see what I was doing. A half Cuban eight is a half loop with a roll out to the upright position on the other side of the half loop. Had I been a little more experienced, or a bit more conservative, I would have concluded that this was a very foolish thing to do . . . and

RIGHT: Flight test division F-8U-2NE (F-8E) on carrier suitability trials, 1958. Note Mark 84 2,000 lb. bomb and ZUNI rockets.

CHAPTER 10: A RECORD OF SORTS

PART III
THE WORLD FAMOUS SEAGOING BOOMERANGS

"Never give in. Never, never, never, never! Never yield in any way, great or small, except to convictions of honor and good sense. Never yield to force and the apparently overwhelming might of the enemy..."

Winston Churchill

VFP-62 RFA from U.S.S. Saratoga (CVA-60) in Caribbean Sea, 1962. (Photo by Tad Riley)

We called ourselves the "World-Famous Seagoing Boomerangs" of Fighter Squadron Sixty-Two. We were one of the two fighter squadrons in Air Wing Ten, embarked in U.S.S. *Shangri La*, (CVA-38). There was a wry irony associated with the epithet because, on the deployment to the Mediterranean Sea, before I reported to the squadron, the squadron had been off-loaded to Naval Station, Rota, Spain for the entire tour inside the Strait of Gibraltar.

Naturally, the remainder of the air wing, the ones who stayed on the ship, took umbrage with the "Seagoing" part of our name. Naturally, in grim retaliation, the squadron pilots "complained bitterly" over being left ashore with all that wine and pretty Spanish ladies. If the truth were known, it was really a bitter pill for the squadron to swallow . . . to be left off of the ship-air wing team.

Fortunately, when I arrived, the tour ashore in Rota was history. Anyone who tried to rib me about that aspect of the squadron's history was told unceremoniously to "stuff it." Soon, we stopped hearing that kind of crap from the rest of the squadrons in the air wing. But, there was a lesson in all of that which I did not forget!

Notwithstanding the fact that I cut my teeth in the Crusader at the test center, it was not until I joined the Boomerangs that I really began to learn how to fly the airplane.

Ten years later, when as a wing commander I was approached with a similar suggestion for our deployment to the Mediterranean, the idea man who came up with that had his head nearly bitten off. His problem was that he never understood why . . . and never even tried to find out why. The lesson was that a good manager, and leader, should never tell a part of his organization that they are not really part of the team. At least, not if the leader ever expects to need them in times of emergency.

Every time my big idea man would suggest off-loading a squadron, I would bite his head off. He had a perfectly flat learning curve!

I suppose, in retrospect, I should have taken the time to explain it to him. But, it was so much fun to see him continue to take that near-lethal dose of gas . . . time after time, that I allowed it to become one of my more innocent forms of entertainment . . . on an otherwise frustrating Mediterranean Deployment. Some naval authority once said: "When principle is involved, be deaf to expediency."

So much for expediency!

11
"Gator"

The Chance Vought F-8 Crusader was a truly unusual airplane. Its design concept was unique and it won "hands down" in a competition with seven other airframe manufacturer's designs in the spring of 1953. In May of that year a development contract was awarded to Ling Temco Vought Aerospace Corporation's Vought Aeronautics Division. In less than two years, on 25 March, 1955, a prototype XF-8A flew its first flight and exceeded the speed of sound in level flight. Six months later, the first production airplane was flying. The Navy had itself a winner. The uniqueness of the airplane lay in its two position, variable incidence wing. The leading edge of the wing, normally flush with the top of the fuselage was raised to a landing incidence angle by a hydraulic cylinder and pivoting on a hinge at the trailing edge. This kept the long, slender fuselage of the airplane roughly parallel to the ground, affording the pilot good visibility over the nose for carrier landings. The unique wing permitted a longer fuselage and plenty of internal fuel capacity.

On 21 August, 1956 an F-8 set the first U.S. national speed record of 1,015 knots over a closed course. On 16 July, 1957 an RF-8A photo reconnaissance version set the first supersonic transcontinental record by flying the two thousand four hundred and forty-six mile route from Los Angeles to New York in three hours and twenty-two minutes. for an average speed of seven hundred twenty-four miles per hour.

Although the F-8 was bought for its performance, it also had its drawbacks. The visibility from within the cockpit was not very good. But, that was also the case for just about every U.S. tactical airplane during that period. Engine technology was just not there yet to produce the thrust that would permit designers the luxury of accepting the drag penalty of a high visibility bubble canopy.

The F-8 had three other flaws. Inherent with the two position wing came serious speed instability in the landing configuration. As a result, speed control in that configuration demanded an inordinate amount of pilot attention. A good F-8 carrier landing required

Author flying VF-62 F-8E (U.S.S. Shangri La) over the Tyrrhenian Sea, 1963. (Author photo)

73

my kind of guy; cheerful, light-hearted, with no affectations and funny as hell.

After Whiting Field our flying careers took different paths. They didn't cross again until eight years later at NAS Cecil Field near Jacksonville, Florida. Tad had become what we called a "photo beanie", a photographic reconnaissance pilot. He was assigned to Light Photographic Reconnaissance Squadron Sixty-Two, (VFP-62), permanently based at Cecil. The squadron trained pilots in the RF-8A photo reconnaissance version of the F-8 Crusader; and provided four-plane detachments for all of the small-deck aircraft carriers in the Atlantic Fleet. (The big-deck carriers operated the larger, and more sophisticated reconnaissance airplane the RA-5C Vigilante).

Officially designated the F-8U-1P at the time, the RF-8A contained about fifteen hundred pounds of additional fuel (over the fighter version of the F-8) and contained a reconnaissance package where the four twenty millimeter guns and ammo cans used to be. Because of its cleaner external configuration, the F-8U-1P could easily outrun the fighter version. This impressive speed capability undoubtedly saved Tad's life.

In October 1962, tensions between President Kennedy and Soviet Chairman Khrushchev had been heightened by a growing body of evidence that Soviet intercontinental ballistic missiles (ICBMs) had been sequestered into Cuba. Regular reconnaissance flights were being flown over Cuba by VFP-62 pilots to produce photographic evidence of this dangerous example of Soviet adventurism.

It was 10:30 on a clear winter morning, the fifth of November, when Lieutenant Tad Riley and his wingman, U.S. Marine Corps Captain Fred Carolan took off from Naval Air Station, Key West on a reconnaissance mission code named "Blue Moon." VFP-62 operated a joint reconnaissance detachment along with Marine Aerial Reconnaissance Squadron Two (VMCJ-2) at this tiny airstrip located at the southern tip of the Florida Keys, just ninety miles north of Havana, Cuba. The detachment pilots and duty officer operated out of the former VIP lounge in the airfield's operations building, which also contained the control tower. The launch that morning comprised six RF-8As in three two-plane sections, taking off at ten minute intervals.

As Tad conducted the pre-flight inspection of his airplane, he reflected on the momentous events of the last four weeks. He remembered very well the high pitch of excitement as October 22nd, or "Black Saturday", approached. That was the day that the U.S. intelligence community estimated all Cuban intercontinental ballistic missile sites would become operational. It was indeed a day of judgement, Apocalypse and Armageddon all rolled up into one!

Tad remembered the terrible confrontation between President Kennedy and Chairman Kruzchev. Then came the collective sigh of relief of two hundred million Americans when the Kremlin agreed

Lt.Cdr. Tad Riley "Reconnaissance Pilot Extraordinaire and Honorary Mayor of Sagua La Grande, Cuba." (U.S. Navy photo)

to remove the missiles from Cuban soil, break up the concrete launch slabs and bulldoze their remains into history.

"What high dreams", he mused; and, "What a great time in U.S. history to be a Navy reconnaissance pilot!"

Tad had looked forward to today's mission. He knew the territory by heart and the flight profile would be fairly simple. He had been to the missile site at Sagua La Grande in northern Cuba a number of times since construction on it first showed up on a high altitude satellite photograph. This was his turf and he knew it well. The mission was two-fold: first, he would photograph the missile launch pad to record the progress of its destruction. Then, he would retrace the route that the eighty foot long missile, loaded on its transporter, had been following on its trip to Cienfuegos, a port on the southern coast of the island. From there it would be loaded on a waiting Soviet transport vessel and begin its long return voyage to Mother Russia.

Almost daily, Tad had flown along the winding dirt road and found his missile inching along. Because of the missile's size and the narrow, winding road it could only advance twelve to fifteen miles per day. Today, he was able to guess, with high confidence,

CHAPTER 12: "BLUE MOON"

just about where he would find and photograph his prey. He didn't worry too much about the predictability of his periodic incursions over hostile territory. In fact, it would make intercept by Cuban MiGs a simple matter once they became aware of his presence over their island. To minimize that likelihood, the photo detachment had adopted a few precautionary operational procedures. The first procedure was to make no radio transmissions until they were "feet dry" over Cuba. Even then, the use of the radio would be restricted to combat circumstances.

As a consequence, the departure from Key West was accomplished in total radio silence. Clearance to taxi and take-off was given by light signals from the tower. The individual sections of RF-8s flew their approach legs to Cuba at an altitude of fifty feet and, in Tad's words, "at the speed of heat," (about 600 knots). Once past the coast-in point, the real part of the reconnaissance mission would be flown at an altitude of 500 feet and an airspeed of 500 knots.

The field of view from an altitude of 500 feet permitted the pilot to look around, generally following the road and, upon spotting the missile transporter, to plan the photo pass for best coverage. The cameras were turned on just prior to crossing the beach and were left on for the entire time over land. There was plenty of film in the cameras for the ten minutes or so that Tad would be over Cuba.

Once the missile was photographed, Tad would fly the shortest route to the Florida Straits, traversing that route, again, at 600 knots and fifty feet over the ground. "It was really a fairly simple mission", Tad recounted to me.

As he climbed into his Crusader he reminded himself of the words of warning from the intelligence officer who briefed the mission. "If you fly the mission properly, the principal threat to you will be air-to-air; in other words MiGs," he had admonished. "However," he had added, "The MiGs you will encounter are not a serious threat because they are mostly MiG-15s, a few MiG-17s and maybe a couple of MiG-19s. They all have serious flight control limitations at high speed and low altitude. That is where your Crusaders have the edge. Just thank your lucky stars the Cubans don't have any MiG-21s because they don't have those limitations." The intelligence officer's voice had been reassuring and confident.

Tad also recalled thinking, it's easy for you to be reassuring and confident, you jerk. You are not going to be stooging around Cuba in a few minutes. But, I am! Tad and Fred took off in a section. Picking up their wheels and lowering their wings, they eased down to fifty feet over the water and turned to a magnetic compass heading that would put them at their coast-in point ten minutes later.

Tad had flown this particular route so often that he had gotten fairly consistent at "dead reckoning" his way to the sugar plantation. He spotted it from several miles away by the smoke from the chimneys. By precise speed control, he crossed over the landmark exactly as the sweep second hand of his eight day clock passed the twelve o'clock position. The plantation was about forty miles east of Havana. Tad then turned his flight to the precise course for the nearest surface-to-air missile (SAM) site and shortly raced across the top of it at a height of fifty feet and 600 knots. It was the safest way to get by a SAM site. They never got any warning and never got off a shot.

"Good morning, fellas," he muttered aloud as the two Crusaders blew up a cloud of dust around the missile site. The next leg, another time and distance event, brought them over the ICBM site. By now they had slowed to 500 knots and climbed to an altitude of 500 feet. Tad observed the bulldozers hard at work. The cameras whirred, recording Soviet compliance to Nikita Khrushchev's bitterest pill.

Tad then turned his section of Crusaders to a southwesterly heading toward the last recorded position of the ICBM transporter. Now, he could see for miles. There was a thin, scattered to broken layer of clouds above them at about fifteen hundred feet. Every few seconds they flashed under a small hole in the cloud deck. They could see blue sky through those breaks. They had begun following the road only for a minute or so when Tad heard a radio transmission which froze his blood!

"MiGs at nine o'clock." The voice didn't bother to identify itself but it sounded familiar . . . like Fred's voice; only more excited than his wingman had ever sounded before. Tad's gaze snapped over his left shoulder.

"Christ," he heard himself saying out loud. "It's a frigging MiG-21!" Those intelligence bastards had told him there weren't any of those deadly airplanes in the Cuban air order of battle (AOB). Instinctively Tad stomped on the left rudder, yanked back hard on the stick and ran the throttle to full military power while simultaneously rolling into a shuddering, near vertical banked turn to the left. He felt the painful grip of six and one-half "g"s as the Crusader "broke" into the approaching MiG.

The giant grip of acceleration squeezed the bladders in his anti-g suit so hard that it caused pain in his thighs. Tad, still looking over his left shoulder at the MiG, watched for the effect of his turn on the MiG's trajectory. A sick feeling grew in his viscera when he saw the plane closing slowly in, drawing lead angle on him for a gun kill. Tad began to see a little more of the MiG's belly as the distance between them slowly, inescapably, inexorably closed. There was a halo of vapor surrounding the trailing edges of the MiG's wingtips. "This guy is really good," he heard himself mutter. "And in another few seconds he is going to kill me!" Tad slammed the throttle all the way to the full after-burner detent and hauled back even harder on the stick. To hell with the six and one-half "g" limit load factor. This was country hardball!

The Crusader shuddered under the eight "g" turn as Tad felt the heart-warming thump of the after-burner light in the small of

PART III: THE WORLD FAMOUS SEAGOING BOOMERANGS

VFP-62 RFA from U.S.S. Saratoga (CVA-60) in Caribbean Sea, 1962. (Photo courtesy of Tad Riley)

his back. The MiG apparently couldn't match this desperation maneuver and began to slowly drift across his tail to the outside of his turn. This was exactly what Tad had hoped for. Tad was now watching his adversary through the rear view mirror on the canopy rail directly above his head. He observed this development with a sense of relief. He had bought himself a few seconds of life. From the intelligence officer's brief he knew that his respite would be short-lived unless he capitalized upon it. Even with full after-burner he couldn't sustain eight "g"s for more than a few seconds. His only escape was to head for the haven of the tall grass . . . high speed and extremely low altitude . . . and he only had a few seconds to get there!

Still watching the MiG through his rear view mirror, Tad calculated that his best chance to head for the tall grass was about to present itself. It was now or never! The MiG's slightly larger turn radius had caused him to drift to the right far enough to permit Tad to execute a reversal into him and then to dump and run for the tall grass.

Leaving the throttle at full after-burner, he reversed his turn rather violently and clumsily dumped the stick forward into a shallow descent toward the verdant vegetation of the Cuban countryside. Through the rear view mirror he watched the MiG attempt to follow his maneuver, lagging a little in the process . . . exactly as Tad had hoped would happen. He leveled his Crusader dangerously close to the tree tops and noted, with some small comfort, that he was headed roughly in the direction of the Florida Straits.

The MiG had leveled out behind and a few hundred feet above him. Tad wondered fleetingly where Fred had gone . . . but knew that there would be no help from him nor could he have offered any in return. At this point in Riley's life it was every man for himself! He was sure Fred would understand, wherever he was.

But, Tad had other problems, of perhaps a more serious nature . . . if that were possible! That marvelous Crusader began to do what it did so well. With its leading edge droop in the "up" position (for high speed), the engine nudged them, Tad and his Crusader, into the silent world of supersonic flight. The needle on the

CHAPTER 12: "BLUE MOON"

machmeter moved slowly up to 1.1 times the speed of sound and the airspeed indicator was reading 740 knots.

Unfortunately, Tad couldn't afford the luxury of looking at the instrument panel. He knew if he took his attention from the wild blur of the tree tops at eye level even for a fraction of a second he might sag a few feet and suddenly become a huge ball of fire. Besides, the Crusader was encountering fairly heavy turbulence and was beginning to act "squirrely." The directional stability of the Crusader under these extreme circumstances was questionable. Nevertheless, Tad chanced a quick glance in his rear view mirror and noted, with a little satisfaction, that the MiG had dropped back about a quarter of a mile. But, he was keeping up with him! Apparently the MiG had been unable to lower his nose, at these extreme altitudes, for a gun kill. Now, he was out of gun range, but far from out of trouble. However, Tad remembered, the Soviet version of the U.S. heat seeking Sidewinder missile tended to droop a little after launch. They were probably below the minimum altitude for a successful missile launch. That slight droop in initial trajectory would put it into the trees.

Suddenly Tad decided that the MiG pilot might not kill him after all . . . but the Crusader might very easily do the job for him. At supersonic speeds and very low altitude flight, the directional stability of these early versions of the photo Crusader approached zero. This meant that any bump from turbulence (and there was plenty of that right now) might overpower the automatic stability efforts of the yaw dampener and the Crusader would simply swap ends. A fraction of a second later, there would be a large ball of fire and Tad Riley would be history.

His Crusader was bouncing around in the turbulence normally found on a hot day at low altitude over changing terrain. At modest airspeeds the effect could be annoying. At 740 knots it was terrifying. With each bump Tad's flight helmet would smash up against the top of the canopy and the "g" meter would peg at plus ten and minus four "g"s. Tad knew that a couple of seconds more of this kind of abuse would probably destroy both him and his airplane; so he came out of after-burner. The terrain below him had changed to a series of low ridges oriented at right angles to his route of flight.

The MiG was still back there falling gradually farther behind. He noted that only one more ridge lay between him and the inviting blue water of the Florida Straits. In a matter of seconds he would be "feet wet" and he would live to see another sunrise! It was precisely at this moment that Tad noted a break in the line of trees that ran along the crest of that ridge. It was a gap of a few hundred feet where there were no trees, just the bare ridge. Tad pointed the nose of the Crusader toward that gap with the intent of crossing the ridge below the level of the tops of those trees. As he closed the distance to that gap, he saw two figures in its center; one standing on the ground and the other sitting astride a burro. Both figures were facing away from him. Since the three airplanes; Tad's, Fred's and the pursuing MiG were supersonic, there was not a whisper of warning to the two figures of the impending arrival of an explosion beyond their wildest nightmares.

The man on the ground had adopted the unmistakable stance of a male relieving a distended bladder. Both hands were probably occupied in the process, Tad guessed, of directing the stream of warm effluent upon some undeserving anthill. The irony of the whole tableau, the airplanes, the Cubans and the burro, caught Tad's fancy.

He forgot all about the pursuing MiG, the wild excitement of the chase, the violent turbulence. Instead he concentrated all of his piloting skills to easing the nose of the careening Crusader down just a few feet more . . . just a trifle . . . so that he could pass directly over the head of the happy Cuban by as little as a few feet. It took all of the experience Tad had accumulated over the years to do this just right. He applied this experience rigorously to the task at hand. When the figure of the peon disappeared below the bottom of the windscreen, Tad estimated that no more than ten feet separated the belly of his Crusader and the broad-brimmed sombrero. It was as close as he could shave it!

He chuckled at the notion of the man enjoying that exquisite pleasure of an emptying bladder in the peaceful quiet of a warm, sunny Cuban morning. The poor man was totally unaware that in a split second, a thirty thousand pound airplane, traveling still in excess of 700 knots, pushed by a giant blowtorch and dragging a titanic shock wave would pass a few feet over his head. Tad often wondered in later years what had happened to those two unsuspecting Cubans. Perhaps in the reflex action as the exploding shock wave ruptured his ear drums the poor man afoot may have involuntarily torn off a critical part of his anatomy!

Our two Crusader pilots cleared the ridge line, slanted down the slope to the water's edge and went "feet wet." They left their throttles at full military power and stayed at fifty feet over the water until the MiG broke off the pursuit and turned back toward Cuba. Like most single engine tactical pilots, the Cuban didn't like flying very far over salt water. In accordance with their briefing the two reconnaissance pilots turned their airplanes toward Jacksonville, Florida and climbed to cruising altitude.

Waiting for them at Naval Air Station, Jacksonville was a team of photo interpreters and a jeep with the engine idling. The Crusader's camera bays were quickly unloaded and the film rushed to a nearby developing room. Moments later the waiting jeep returned to the flight line where an A-4 Skyhawk waited with its engine also idling. The Skyhawk pilot took the film can, lowered his canopy and taxied to the duty runway with priority clearance for take-off. An hour and a half later the Skyhawk landed at Andrews Air Force Base in Camp Springs, Maryland. Again, waiting, was a helicopter with its engine idling. The film can was passed to a helicopter crewman and the aircraft lifted off headed for the White House. Fifteen minutes later the pictures were being viewed by the President of the United States!

Meanwhile, back at Cecil Field, the VFP-62 maintenance chief petty officer couldn't believe his eyes. He was standing alongside Tad's airplane, a look of incredulity on his face. The paint was completely gone from the radome. It was also completely stripped off the leading edge of the wing, top and bottom, for a distance aft of several feet, wing tip to fuselage. In the belly of the airplane was a gash about two feet long running fore and aft. Around the wound was the distinct color or green vegetation. The statistical accelerometer in the main wheel well had recorded so many overstressing gust loads that the airplane was grounded pending the results of a complete structural inspection.

Tad asked Fred for his account of the event and Fred admitted to having no difficulty following Tad's airplane after the disengagement maneuver put him in a hopeless quarter mile trail. "I had no trouble keeping you in sight", he explained. "I just followed the explosions." When asked what he meant, Fred explained that as the flight passed from forest vegetation to plowed fields after harvest, the shock wave created a small cloud of dust . . . a sort of explosion. Tad had left a distinct trail of small explosions.

Tad recalled a few weeks later, an event at the supper table in his modest home in Jacksonville. The family was watching the evening television news broadcast transfixed by the scene of U.S. Ambassador to the United Nations, Adlai Stevenson berating a stony-eyed Russian Ambassador. Stevenson held a pointer in his hand and was using it to point to a large blown up photograph on an easel. The photograph was an aerial shot of a Russian ICBM site in Cuba. Tad quickly recognized it as one he had recently taken of the site at Sagua La Grande. "Blue Moon", at the time, was a highly classified project. His family was totally unaware of his flying activities.

"How did your day go, Honey," she asked somewhat detachedly, as she served up the plates and sat down to eat.

Tad answered, his eyes glued to the television screen, "Oh, not much, really. Same old stuff!"

13
"Beaver"

Lieutenant James H. "Beaver" Heiss was one of the most unforgettable characters I have ever known. He stood about five feet ten inches tall, was of medium build with brown hair, the palest of blue eyes and usually had a pleasant grin on his face. He had a laid back approach to life in general and was the most pleasant, most easy-going young pilot in Fighter Squadron Sixty-Two.

When I first met "Beaver", he was the squadron training officer and reputedly, the best "dog fighter" in the organization. New arrivals in the squadron were quickly scheduled for a one versus one tactics training flight against "Beaver." It was intended to be a lesson in humility. It also served as a measure of the newcomer's ability as to how long he lasted in the first engagement before "Beaver" slid into a gun firing position and transmitted, over the radio, "guns, guns, guns."

The rest of the squadron pilots, gathered around the squadron base radio in the ready room, would grin with amusement and return to their duties, comfortable in the knowledge of how good, or how bad, a pilot the new guy was. "Beaver's" skill in aerial combat gradually became legendary.

When he finally left the "Boomerangs", he was assigned as a tactics instructor in the F-8 fleet replacement squadron, VF-174, at Cecil Field. He developed his tactics skills in his two years as an instructor to the point where he was generally acknowledged to be the best F-8 aerial tactician in the Navy . . . nay, in the world.

But, it was this very talent that lead to a bizarre accident in which he was literally crushed to death, in an instant, in the cockpit of a Crusader.

When I reported to Fighter Squadron Sixty-Two, I was a brand new Lieutenant Commander fresh from the training squadron with a total flight time of sixty-eight hours in the crusader. The squadron commanding officer, Commander Joe Moorer, assigned me to be the squadron standardization officer. It was a newly created billet.

As the "Stan Man" I was to enforce the new standardization manual under which all fleet operations would hereinafter be conducted. This standardized approach to flight operations was intended to enhance combat readiness and, at the same time, improve aviation safety. The concept was one which had been adopted earlier by the U.S. Air Force with singular success in improving their safety record.

The acronym NATOPS (Naval Aviation Training and Operations Procedures System) was greeted with mixed emotions by the squadron pilots. Although grudgingly understood to be a safety measure, it appeared to impose far too many silly restrictions on all of us. For example, there was a section in the manual which spelled out, in nauseating detail, all of the rules which would be rigidly imposed on the conduct of tactics training, or air combat maneuvering (ACM), as the new buzzword described it.

As the standardization officer I was, by definition, an unpopular guy. I was the one who would conduct an annual evaluation, or standardization check, on every pilot in the squadron, including the C.O. The evaluation included an open and closed book examination, a simulator check flight and a flight check. Although Joe Moorer fully supported my efforts to impose the new system, I was viewed by the junior pilots as a royal pain in the neck . . . someone to be endured.

And so, when the day arrived for my one versus one tactics lesson from "Beaver", an unusual number of pilots appeared to have business in the ready room coincidental with my pre-flight briefing and subsequent tactics mission. They would listen to our briefing; and then they would listen to our fight over the radio in the ready room. I knew that I was double damned.

The fleet pilots looked upon Patuxent River pilots as prima donnas. Here I was a Patuxent River pilot and also a standardization officer about to get his come-uppance.

"Beaver", I said in a loud voice, as I began the briefing, "Anyone can bend an airplane at twenty thousand feet (the altitude where most tactics training was conducted). But, the real test of airmanship is how well you can bend it at forty thousand feet."

A hush fell over the low buzz of conversation in the ready room. I knew they were all astonished at my audacity. None of them would even think of saying such a thing to "Beaver" Heiss.

83

They were all thinking to themselves, with grim satisfaction, that now I was going to get really beaten, and badly! "Beaver" merely shrugged at my statement. I knew that he was such a natural aviator that he really couldn't have cared less at what altitude we fought.

Finally, the briefing was completed and we left the ready room to man our airplanes on schedule. The conduct of the flight called for a standard running rendezvous as we climbed to altitude enroute to the tactics training area southwest of the field. The wingman would check out his weapons system during the rendezvous, as he closed on his leader. He would acquire the leader with his radar, lock on, check the gun sight symbology, the radar break-away tones etcetera. He would then pass by the leader and temporarily take over the lead while the leader checked-out his system on his wingman. It was standard procedure. Once the leader completed the check-out of his system, he would join up and resume the lead of the flight.

"Beaver" and I completed this process with one significant addition. While I was behind "Beaver" in his blind zone I hit the emergency fuel dump switch and dumped four thousand pounds of fuel from my wingtip dump valves! "Beaver" never suspected I would do such a thing! Unbeknownst to him, he would now be fighting an F-8 which weighed two tons less than his airplane. It would therefore, be much more agile since its thrust-to-weight ratio and its wing loading would be dramatically improved. I knew I now had a distinct advantage over "Beaver" of which he was unaware. Of course, I now only had enough fuel for one engagement. But, that was all I really planned on.

The standard way to begin a one-versus-one engagement was in a head-on pass in full afterburner at twenty thousand feet. To arrange for this to happen, the flight leader would direct each airplane, during the final stages of the climb to altitude, to turn away from the other exactly thirty degrees. Then, when the two airplanes were sufficiently far apart, the leader would call for the two to turn into each other. The head-on pass ensued. When the two airplanes passed each other in opposite directions, close aboard (a hundred feet apart, he would call, "The fight's on."

At this juncture, each airplane was free to do whatever it could do to win. The first move, of course, was a maximum power turn

BELOW: VF-62 F-8E at NAS Cecil Field, Florida, 1962. (Photo courtesy of Tad Riley)

CHAPTER 13: "BEAVER"

into the other airplane. After that it was up to the pilot to maneuver into his opponent's rear hemisphere for a Sidewinder shot or a gun kill.

But, this time we were climbing to forty thousand feet. The F-8 (or any airplane for that matter) lumbered along heavily in the thin air and didn't turn nearly as well at forty thousand feet as it did at twenty thousand. We started our separation turn, and I almost lost sight of "Beaver's" airplane before I felt I could call a turn in. When I did I saw his wings bank steeply toward me, the after burner light and his airplane begin a pronounced descent. "Smart guy", I said to myself as I did likewise.

We were descending through thirty-five thousand feet, almost supersonic and on opposite courses when "Beaver's" airplane flashed by me surprisingly close abeam. Our closure rate must have been a thousand knots when I wrapped the F-8 up into a hard climbing turn into his airplane. The initial "g" force of the turn was the limit load factor of six and one half "g"s; but I immediately felt the airplane go into heavy airframe buffet and slacked off a bit. The airplane turned best in light buffet. I settled on a turn with slight buffet making the airplane quiver.

My head was twisted around as far to my left as I could turn it watching "Beaver's" airplane slip momentarily out of sight down my left side. For the next forty degrees of turn "Beaver's" airplane was in my blind zone; and my airplane, presumably in his.

He reappeared where I expected him, at eight o'clock low about two miles away. I had a bad feeling when I realized that I had not gained nearly as much angle on him as the two ton difference in weight should have produced. However, I was able, at that point, to roll into a 120 degree bank angle and pull my airplane down toward his . . . coming downhill at him.

Theoretically, if each airplane performed perfectly, they would continue to meet in a series of head-on encounters until the fight was called off because of time, fuel or altitude. But, fortunately no two pilots perform perfectly against each other. That, after all, is what aerial combat is all about!

At the second pass I noted with satisfaction that I had gained a good twenty degrees, and a thousand feet of altitude on him. Furthermore, I was continuing to gain angle off. He was forced to come up hill at me, losing some energy in the process. My next turn was a level one and it ended with my now having a good thirty degrees on him. But, I knew I couldn't make any mistakes. Even with the weight advantage in my favor, "Beaver" was a formidable opponent.

I saw "Beaver's" airplane go into the vertical plane. He must have realized he was getting into trouble and was throwing a little more sophistication into the tactical situation. With my weight advantage, I knew that the vertical fight would work to my advantage and I matched his move. We were now turning in the purely vertical plane and running out of energy fast. I didn't notice the ache in my `neck muscles or my right arm from the monster forces of gravity and stick forces. This was real "country hard ball, kick-ass" aerial combat and the adrenaline was pumping!

"Beaver's" heavier airplane got slow sooner than mine did, and he started downhill to regain energy. At the fourth pass, I knew I had him and I think he sensed it too. So far, I hadn't made any mistakes. But, "Beaver" Heiss sure as hell was not giving up. I cautioned myself not to relax. "Beaver" was a dangerous opponent, weight difference or not.

Still nibbling at the edge of buffet, my F-8 was now in a downhill, six and one half "g" turn doing about nine tenths of the speed of sound. I was closing on "Beaver's" airplane with about forty-five degrees angle off and a range of about two thousand feet. There was no way I could simulate firing a Sidewinder, so my only recourse was to close to gun range. I was at about fifteen hundred feet when "Beaver's" airplane started back up into a vertical turn. "Jesus", I shouted to myself. Glancing at my own "g" meter I saw seven "g"s and I was just matching his turn.

He must be pulling eight "g"s to match my turn! (six and one half was the limit allowed). He had pulled up into the sun to cause me to lose sight and break off. By the rules I was supposed to call losing sight of him and turn away to avoid collision. The pull up into the sun was a standard last ditch maneuver. "Beaver" must have been counting on me, as the standardization officer, to follow the rules and break off.

"Bull crap", I muttered to myself. "I'm not slacking off one God damned bit. I don't give a damn if I hit him." My eyes were watering from looking directly into the sun, but I kept them painfully open . . . sweat running into them and stinging.

Suddenly, "Beaver's" airplane appeared out of the blazing sun. I was extremely close and almost in a firing position. "It's now or never," I shouted into my mask. "I've got you, 'Beaver', you son of a bitch."

As "Beaver's" Crusader descended into a near vertical dive, it slid right into my gunsight. The "g" was starting to build rapidly, so I knew I had only a few seconds to get a firing solution. The sweat was running down my forehead and into my eyes burning the hell out of them. The range closed to about six hundred feet and the pipper hung on his canopy for the requisite two seconds so I called.

"Two, this is One. Guns, guns guns. Okay, let's knock it off." I wondered as I did so, what the expressions were on the faces of all the pilots gathered in the ready room, listening. Chuckling into my mask, I shouted into it. "How do you bastards like them apples?"

I was snapped back to reality as though a dash of cold water had splashed down my spine. The low fuel level warning light was ON and I had no recollection of how long it had been that way! I yanked the throttle out of afterburner and all the way to the idle stop. My fuel gauge read nine hundred pounds! Squadron operating procedures called for being on deck before the light came on which was supposed to happen at 1,050 pounds. Here I was, thirty miles from the field at twenty thousand feet with only nine hundred pounds. Some standardization officer I am, I remonstrated.

I bobbled the nose up and down a few times as I headed back toward the field. The was the signal for "Beaver" to join up on me. When he slid up onto my left wing I gave him the no radio signal. Pointing at my mask I followed with a thumbs down. Next I pointed at my helmet and gave a thumbs up signal. That was intended to tell him that my radio transmitter was not working but my receiver was okay.

"Are you okay?", his voice inquired over the radio. I gave him a thumbs up and then gave him a landing signal and a wave goodbye. He nodded and announced. "I'll call the tower and tell them to expect a NORDO (no radio) in the break. Do you mind if I stay out here and burn down a bit?" I gave him a thumbs up and turned my full attention to getting my airplane on deck at Cecil before it ran out of gas.

I entered the break at Cecil rocking my wings and got a green light from the tower. The landing was uneventful. But, I had to continue the charade that my radio transmitter was not working when it was really perfectly alright. It was my excuse for returning to base early. Aerial combat is not permitted without a working radio.

When I walked into the ready room, helmet in hand and flight suit soaked with sweat, there was a subdued air of curiosity and thereafter I was treated with a great deal more respect than I deserved. What I had done was a dirty trick. I was genuinely ashamed of my duplicity. Many times thereafter, "Beaver" beat me soundly in aerial combat . . . but I was getting better with each go. Nevertheless, among the other pilots, there was always an air of respect because they never forgot my first engagement with "Beaver" Heiss.

It was a good twenty years before I told this story to anybody. I wish now, that I had confessed my trick to "Beaver." I always intended to do so, but the occasion never seemed right. It will always remain one of my deep regrets . . . because we were real friends. "Beaver" would have laughed it off over a beer, I'm sure. But, he died before I ever got around to telling the story on myself.

It was 14 February, 1968 when "Beaver" Heiss got a set of orders which ended his tour of duty with the "Boomerangs." He was assigned to be an instructor pilot in the F-8 fleet replacement squadron at Cecil Field. For "Beaver" it was the best possible set of orders he could have gotten. It kept him in the same area which he

CHAPTER 13: "BEAVER"

and his wife Jan had grown to love. She had a good job as a registered nurse. And, best of all, he would continue to fly the F-8, and as a tactics instructor, no less.

It went without saying that VF-174 would assign "Beaver" as one of its tactics instructors. It was the choicest assignment in the squadron; one reserved for only the very best Crusader pilots.

The "Boomerangs" held a hail and farewell party for the incoming new pilots and the outgoing old ones. The party was, as always, a happy event at the officer's recreational facility (ORF) at Cecil Field. Most of us drank more than we should have. We laughed, told sea stories and plied "Beaver" and his attractive wife, Jan, with drinks. Finally, the CO, Joe Moorer stood up, said a few kind things about "Beaver" and gave him the traditional "Boomerang" plaque. "Beaver" accepted it, and the Skipper's gentle ribbing with good grace and said the usual good-byes. The party ended and "Beaver" and Jan drove off into the night to bigger and better things.

The eighth of April, 1963 found "Beaver" finishing up his period of training as an instructor under instruction (IUT) at VF-174. The final training flight was a knock-down-drag-out one-versus-one aerial battle between "Beaver" and his instructor mentor. "Beaver" had done well. holding his own and more. Each had won one engagement with the third and final one ending in a stand-off. The instructor call a final "knock it off" and the two headed back toward Cecil Field.

There was a club at Cecil Field, with an unwritten charter, and no annual dues. The fleet pilots called it the "close abeam club." Its members were an elite group from the senior ranks of the training squadron, VF-174. The club was started in the early 1960s by a Crusader plank owner when he was a squadron tactics instructor.

To be a member in good standing one had to be a very good stick-and-throttle man, have a cavalier disregard for the standard rules which governed the way mere mortal F-8 pilots flew the airplane and, finally, possess a total disregard for one's own well-being!

While normal pilots flew a reasonably safe 350 knot entry into the break (with an occasional 450 knotter for special days), the "close abeam club" did otherwise. When they approached the field the first thing they did was manually open the oil cooler doors. This allowed air to pass through the louvers of an oil heat exchanger which was normally left in automatic. (the feature automatically cut in at some supersonic speed where engine oil temperature became a factor). This air flow through the louvers produced a low pitched, moaning sound whose loudness was directly proportional to the speed of the airplane.

Next, the club members would kick the speed up to six hundred knots thereby announcing their approach to anyone who was out of doors at Cecil Field. It was a kind of heralding for the return of the conquering hero . . . very macho! Then, at a point directly over the point of intended landing, the club member would roll into an eighty-five degree bank angle and pull well in excess of the allowable (6.5 "g") limit load factor in order to kill off excess airspeed for purposes of putting down wheels and raising the wing.

This maneuver usually put the airplane at a position abeam the point of intended landing which was far too close for a prudent landing approach (the origin of the name "close abeam"). Usually their speed was still too great to lower the wheels safely so that detail was delayed until later in this highly non-standard landing attempt. Since a landing at the completion of this evolution was a requirement, great airmanship was often demonstrated to avoid overshooting the final approach centerline. More often than not, the throttles remained at IDLE for the entire approach while airspeed was being "killed off." Then, at the last moment, a large, uncontrolled burst of power was necessary to prevent the airplane from falling out of the sky "on final."

I had observed "close abeam" landings in which the pilot never got below a bank angle of sixty degrees until rolling out on final. Lest I be misunderstood, I never denigrated the skill needed to successfully complete a "close abeam" landing. It was considerable! What the majority of the fleet pilots, who observed these ridiculous demonstrations, did was question the judgement of the pilots.

The club was a small one, thank goodness. Maybe it had half a dozen members at any given time. Members came and went. But the club had dues! The price was high! Fortunately (or perhaps unfortunately) nobody paid his dues until "Beaver" Heiss came along. Everyone at Cecil Field knew who the club members were. Everyone at Cecil Field also knew that "Beaver" Heiss paid their dues with his life!

Dick Oliver and I watched a "close abeam" landing one day. Dick, later a Blue Angel soloist, was one of the most accomplished aviators I have known and a good friend. We heard the moan of the oil cooler doors and stopped to watch the resultant landing. It was a classic "close abeam" landing. When it was over, we resumed walking to my car and Dick made a very trenchant, and prophetic, observation.

"Paul," he observed, "One of these days, one of those idiots is going to kill himself!" Dick was right!

At about 10 o'clock AM "Beaver" and his instructor mentor opened their oil cooler doors! The resultant banshee howl told two observers; John Nichols and Phil Craven, also in the training squadron, that this was going to be a "close abeam" landing. They watched!

The two Crusaders appeared in the break easily in excess of six hundred knots and the instructor mentor broke to the downwind leg, followed closely by "Beaver." Two seconds later "Beaver" Heiss was dead!

The long thin fuselage and the large wings of the Crusader flexed considerably under heavy "g" loads. The heavier the load the greater the flexion. During the stressing, the wing's outer wing panels would flex upward as the wing strove to generate the lift necessary to support the weight of the airplane multiplied by the number of "g" being pulled. The resultant lifting force vector of both wings was generated at a point (called the center of lift) in the fuselage centerline over twenty feet behind the pilot's seat.

During high "g" maneuvering the forward fuselage flexes downward much as though the pilot were sitting on a diving board twenty feet long. The higher the "g" load, the farther down the diving board flexes. This is exactly what "Beaver's" fuselage was doing when the wing attachment structure failed catastrophically! "Beaver" quite literally pulled the wings off his airplane.

The structures engineers from the plant in Dallas had varying opinions as to how far down the seat supporting "Beaver" had flexed. They also had varying opinions on what exactly was the precise coefficient of restitution of the fuselage at the moment of failure. They also had varying opinions as to how many times the wing had been overstressed before the accident. And, finally, their calculations also varied over what the resultant "g" load was which was exerted upon "Beaver's" spinal column that caused it to be crushed so devastatingly! The most conservative engineer said fifty "g"s! The most liberal engineer said one hundred fifty "g"s! The question is really moot! Either would have killed him just as instantly!

John Nichols and Phil Craven, watching, both heard the "popping" sound and saw the fuselage separate from the wing with a small puff of white smoke. Simultaneously they heard and saw the engine go into afterburner, and the fuselage, pitching up nose high, continued roughly up the runway line, finally disappearing from view behind the stand of loblolly pines near one of the field's picnic areas along the runway. The wing fluttered to earth vertically, like a falling leaf.

Almost gently, the fuselage descended onto the ball diamond at the picnic area. It came to rest in an almost upright position, its afterburner in full throated power until contact with the ground cause it to extinguish . . . much like "Beaver's" life had done seconds earlier!

John and Phil were quick to act, and, having commandeered the nearest operating vehicle, were the first to arrive at the scene, just moments later! They opened the canopy and inserted the safety pins in "Beaver's" ejection seat.

Inside the cockpit, they found "Beaver" looking relatively unscathed, his head slumped forward in death. The autopsy revealed that his spinal column had been crushed as if by a giant hammer blow! The accident board, employing some engineers from the plant at Dallas investigated the event as meticulously as they could. It was a very high interest accident in Washington. D.C. because of its unusual nature.

Among the conclusions of the board members was the fact that the wing-fuselage separation occurred as a result of excessive "g" forces induced by the pilot. The forward fuselage was flexed downward by some substantial amount when wing separation occurred. Thereafter, the fuselage restored itself to an unstressed condition so rapidly that it imposed enormous "g" loads on the cockpit area producing loads that "Beaver's" spinal column was unable to stand.

Death had been instantaneous! The "close abeam club" evaporated into thin air. Sadly, it took the loss of a magnificent aviator to bring sanity back to a small group of naval aviators who had become too enamored with their own prowess!

14
"Gitmo"

Guantanamo Bay, Cuba is one of the most strategically located bases in the Caribbean Basin. It is a natural, deep water anchorage located at the southeastern tip of the island of Cuba. Today the bay is the site of a large U.S. Naval Base including two airfields, weapons storage facilities, refueling piers, a joint reconnaissance center and personnel support facilities necessary to keep the four thousand or so military and civilian personnel adequately provisioned with no help from the parent owner of the island. Of course, a small labor force of Cuban nationals comes onto the base and leaves every night, hopefully, to do menial housekeeping chores. All in all, it is both a unique and odd arrangement with the Cuban government which has been, for over thirty-four years, hostile to U.S. interests.

My first visit to "Gitmo" was as a third class midshipman on the battleship U.S.S. *Missouri* in the summer of 1949. At that time the only air facility was the short airstrip on the eastern side of the bay. McCalla Field, as it was called, was only 4,800 feet long and could handle only small propeller driven cargo planes. At both ends of the small runway, the terrain fell away steeply into the water. In the bay itself was a seaplane facility.

On my first visit I took little notice of such things as air facilities. The intense Caribbean sun had turned our berthing quarters in *Missouri* into a sweat box. But, I was eager to do some sightseeing. Relations at the time between the U.S. government and the regime of President Fulgencio Battista were reasonably good so the crew of *Missouri* were free to take the short walk from the main gate of the base to the metropolis of Caimanera where we were able to cool off with a mug of beer.

The town, like many other towns outside naval bases, catered to the more prurient interests of its tourists and had, therefore, its share of brothels and bars.

The sights were not too elevating but the beer was cool and delicious . . . well worth the walk.

Twelve years later I paid my second visit to "Gitmo." Things were far different; and the events surrounding my visit were to become a nightmare for the Skipper of the Seagoing Boomerangs of Fighter Squadron Sixty-Two!

Commander Joe Moorer, our colorful Skipper, had just been relieved of command of the Boomerangs and had assumed command of the air group, Carrier Air Group Ten, embarked in *Shangri La*. His relief was an interesting change of pace. The new Skipper was a small, energetic and intense man, bubbling with enthusiasm and imbued with boundless energy.

He was bright, energetic and vigorous; but he was not a leader. His performance at a previous staff assignment had been so brilliant that he "screened" for a fleet squadron. The assumption must have been; if he was a good staff man, he would make a good commanding officer. Not so!

Unfortunately, our new Skipper had two strikes against him from the outset that were to prove his eventual undoing. He had very little fleet experience and he was a weak aviator. His weakness as a carrier pilot must, in retrospect, have been a severe burden to him. Certainly, it inhibited his ability to lead men.

It had become immediately apparent to all of us that he couldn't fly the Crusader very well. Even the most inexperienced of the young pilots soon found that they could "clean his clock" with ease in simulated aerial combat.

The decision had been made to deploy the air wing without a day fighter squadron. In its place another light attack squadron was substituted as an experiment. Faced with a six month hiatus "on the beach", the new C.O. decided that the squadron ought to deploy on an aerial gunnery detachment to "Gitmo."

As the squadron operations officer, I was given the responsibility for planning and execution of this elaborate training evolution. Several days of extensive planning and writing culminated in the production of an Operations Order (OPORDER) which outlined, in great detail, the schedule of events and specific duties of everyone involved. Details, such as personnel support, airlift, logistics, ordnance, targets, and barracks availability were worked out. An eight plane detachment of Crusaders would fly to Cuba in

August 1962. Maintenance and ordnance personnel, comprising the bulk of the personnel in the squadron detachment, would be moved to Naval Air Station, Leeward Point, Guantanamo Bay ahead of the actual movement of the eight airplanes to prepare for their arrival. All of the preparatory actions were initiated and the day for the flight to Cuba arrived.

The C.O. was scheduled to take-off from Cecil Field at 8:30 AM on the 16th of August leading the first, four plane division. One half hour later, I would follow leading the second four plane division. I made sure that the most experienced second tour pilot, Lieutenant Dick Oliver, was flying with the C.O. as his section leader. This was intended to lend some flight experience to the division in the event anything went wrong. I wasn't expecting trouble, but then one had to be careful when flying to Cuba.

The flight was a simple 875 nautical mile cross-country flight with a good deal of it over open water. The tricky part was to avoid, at all costs, flying into Cuban airspace. The Cuban Air Force, equipped with the latest export model of the MiG fighter, could be expected to behave in a hostile manner if U.S. tactical airplanes ventured over the Cuban land mass. Their combination of new fighters and a very sophisticated network of surface-to-air missiles and anti-aircraft artillery argued for a careful skirting of the island well to the east . . . then turning westward to fly directly into "Gitmo." I hand-picked every pilot to fly with the Skipper on this trip. His wingman, Lieutenant (junior grade) Ben Walker was a very likeable, capable and hard-charging young pilot; and the newest "nugget" in the squadron. Lieutenant Dick Oliver was the second section leader. His wingman, Lieutenant (junior grade) Tom Malloy was every bit as good an aviator as Ben, and a very volatile and droll young man. The four of them, I thought, ought to be able to handle this relatively routine flight with no hitches. Little did I suspect that, in short order, it would turn into a squadron operations officer's worst nightmare!

As leader of the second division, I chose, for my wingman, Lieutenant (junior grade) Don Ressel, a very capable second tour pilot, and very experienced. My section leader, Lieutenant Al Wattay would have, as his wingman our squadron assistant landing signal officer, Lieutenant John Nichols, also the squadron safety officer.

The C.O.'s flight briefed, manned aircraft and launched on schedule. My division briefed one hour behind the C.O.'s, manned up and was preparing to launch when the first bit of trouble struck. As I was preparing to taxi out of the flight line my airplane developed a massive hydraulic leak in one of the two flight control systems. The spare airplane had just gone "down" for a discrepancy and there were no spare airplanes. I hand signalled for the section leader to come up on the squadron common radio frequency. When he did so, I told him my airplane was "down" and directed him to take the flight to "Gitmo", and that I would follow as soon as I could get an airplane. They taxied out and I returned, disgruntled to the ready room. His crisp, "This is Three. Wilco. Out," told me that the flight was in good hands.

Five minutes later I was sitting in the squadron ready room reading the message board when the duty officer's telephone rang. The duty officer's voice took on such a serious tone that all of us in earshot stopped talking, and listened. Lieutenant (junior grade) Tom Napoli's face paled as he listened. Finally, he looked at me, covering the telephone receiver with his hand and said, "The Skipper is down."

For the next several hours I attempted to piece together what had happened. The voice of the air controller on the phone from Miami Air Route Traffic Control Center (ARTCC) informed me that a MAYDAY report had been received from Silverstep Two Zero Nine (Dick Oliver's airplane) that Two Zero Zero (the C.O.'s call sign) indicated he was ejecting shortly after entering a large cumulonimbus cloud over Andros Island. The Center also received the emergency beeper signal, presumably from the Skipper's airplane until it crashed. Then it stopped. They then reported another emergency beeper signal, assumed to be Tom Malloy's airplane which also shut off abruptly a short time later.

Dick Oliver's report gave a position relative to the TACAN station at Homestead Air Force Base south of Miami. Dick further reported that his airplane had been damaged by several lightning strikes. He was unable to locate the C.O. because of torrential rain squalls; but, both he and the controller were able to hear what they thought was the C.O.'s emergency radio beeper. Dick announced further, that his fuel reserves were low and he was continuing on to Guantanamo.

A hush settled over all of us in the ready room as the duty officer began monitoring the Coast Guard search and rescue effort which was now underway from the Miami station. A cutter was gotten underway. A P-3 Orion patrol plane was launched from Naval Air Station Jacksonville and vectored to the scene to take over as on-scene commander of the search effort. A Coast Guard helicopter was launched from Miami to effect a rescue.

Hopes of finding the Skipper before darkness set in dimmed when the patrol plane reported visibility near zero in heavy rain underneath a five hundred foot overcast. The on-scene commander also reported huge swells and high surface winds. The silent vigil began with a grim note.

The duty officer's telephone broke the silence like a shattering piece of china. "Yes, Sir." He said excitedly. "I've got it, Sir. Thank you, Sir," and hung up the telephone. He announced to me excitedly. "Ben Walker just landed at 'Gitmo' but he doesn't know anything about anybody else." Another tense fifteen minutes ticked by while we set up for the squadron executive officer and his wife to go to the Skipper's home, tell his wife and stay with her for the moment. The appropriate messages were sent out in accordance with current directives. Then we continued to wait.

thought about the implications of what I knew my answer would have to be . . . loyalty set aside. Finally, with the Skipper looking directly at me with those intense dark eyes I answered, "No!" To this day, I believe that my answer probably had little to do with the outcome of the court. However, the hurt expression on the C.O.'s face has haunted me ever since. He looked at me as though I had thrust a knife into his back! I found myself wondering whether he would hold it against me in his next fitness report.

What happened next still boggles my mind when I recall it. The wing commander proceeded to question me on why I had grounded my airplane and passed the lead, for the second flight to Cuba, on to my section leader, Al Wattay. After I explained that my airplane had experienced a flight control system hydraulic failure and the spare airplane had gone "down", he smirked. Then he insinuatingly asked if it wasn't true that I had "downed" my airplane deliberately because I knew about that bad weather and was afraid to launch on the mission.

I was so stunned for a moment at the enormous personal insult to my integrity, that I didn't know what to do. Then I was on my feet leaning across the table, my nose inches from his (he had jumped up in startled reaction), shouting, "Retract that question, Commander!"

"Bullmoose" Woerner had also jumped up when I did and shouted, even louder than I, "Gentlemen, sit down, God damn it!", in a booming voice. Then he addressed me, "Paul, don't even bother to answer that question." He looked scathingly at the commander, who had turned pale as a sheet. He must have realized how close he had come to having his teeth rearranged.

Captain Woerner then turned to the court recorder, the only other person in the room, and ordered, "I want that last question deleted from the record of these proceedings, is that clear?" The recorder nodded. I sat back down still seething in a rage.

In due course, the court of inquiry completed its deliberations and found the C.O. guilty of bad judgement and dereliction of duty for having blatantly violated directives as a squadron commanding officer and thereby having contributed to the loss of a life and two airplanes.

He was issued a letter of caution and was summarily relieved of his command. He never returned. Sometime later I learned that he had been promoted to Captain based on a superb performance as a staff officer. Subsequently, he retired. Fifteen years later, I met him briefly at CINCLANTFLEET headquarters. We chatted for a moment. He told me that as a civilian he was doing some consulting work for the Navy. When we said our good-byes I noted that same haunted look in his eyes. I never saw him again. Sadly, I learned a few years ago, that the Skipper had suddenly fallen ill and died of a heart attack.

I don't believe he ever got over that terrible experience. But, the nightmare had finally ended for him.

Of the other three aviators only one is alive today. Tom Malloy died a violent death that day, but no one will ever know exactly how! Lieutenant Oliver had a distinguished career which culminated in a violent death. As a solo pilot with the Blue Angels he flew into the water at five hundred knots during an air show at Niagara Falls. Ben Walker is the only one alive . . . and he, when last heard from, was happily employed as an airline pilot for Delta Airlines stationed out of Los Angeles International Airport.

The Skipper was a good man. a good naval officer and a superb staff man. But, he was not a good aviator nor was he a leader of men. I personally liked him. But, those last two facets of his make-up cost him his flying career.

15
SHARK BAIT

Lieutenant Commander Hamilton "Ham" Love, USN was one of the most experienced junior officers in Light Photographic Reconnaissance Squadron Sixty-Two (VFP-62). Nevertheless, as experienced and professional an aviator and photo pilot as he was . . . he was also, in a sense, a scofflaw!

Although the U.S. Navy began its transition to jet propulsion way back in the early 1950s; it did so in an incremental way. Of the 6,000 or so airplanes in the Navy's inventory, only a small percentage had transitioned to jets by the time the Crusader photo reconnaissance version hit the fleet. There were still a great number of aviators flying propeller-driven airplanes . . . and habits are hard to break. One of those habits was smoking in tactical airplanes.

Mind you, it was never a good idea for tactical aviators to smoke while flying. Fire and airplanes don't mix. They are an often fatal combination. Nevertheless, there were countless pilots of single-seat tactical airplanes who still broke out the pack of cigarettes and lit up when things got slow and time hung heavy on the aviator's hands. By the mid to late 1950s there were still many "Spad" (Skyraider) squadrons in the Navy. Their pilots would routinely go off on eight and nine hour long-range, low-level missions. As dangerous as certain portions of those flights were (catapults, deck launches, arrestments and strikes), there was still a great deal of time spent just cruising along at 180 knots with not much to do.

However, with the arrival of jet propulsion, came the requirement to fly at high altitudes and therefore the mandate to wear oxygen masks from take-off to landing. Certainly, the "Spads" were equipped with oxygen tanks and Skyraider pilots used masks. But, it was only on those occasions when the airplane was above an altitude of say, ten thousand feet, for any length of time. Those were rare occasions for "Spad" pilots. So, most of the Skyraider pilots who smoked, did so in the air; mandates to the contrary notwithstanding.

A much smaller number of jet pilots smoked in their airplanes, although they did so much more surreptitiously. "Ham" Love was one of these few. He did it on long range missions whenever things got quiet. He was never blatant about it. But, everyone knew he did it, nonetheless. After all, he rationalized, all a careful pilot had to do was ensure he was not off the oxygen for a long time at altitude. He also, had to make sure he turned off the flow of oxygen to his mask when he disconnected one side of it and let it dangle while he smoked. Obviously, everyone knew how flammable pure oxygen was.

So, "Ham" Love indulged himself whenever the situation permitted. None of his squadron mates said anything about it. After all, "Ham" was one of the most senior and most experienced of the pilots in VFP-62 and he never let his habit interfere with the professional performance of his flying duties . . . almost never.

Sometime during the summer of 1962, when the Cuban missile crisis was heating up and the utilization of all U.S. photo-reconnaissance assets rose in priority, VFP-62 had the need to send an airplane to the Caribbean. The parent squadron, VFP-62, operated three and four-plane detachments from all Atlantic Fleet carriers. One of the carriers, on a training work-up deployment to the Caribbean, put in to port for a weekend in Puerto Rico. They had damaged one of their photo Crusaders and needed a replacement. The requisite messages had been sent and, in short order, Lieutenant Commander "Ham" Love took off from his home base at Naval Air Station, Cecil Field, Florida for the nine hundred fifty mile flight to Naval Air Station, Roosevelt Roads, Puerto Rico.

It was not a particularly difficult flight. The photo Crusader had plenty of fuel for the trip and the weather was good, so it was decided to send the airplane alone. There were no other airplanes going to Roosy Roads on that particular day so the decision to send "Ham" Love alone was not an unusual one. Of course, the squadron would not have sent an inexperienced young pilot . . . but "Ham" was an old pro! The only caution for the flight was the absolute requirement not to stray over the island of Cuba. Russian-built Cuban MiGs could be expected to try to shoot down U.S. pilots who wandered over their turf and into their sovereign airspace.

The leg from the southern tip of Andros Island to Puerto Rico was specifically laid out because it carried the airplane traversing

CHAPTER 15: SHARK BAIT

it, just to the east of the eastern tip of Cuba. But, it was a long leg.

It was while on this leg that Lieutenant Commander "Ham" Love engaged his automatic pilot, noted that it was holding heading and altitude properly, and proceeded to have a smoke. "Ham" claims to this day that he shut off the oxygen valve on the left hand side console before he unsnapped his oxygen mask letting it hang on the left connector while he drew out the cigarettes and the lighter. There was the anticipatory pleasure experienced as he tapped the cigarette on the face of his wrist watch to pack down the tobacco fragments. Then he put the cigarette in his mouth and flicked the lighter!

"POOF!" There was a loud explosion and the next thing "Ham" Love knew he was hanging in his parachute descending quietly toward the green waters of the Caribbean Sea!

Crusaders normally cruise at 35,000 feet and higher for long range missions. That's where they get their best cruise performance. The adiabatic lapse rate for upper air temperature on a standard day is two and one-half degrees per thousand feet. Therefore, off the end of the Island of Cuba, whose surface temperature in October at mid-day is normally in the high sixties, the temperature at 35,000 feet ought to be about thirty degrees below zero. It is for this reason that "Ham" Love's ejection seat was designed to free-fall all the way to ten thousand feet before the barometric trigger automatically opened the main parachute canopy. To stabilize the seat during the 90 second free-fall, a small drogue parachute deployed from the seat's headrest 1.2 seconds after the ejection seat fired.

"Ham" never fired the ejection seat himself. The cigarette lighter took care of all of that for him . . . free of charge (pardon the pun). Somehow a flammable mixture of pure oxygen and ambient air existed inside the small cockpit when he "flicked his Bic!" The explosion shattered the plexiglass canopy into thousands of small, sharp shards. Since the cockpit pressure inside the pressurized cockpit was higher than the outside air pressure, most of those shards exploded outward and away from his exposed facial skin. But three hundred knots of airspeed (indicated airspeed) swirling around the canopy did two very bad things to "Ham" Love that day. The first bad thing it did was to blow many of those sharp fragments of plexiglass into his body, inflicting dozens of small wounds on his exposed face, hands and wrists.

The second bad thing the errant gale did was to grab the ejection seat face curtain handle, pull it out of its retaining brackets behind and above the pilot's head and initiate the ejection seat firing sequence. All of this happened in fractions of seconds. "Ham" Love remembered nothing of being blasted out into the three hundred mile per hour gale of 30 degree below zero air. Nor does he, remember the drogue gun firing as the seat moved up the rails and out into the slipstream. Neither does he remember the drogue canopy deploying and stabilizing him in his seat in a face down attitude while the assembly plummeted earthward for a full minute and one half. Nor does he even remember the seat automatically separating from him and the enormous deceleration force of the main parachute canopy opening.

What he does remember, however, is becoming aware (still in a state of shock) that he was hanging in a parachute in the deathly quiet, looking at the green rippled surface of the Caribbean Sea two miles below him. Somewhere far below him there was the roar of a jet engine, then the sound of an explosion, then quiet.

Something was wrong . . . terribly wrong. His face was wet . . . and so also were his hands. When he opened his eyes and focused on his hands, he saw that they were covered with blood. So also was his flight suit. He touched his face, not feeling any pain whatsoever, and his hands came away covered with even more blood. On his wrists and hands he saw numerous small cuts . . . all of them bleeding profusely. Jesus! What in the world had happened? How did he get here? Then he remembered. The last image recorded by his retina was of the small flame at the end of the cigarette lighter in his hand!

"Ham" Love's guardian angel was there on duty, thank God! A lookout on a Polish freighter bound from Cuba's southern port of Cienfuegos and declared for Gdansk spotted the Crusader and had been watching its passage overhead. Then he saw the flash, the plane crash in the ocean and the tiny white speck of a parachute. It took several calls to the bridge before the Captain was convinced that he should alter course and head towards the descending pilot. But, he had plenty of time. The parachute descent took almost ten minutes.

It was during that ten minutes that "Ham" Love took stock of his situation, deployed his two-man life raft on a lanyard suspended below him, ignited a day flare to get the attention of the freighter, and inflated his flotation gear in preparation for water entry.

With the exception of the blood which, he had concluded, must look worse than it really was, he was really in pretty good shape. All he had to do after water entry, was to get rid of his parachute, climb into his raft and await the arrival of the freighter which, he had already noted, had obviously seen and turned toward him. Piece o' cake!

It was only after he felt the bump under the raft that the sudden realization came over "Ham" Love that he was not alone in this particular stretch of the Caribbean. Looking up he saw the dorsal fin immediately followed by another bump and another dorsal fin. Looking down, into the raft, he saw that the accumulated water in the raft's bottom was dark red. The blood-stained trail of water behind the raft gave him the answer. He had been chumming for the last five minutes without knowing it. Another look around him and he counted at least a dozen more dorsal fins . . . then came another bump. But, the last bump almost overturned his fragile raft! The shark, a good eight feet long, rolled on its side and he had the horrible view of a huge gaping maw, rows of sharp serrated teeth and a single black eye, the color of dull obsidian! Fear, raw, animate, all-encompassing fear seized him as it had never done before . . . never in his whole life.

Now, the position of the freighter became all important to him. It was still a mile away and slowing down to put a small boat in the water. He heard a harsh, raw, gut-wrenching scream before he recognized his own voice. "HURRY, SHARKS, HURRY, FOR GOD'S SAKE!"

The .38 caliber revolver appeared in his hand as if conjured there. He fired five times at the nearest dorsal fin before reason got control once again. Jesus, don't shoot a hole in your own raft! Calm down. Take several deep breaths while you reload another six rounds from your bandolier. Then. "Ham" Love systematically fired at the nearest dorsal fin, slightly below its base in the hope of doing some damage. The gun made a great deal of noise, but the sharks ignored it. There was another bump, this time a big one which nearly overturned his raft. He panicked and grabbed the gunwales of the raft in an almost frantic effort to right it. The gun fell into the water but came to rest at the end of its lanyard. Thank God for the lanyard! "Oh, Jesus, don't let them get me!"

He recognized the voice as his own . . . and he was sobbing in abject terror. Then, he remembered the gun as he began loading it for the second time. SAVE ONE BULLET! SAVE ONE BULLET! A calmness came over him as he made his resolve. The oars were flashing in the small boat as it approached him . . . still a quarter of a mile away.

How long, he asked himself would it take for them to get to him . . . two minutes, three minutes . . . maybe even four. Oh God, another bump. He was sobbing again. SAVE ONE BULLET!

He resolved as he watched transfixed, another dorsal fin approaching. If I get over turned, they will get me before I can get back into the raft. There are too many and too close to hope to get back in without losing a leg or an arm or half a torso. OH, JESUS, HURRY . . . HURRY! I will save one bullet, and when I get knocked into the water I will put the barrel of the revolver into my mouth before those teeth get me . . . and pull the trigger!

"Ham" Love went on to become one of the more professional photo reconnaissance pilots in the U.S. Navy.

"Ham" Love also stopped smoking in airplanes!

16

SCRAMBLE

Of the many hurried launches I've made, the one I remember best occurred on New Year's Day in 1963 at NAS Key West, Florida. I assumed charge of the Fighter Squadron Sixty-Two air defense alert detachment for a five day period beginning 28 December 1962. My wingman, Lieutenant (junior grade) Ben Walker and I were to begin our four hour stint at six o'clock that morning. The evening before we had gone on liberty in the town of Key West to ring in the new year.

One thing led to another and, before we knew it the time was two o'clock. We headed for the base. Enroute, we argued the pros and cons of going to bed at all. Finally, we decided that it made the best sense to go directly to the alert office and assume our duties. Then, we could turn in and not have to get up, shave, shower etc. The only down side to our plan was that we would have to get completely dressed in our flight gear, all forty pounds of it, and then crawl into the alert pad bunks. It wasn't the most comfortable way to sleep . . . but it would be, hopefully, an uninterrupted sleep.

It had seemed like such a great idea on the ride back from Key West. Now, as I lay in the top tier of my metal bunk, tossing and turning, I was beginning to wonder about it. Finally, at about five o'clock, I rolled out of the bunk, poured a cup of coffee, grabbed a doughnut (one of those store bought ones covered with white powdered sugar) and began to shave. The doughnut was the kind that congeals in your mouth and sticks to the palate like wallpaper paste. It wasn't the most pleasant of my moods.

Since my DOP kit was back in my room at the bachelor officer's quarters I borrowed the one belonging to the detachment duty officer. I noticed that the gentleman used one of those aerosol cans of shaving cream, and the cream came of the can colored light blue . . . the same color as the hair of old ladies in nursing homes. Looking dolefully at Dorian Gray in the mirror, I drew the blade through the blue foam and realized it was one of those throw away razors that the owner should have thrown away months ago. Christ, I felt lousy!

I took a sip of very old, but scalding, coffee from the styrofoam cup. Even though it burned the inside of my mouth, it tended to peel the congealed doughnut off of the roof of my mouth. My commentary on the quality of my New Year's Day breakfast brought a chortle from the duty officer.

"Don't worry", he assured me. "I sent the duty yeoman to the canteen for a thermos of fresh coffee and a dozen of the fresh bakery doughnuts you love. He should be walking in at any moment."

That was when the klaxon went off with an ear-splitting roar!

I dropped the razor and sprinted for the door, grabbing my kneeboard as I went by my bunk. There was a bat-wing door leading out onto the second floor landing. I hit the doors hard with the weight of my body . . . like a linebacker. They flew open like an explosion catching the yeoman carrying the tray of coffee and doughnuts directly in the midriff. The two of us went down in a crash. Scalding coffee and doughnuts flew everywhere. I remember a terrible burning sensation on top of my head where a dollop of steaming coffee landed . . . right in the bald spot.

"Christ," the yeoman screamed as I rolled to my feet and went over the top of the stairway like a waterfall. All of the cards from my kneeboard had come out and they were cascading down the stairwell while I scampered down the steps three at a time trying to beat them to the main floor of the building. I went out of the building about twenty feet behind Ben and began the seventy-five yard sprint across the tarmac to the airplane. I caught sight of the ground crews starting up the engines of the jet starting carts and waiting for us. Ben was puffing mightily as he slowly opened the distance between us.

A piece of that confounded doughnut had gotten caught in my windpipe and I was gagging over it as I put my boot on the bottom step of the Crusader's boarding ladder. The ground crew was giving me strange looks. I was not a bit surprised at their astonished glances. Here I was, running at full tilt, coughing up small pieces of doughnut, my face covered with blue foam and my head and flight suit soaked in hot coffee from which the steam was still emanating. What a sight!

I brought the throttle around the horn and into the idle detent before starting to strap in. While the engine was winding up I was hooking up my shoulder harness and lap fittings. When I jammed the flight helmet onto my head I felt the blue foam shaving cream ooze out around the edges and into my oxygen mask.

"Are you okay, Sir?", the plane captain asked as he helped me strap in.

"Jesus, I feel awful," was all I could think to say to the startled young man as I nodded for him to jump clear and disconnect my electrical power. I slammed and locked the canopy and, as I poured the coal to the throttle, looking up, saw the duty officer standing on the ramp next to my airplane holding up the plexiglass data board. On it was hastily printed in large grease pencil letters:

"VECTOR 185 DEGREES, GATE,
ANGELS SIX. 284.6 HAPPY NEW YEAR!"

The information had come over the telephone to him as we ran from the building. It told me all I needed to know for the moment. Our initial compass heading for the scramble was one eight five degrees magnetic. "Gate" meant use full afterburner for the intercept. Angels six meant for us to go out at an altitude of six thousand feet. Our assigned radio frequency was two hundred eighty-four point six megacycles . . . The season's greetings was gratuitous!

There was even a smiling face cartooned at the bottom of the board. I didn't feel like smiling! My head burned on the outside, ached on the inside. My mouth felt as though I had eaten wallpaper paste. All I could think of was to ask myself, why am I doing this?

We didn't set a scramble record that day; but we still beat two minutes. As soon as my Crusader lifted off the runway I rolled it into a 75 degree bank angle to get it pointed south in a hurry. The word "GATE" in our vector instructions (full afterburner) gave an adequate indication, I thought, of the urgency of the scramble order. At about three hundred feet I entered a solid cloud bank and realized that Ben and I needed vertical separation. Keying the microphone button, I gave Ben his instructions. He was behind me by a few seconds.

"Two, this is One. Take angels five. Over." I heard two radio clicks and was satisfied that, as a minimum, he and I wouldn't run into each other in this crap. The mental note was also made that getting back into Key West, with a three hundred foot overcast was not going to be easy. We needed, I decided, to be sure we came back from this caper with plenty of fuel.

Something also had to be done about our vector instructions. We were given directions to go in full afterburner and at six thousand feet. It needn't take a rocket scientist to figure out that once we leveled at six thousand feet, our Crusaders were going to accelerate to well beyond the speed of sound in short order. Did we need this? Of course not!

I decided to query our scramble authority to the Joint Aerial Reconnaissance Control Center (JARCC).

"Fandance Control, this is Silverstep One. Request bogie dope, Over." Then I waited. Ben and I were passing six hundred seventy-five knots and would be, in a matter of minutes, over downtown Havana, Cuba inside half a dozen surface-to-air missile envelopes.

"Silverstep One, this is Fandance. Wait one. Out." Jesus, how silly can you get?, I thought. They tell us to use full afterburner then they tell us to wait. I decided I wasn't going to wait any further; and gave Ben further instructions.

"Two from One. Gates in, now." There were two more microphone clicks. Ben was paying attention, thank God! Our controller probably didn't even know that "Gates in", meant to come out of afterburner. My TACAN needle was beginning to spin around. We were too low to keep a lock on the Key West TACAN station. The last distance measuring equipment reading (DME) said seventy-six miles. At full military power our Crusaders were still clicking off six hundred forty knots. Within seconds we would be over the Cuban land mass. These stupid bastards! I couldn't wait any longer; and called Ben again.

"Two from One. One eighty left, and eighty percent, now." There were two distinct radio clicks as I rolled into a hard left reversal of heading and pulled the throttle back to a sedate cruise setting of eighty percent RPM. Paul and Ben weren't going to get our asses shot down over Havana today because some stupid radar controller didn't know what he was doing.

But, by the same token, our controller had to know that we had deviated from our scramble instructions. After all, I knew, there were people in CINCLANT headquarters in Norfolk as well as the Pentagon who were watching the progress of our intercept with great interest. If anything, I had to be super professional.

Who knows, I thought, I may end up standing before a long green table before this is over. I had better have a good reason for everything I do . . . especially whenever I deviate from my controller's directions.

It had finally dawned on me, as well as those observers in Norfolk and the Pentagon, that this was a false scramble. There was no incoming bogey; and the people at JARCC were probably standing around the radar scope picking their noses and wondering where that funny blip was, that they had seen so clearly a few moments ago.

"Fandance. This is Silverstep One. We have reversed course and reduced speed. Awaiting further instructions. Over."

"Silverstep, this is Fandance. Your signal is RTB. Over."

This was a new and much more authoritative voice which now spoke. RTB meant return to base. I would have appreciated some sort of explanation . . . something like, "sorry we screwed up, fellows, and nearly got you killed in the process. Come on home." But, no such luck! But, enough of this. We would have our hands full just getting back into Key West.

A few minutes later we were given individual ground controlled approach (GCA) radio frequencies and made separate approaches

CHAPTER 16: SCRAMBLE

and final landings at Key West. I never saw Ben for the entire forty five minutes we were in the air. On final approach, my Crusader broke out of the clouds at exactly two hundred fifty feet (the allowable minimum for a single piloted Navy airplane). As I taxied into the alert pad and shut down the engine I felt really wrung out.

Several days later, on our arrival back at Naval Air Station, Cecil Field, the squadron executive officer, Hal Terry told me with a twinkle in his eye, "I heard you and Ben were colorful on New Year's morning!" That, I thought, was putting it mildly!

17
LOST OPPORTUNITY

If there were a category in the annals of Naval Aviation titled Lost Opportunity, it would have to mention Howie Bullman as one of its principal characters. In August 1962 Fighter Squadron Sixty-Two was assigned "hot pad" duty at NAS, Key West. The officer in charge of that detachment was the squadron maintenance officer, Lieutenant Howie Bullman . . . call sign "Kickstand." Howie was one of those unforgettable characters who always evokes smiles when his name is raised in bar talk and sea stories. He derived his call sign from a peculiar habit he had at social events when the third martini began to take effect. His eyelids would droop to half mast and his body would lean several degrees off the vertical. He could stand for hours tilted as though his body were supported by some invisible kickstand. Howie stood about 6 feet 1 inches tall, was of medium build with short cropped brown hair, gray eyes and an absolutely wonderful sense of humor. He was also a very dedicated naval officer, a very skillful pilot and an extremely hard-working maintenance officer. I felt very comfortable knowing that VF-62's Key West "hot pad" detachment was in such capable hands.

The alert assignment for the Cecil Field F-8 squadrons was sort of a second team arrangement. The U.S. Air Force invested some military construction money and built a proper alert installation with air conditioned trailers mounted on concrete slabs alongside alert ramps set adjacent to a taxi way leading directly to the main runway. Whenever the Air Defense Command set a high enough defense condition it deployed Air Force F-4 Phantoms to Key West from Homestead Air Force base.

When they were not there but when the Joint Aerial Reconnaissance Control Center wanted coverage during surveillance flights the F-8's took the second string duty. However, the F-8s were not allowed to use the Air Force's unoccupied facilities. As a consequence, they were forced to park their F-8s on the regular parking ramp, use borrowed ground support equipment and operate out of unoccupied spaces in a hangar nearby. We jury-rigged an acceptable way of carrying out an alert commitment but it was far from satisfactory.

Our two alert pilots slept in an old double decker metal barracks bunk in an un-airconditioned room on the second story of the old hangar. The alert procedures called for the two pilots standing four hour watches, fully dressed in flight gear. The beginning of each four hour watch involved the pilots inspecting their Crusaders, climbing in, starting up the engine, going through all the pre-taxi and pre-take-off procedures then just before shutting down the engine, the F-8 was configured for take-off with the wing up, trim tabs set for take off, helmet plugged in and electric power starting cables connected.

A duty officer sat in a room next to the bunkroom with a klaxon bell mounted in the window. When an alert call came in on the telephone, he would push the flaxon button then copy down the vector, speed, altitude and radio frequency in grease pencil on a large piece of plexiglass. While he was doing this the alert pilots rolled out of their bunks, sprinted down a flight of stairs and across the concrete ramp to the alert F-8, a distance of about seventy-five yards.

Meanwhile the ground crew, having been alerted by the klaxon, started up the electrical power carts and helped the puffing pilots get strapped into the cockpit while they were starting their engines. About this time the duty officer would arrive and hold the plastic board up for the lead pilot to read. Then he ran to the other F-8 and did the same. This was because the F-8 radios would come on line about the time the F-8 was breaking ground on its take-off roll. It was a ridiculous arrangement; so ridiculous, that we did our damnedest to beat the F-4's scramble times . . . and succeeded! The best recorded VF-62 time was one minute and five seconds from klaxon to lift-off. With my stiff knee, my personal best was 1 minute and twenty-four seconds. But we all watched each F-4 scramble with great amusement. It had two engines to start and a more complicated weapons system . . . but I never saw one beat two minutes!

Howie's detachment had been in place only a couple of hours when he and his wingman, Tom Napoli, were scrambled. The JARCC command center was monitoring a surveillance flight by a U.S. Navy P-2 patrol plane which had been flying along the north-

CHAPTER 17: LOST OPPORTUNITY

ern coast of Cuba, presumably outside the twelve mile limit, on an electronic intelligence mission. The P-2 radioed that it was being harassed in international air space by two Cuban MiG-19 fighters.

Of course, Howie didn't know such details as he climbed, huffing and puffing up the boarding ladder of his F-8. All he knew was what the duty officer had written on the plexiglass board which was being held up for his perusal: "Vector 165 degrees", Buster, Angels 5, 284.6 mc!"

This told him to fly out on a magnetic heading of one hundred sixty-five degrees magnetic at full military (not afterburner) power at 5,000 altitude and that his controller's radio frequency was two eight four point six megacycles.

It turned out that the pilot of the patrol plane was a Naval Academy classmate of mine. His account later confirmed Howie Bullman's observations. The two Cuban fighters showed up while the patrol plane was flying a track at ten thousand feet altitude on a course that paralleled the northern coast of the island just north of Havana. The Lockheed built P-2 Neptune had turned back toward Key West and dove his patrol plane to "the deck," firewalled all four engines, the two reciprocating engines and the two turbojets. He apparently thought that flying that low would deter the MiG's from any further harassing passes close aboard.

About the time that Howie and his wingman arrived on the scene, Bill Lancaster sounded pretty excited. He was on the same frequency as Howie and had just informed the JARCC controller that one of the MiG's has nearly killed them all. In fact, the MiG made a high speed pass from directly astern and passed underneath the P-2 pulling up directly in front of the airplane. This would be disconcerting to a pilot at anytime. But when it happened at such a low altitude it scared the daylights out him. He was pretty excited and shouted for JARCC to do something.

Howie Bullman picked up one of the MiG's on his radar and ran an aft quarter intercept on it. He visually acquired the MiG at about four miles and slid into a good shooting position behind him. Howie's MiG was loafing along at about five thousand feet watching the other MiG beating up on the U.S. patrol plane. Its pilot was apparently unaware of the presence of the F-8s as Howie armed his ordnance switches and heard the growl of the seeker head of the Sidewinder missile in his radio headset. Then he made the radio transmission that he lived to regret.

"Fandance, this is Two Zero Eight. In position on one of the MiGs, request permission to fire, over."

"Wait one," was the inevitable response. By the time the controller's supervisor had made contact by telephone with the Admiral who had the authority to grant such a request the Cuban radar controller had long since spotted the two F-8s and alerted the MiGs who then departed the area returning to home base.

When Howie and I discussed it later after consuming several beers and armed with the brilliance of hindsight we had come up with all sorts of alternatives to what Howie did. Most of them were off the wall. Probably the only thing he could have done and avoided a court martial would have been to announce to the controller that he judged the actions of the MiGs to be imminently life threatening to the P-2 and, as on-scene commander, was exercising his authority to defend the threatened U.S. airplane with force. Then he could have pressed the red button on his control stick and smoked the high MiG while his wingman went after the low one.

Whatever he was, the pilot of the harassing MiG had to be a good pilot to execute that last stunt . . . it was a pretty gutsy (and stupid) maneuver. The bar conversations went on and on. "What would I have done?" I've asked myself a thousand times.

Howie Bullman, as a fighter pilot, trained himself to shoot down MiGs his entire flying career. The only opportunity to do so came and went like a wraith. He did the right thing . . . or did he?

18

GUNSIGHT

The gunsight in the Crusader includes a piece of clear glass upon which the gunsight reticle is projected. The glass, of high optical quality, is mounted forward and above the instrument panel between the forward windscreen and the pilot's eye. The gunsight glass, as it was called, was a rectangle about five inches long, three inches wide and a good half inch thick. It weighed about two pounds. The gunsight glass mounting held it standing on its narrow end and tilted back toward the pilot at an angle of 45 degrees. For ease of removal and cleaning, the glass was secured to the metal mounting frame by a pair of black metal clips four inches long; one clip on each side.

It was routine procedure for ordnancemen to unclip the glass, lift it out, clean it with glass cleaner and clip it back in. It was a 30 second operation and frequently done by the pilot. Forward visibility is reduced in a Crusader by the fact that the pilot had to look through a couple of inches of glass, counting the bullet-proof forward windscreen and the gunsight glass. Forward visibility is just as important in the carrier landing pattern as it is in aerial combat, perhaps more so.

I can clearly remember a sense of unease as I taxied forward to *Shangri La*'s port catapult in my F-8 one day in the summer of 1963 in the western Mediterranean. The uneasiness bothered me because, try as I might, I couldn't put my finger on its source.

But, my instincts and forebodings have always been uncannily accurate. I knew there was something wrong and it escaped me.

I went over each item on the pre-catapult take-off check-off list three times; touched each item on the list "Catapult trim set, catapult throttle grip extended, wing incidence set," et cetera. I even repeated it out loud. Still the bad feeling . . . still no apparent thing wrong. By now the nose wheel of the Crusader climbed over the catapult shuttle cover and dropped down again to the flight deck.

Now the flight deck director was inching me forward in small increments until I felt the clunk as the airplane came up against the catapult hold back fitting. It was a bright, sunny day with blue skies and a clear horizon. Nothing to worry about, I decided, and still the nagging doubt . . . the uneasiness, the sense of impending doom engendered by ten years of flight deck experience. Something was wrong! I knew it!

As soon as the clunk of the hold back was felt, the flight deck director gave the shuttle forward signal to the deck edge catapult operator; simultaneously giving me the feet off brakes signal. I ran the throttle to the firewall and grabbed the throttle catapult grip. One more time I went over the check-off list and looked quickly around the cockpit. God damn it, I thought, what the hell is wrong? Where is it? But nothing!

From the corner of my eye I saw the deck edge catapult operator raise both of his hands above his head. Those hands, I knew, would stay right where they were until the catapult officer's hand touched the flight deck. Then, and only then, was he allowed to drop his hands and, with one of them, push the catapult firing button on the console in front of him. From that moment it would take about two seconds for the steam valve to open and the shuttle to start forward; snapping the hold back fitting and sending me and my Crusader off the bow.

The engine was stabilized at military thrust and, one more time I went over the take-off check-off list. Maddeningly, the uneasiness persisted. Oh well, what the hell, I decided; it must be my imagination. Putting my head firmly against the head rest, I gave the catapult officer a snappy salute and returned my right hand to the control stick; positioning it for the stroke. It was at that moment when I shifted my gaze from the catapult officer to the forward windscreen that I saw it! I was horrified!

There, directly in front of my face was the gunsight glass with the two retaining clips peeled forward so that only the forward corner of each clip was holding the glass to the mounting frame. It took a split second eternity of time to understand the implication of what my eyes had taken in!

One of the laws of physics states that a body at rest, tends to remain at rest. At any moment, a fraction of a second perhaps, or would it be an entire second? the catapult was going to fire; and that two pound piece of glass would accelerate to 150 miles per hour while it traversed the thirty inches or so from its mounting to

CHAPTER 18: GUNSIGHT

Left: F-8 gunsight. Right: F-8 gunsight with reticle illuminated. (Photos by Redditt)

the bridge of my nose. Instant death would follow . . . as surely as if I had put the muzzle of a loaded .357 Magnum in my mouth and pulled the trigger. The Crusader would launch, go out of control, crash into the sea and, with me in the cockpit, settle rapidly to the sea floor some six hundred fathoms beneath the surface of the Aegean Sea.

No one would ever know whatever happened to old Paul . . . just another one of those carrier aviation mysteries . . . another unexplained casualty of this risky business. Nano-seconds were creeping by, taking me inexorably closer to the end of my life! My eyes were locked on those two God damned clips. . A signal went from my mind to my central nervous system, traveling at the speed of light. It was transformed to an electrical impulse, also traveling at the speed of light from my central nervous system to my neck muscles. They accelerated my head my head to the left, slamming my flight helmet against the side of the canopy just as I saw the gunsight glass slide up the mounting, the clips snapped shut on air and the glass began accelerating backward, as if in slow motion directly at my eyes. The glass struck the headrest, where my head had been moments earlier and shattered into a thousand tiny shards, half of which seemed to have found their way to the inside of my flight suit collar.

Meanwhile, the Crusader tore down the catapult track with my neck muscles almost tearing in that awkward position . . . semi-supported and leaning against the side of the canopy. But, the important thing was that the Crusader was flying and I was alive. As the itchiness of glass fragments began to abrade the skin around my neck, I made a vow. Never again would I ignore that sixth sense, the persistence, the sense of uneasiness, that important signal that my subconscious mind was desperately sending!

F-8E's, getting ready for a Mediterranean deployment. One of the pilots in VF-62, while on a routine training mission reported on the squadron base radio that the fuel quantity indicator showed that the level of fuel in the tank that fed the engine was dropping at a rapid rate for no explainable reason. The fuel transfer system in the Crusader was complex. There were eight fuel tanks in the airplane; four in the wing and four more in the airplane's long thin fuselage. All of these fuel tanks were pressurized and fed sequentially into one tank, the feed tank. Inside the feed tank were several electrically powered fuel pumps which pumped fuel to the fuel control which, in turn, fed fuel to the nozzles of the engine. Obviously, the level of fuel in the feed tank was critical. Even though the Crusader held over eight thousand pounds of internal fuel, if the feed tank ever went empty, the engine would "flame out." In a single engine jet airplane a flame out was serious. As the saying went, "It could ruin your whole day."

The first incident of the fuel transfer problem ended close to tragedy. The pilot reduced his throttle setting all the way to idle. He was close enough to Cecil Field that he was able to make a straight in gliding descent to a safe landing on the twelve thousand foot long north-south runway. As the Crusader rolled out on the runway the engine flamed-out and a ground crew had to hook up a tractor and tow the airplane back to the VF-62 flight line.

The maintenance technicians were unable to determine what had caused the transfer system to reverse the flow of fuel and pump it backwards out of the feed tank and into the wing fuel tank. There were several more similar instances and the maintenance crew was perplexed. An engine fuel expert was called in from the Ling Temco Vought plant in Dallas. He was unable to locate the culprit. As the date of our impending deployment approached, this unresolved problem became a major worry for the entire squadron. A flurry of messages went back and forth from the squadron to the Naval Air Systems Command, the engine manufacturer, Pratt & Whitney, and the airplane manufacturer in Dallas.

Finally, in desperation, the Naval Air System Command directed a series of tests to be conducted on the next squadron airplane to develop the peculiar transfer anomaly. But, it would be necessary for the pilot to keep the engine running long enough for the maintenance technicians to open up an access panel of the airplane and conduct an inspection of a fuel transfer sequencing valve. This requirement and the procedure was thoroughly and repeatedly briefed to all squadron pilots to ensure that this serious problem could be resolved. Since VF-62 was the only fleet squadron operating that particular block number of the F-8E at this time, all eyes were watching us.

Another recurrence didn't happen. The squadron deployed in *Shangri-La* to the Mediterranean with the fuel transfer problem unsolved. As squadron operations officer I was particularly concerned and attuned to any indication whatsoever of an impending fuel transfer problem. As a consequence, when Stu Harrison told me, that day, of the symptoms his airplane was displaying, I had no hesitation in asking for *Shangri-La* to execute the pull forward. Fortunately, the problem had developed during the daytime and the ship was only thirty miles away. Stu would be able to set his throttle at idle RPM and make a straight-in descent to a safe landing before his engine flamed out. I was holding my breath. If only the ship could complete the pull-forward and declare a ready deck by the time we got there. It would take perfect execution by a lot of people to get Stu safely back on deck.

As we descended from thirty thousand feet I kept checking on the quantity of fuel in Stu's feed tank every couple of minutes. The numbers that Stu read back to me were scary. It looked to me that at the present rate of reverse transfer Stu would barely make it to a landing before his engine quit. Descending through five thousand feet I noted that we were much too high to complete a straight in approach to a carrier landing. The ship was turning into the wind and had about ten more degrees of turn to complete when the Air Boss called on the land/launch radio frequency the heart-warming radio announcement, "Silversteps, you have a ready deck." I rogered his transmission with a sense of real relief. "With a little more luck," I thought, "we're going to make it."

Passing through three thousand feet, I could see that we were still too high to make it. Certainly, I didn't want Stu to have to do a turn around the landing pattern. There wasn't enough fuel in his feed tank to do that. I suggested, as I flew on Stu's left wing, that he "dirty-up." In response Stu put his tailhook down, lowered his landing gear and raised his wing to the landing position. I matched his configuration changes with my own hook, wheels and wing.

Fortunately, Lieutenant John Nichols was the Landing Signal Officer (LSO) on the platform that day. Although John was one of the more junior LSO's in the air wing, he was clearly the best. All I could think when I heard his voice on the radio saying "Paddles is up," was "thank God John is waving."

The optical landing system represents a cone of airspace pointed up and aft from the mirror at an angle of four degrees. Anytime an airplane is inside this cone its pilot can see the "ball" located somewhere on the mirror. The location of the ball relative to the horizontal row of green datum lights tells the pilot where he is vertically in the cone. The cone narrows to a single point as the airplane touches down at the number three wire. If the airplane is above the center of the cone the ball is seen above the datum lites. If the pilot is low the ball appears below the datum lights. If the pilot gets dangerously low the ball also turns red to indicate danger.

We descended into the cone about four miles behind the ship. I was trying to fly on Stu's left wing as close as I could and still monitor my own instrument panel to be of assistance. As we descended into the top of the optical landing cone I glanced at the ship and saw the ball way above the datum lights. I knew that we were "fast as a fox," but when I flicked my eyes to my own airspeed indicator I almost burst into tears.

CHAPTER 20: DEAD STICK

"Jesus Christ, one hundred seventy-five knots," I shouted to myself. "He'll rip the tail right off the airplane if he catches a wire at this speed." We were still at idle power and decelerating but I knew we would never slow down forty-five knots in the next twenty seconds . . . and that was all the time Stu had . . . in his life. A minute ago Stu had announced that his fuel gage indicated zero fuel. He would never be able to execute a go-around. I also knew that in another ten seconds Stu would be outside the envelope for a safe ejection.

Now we were passing five hundred feet, still decelerating when the ball went off the top of the mirror and disappeared. In utter despair I started to add throttle intending to tell Stu to do likewise, level off and eject when I noticed him dropping below and behind me.

"What the hell?," I shouted into my oxygen mask. I looked back over my right shoulder as I leveled off about fifty feet above flight deck level and flew along the left edge of the landing area. To my utter astonishment I saw Stu's airplane rolling to a stop. His tailhook had engaged the number one wire. "How in Christ's name did he do that?," I wondered to myself as I turned my own F-8 onto the downwind leg for my own landing.

"Take it around, Two Zero Six," came John Nichols crisp orders to me as I saw the red wave-off light flashing at me. As I flew by the left side of the landing area I saw Stu's F-8 still sitting in the landing area. The hook runner was swinging his crowbar at the wire tangled around the F-8's hook point. Climbing to six hundred feet I turned down wind and set up for another approach. I had mixed feelings as I turned abeam for my next carrier approach. I was thanking God that my friend Stu Harrison was alive and safely aboard *Shangri-La*. On the other hand I kept wondering how in the hell Stu had managed to fly from a position off the top of the optical landing system cone and thirty knots too fast, to a one wire. It was simply not physically possible for an F-8 to do what I saw it just do. Immediately before I added power, my throttle had been at idle. There was no possible way to kill off thirty knots of excess speed and simultaneously traverse from the top of the landing cone to the bottom of it without splattering the airplane all over the flight deck and ripping the tailhook right out of the keel.

Again John Nichols cool voice waved me off and as I flew by Stu's airplane I saw that it was being towed out of the landing area and across the flight deck's foul line. My final approach was another uneventful carrier landing. As I unstrapped and climbed out of the cockpit I kept muttering to myself, "How, in Christ's name did he do that?"

When I got to the ready room Stu was no where to be found, neither was the CO. When I asked what the hell was going on I got nothing but mystified looks from the several pilots lounging in their ready room seats. The squadron duty officer told me that the C.O. was up in flight deck control and Lieutenant Commander Harrison was in his stateroom.

Fifteen minutes later I sat down on Stu Harrison's bunk in his stateroom. Stu was nursing the second of two medicinal brandies which the flight surgeon had given him as he told me the most incredible story of a carrier landing I had ever heard. Stu recounted seeing me add power on my F-8 as I levelled off. At that moment the ball had just disappeared off the top of the lens and he noted his airspeed indicating one hundred seventy knots. Also, at precisely that same moment Stu's engine flamed out!

Stu said he began programming the control stick aft with his right hand and at the same time reached over and down with his left hand and grabbed the alternate ejection seat firing handle located between his knees. Gripping the ejection handle tightly in his left fist, with his heart in his mouth, he saw the ball reappear on the top of the lens. Stu and I both estimate that he was about one third of a mile from the ship when his engine quit. As the ball began to move down the lens Stu continued moving the stick back, raising the nose of his F-8 as it continued decelerating. Stu said he just stared at the ships steel ramp as he hurtled toward it.

Stu calculated that he had one or two seconds to decide whether or not to eject before he would be outside the safe ejection envelope. A strange choice! Would he die in a fireball on the steel ramp of the U.S.S. *Shangri-La* on a bright sunny day in the Aegean Sea, or would he drown in the ship is turbulent wake knocked unconscious by the violence of his one hundred fifty mile an hour impact with the water? Stu said he decided to "stay with the f#%@ing airplane."

He continued slowly raising the nose of the F-8. The ball began to descend on the Fresnel lens and turned red as Stu saw the airplanes angle of attack indexer go from a green (fast) chevron, to an amber (on speed) donut, to a red (slow) inverted chevron. The ramp was rushing at him as he watched it heave upward in his field of view.

"So this is how it feels to die," was his last thought as the F-8 slammed into the flight deck. The F-8 seemed to accelerate as it roared across the ninety feet from its touch down point right at the edge of the ramp to the number one wire. The tail hook engaged the cross-deck pendant and yanked Stu from eternity to the present. He had just made the first dead engine carrier jet landing in the history of carrier aviation!

The flight deck director, unaware that Stu's engine had flamed out, gave him the hook up and taxi forward signal. When he got no response he repeated the signals, annoyance clearly evident by the expression on his face.

Meanwhile, the squadron C.O., having been apprised of the fuel transfer problem, rushed to flight deck control and donned a brown colored flight deck jersey, helmet and goggles. He was going to personally see to it that he and the maintenance chief petty officer got to the bottom of this mystery. As Stu Harrison's F-8 trapped on board, Joe Simon and the maintenance chief ran out to the airplane stopped in the landing area. Joe Simon stood on the right side of Stu's airplane frantically giving him the two-finger circular signal meaning "keep your engine running."

Having just barely escaped death, Stu watched the flight deck director and his own squadron commanding officer signalling him to do things he couldn't do, with no small amount of disgust. Finally, after shaking his head several times he flipped them both the bird, popped open the canopy and climbed out of the cockpit.

Stu's cavalier obscene gesture sent Joe Simon into a towering rage. Running around to the left side of the cockpit, the C.O. caught Stu Harrison just as his foot touched the flight deck. Almost overcome with anger, Joe Simon screamed at the top of his lungs, "God damn you, Harrison, I told you to keep that engine running. Who the hell do you think you are giving me the finger? I'm your commanding officer and I'm throwing your ass in hack (confinement) for the rest of this cruise."

Stu's response was equally strident. "God damn it, Skipper, the reason why I didn't keep the engine running was because the f#%@ing thing quit while I was in the groove." With that Stu turned on his heel and walked into the history book.

21
CRASH LANDINGS

U.S.S. *Shangri-La* departed Naval Station, Mayport, Florida on the morning tide 8 April, 1963 headed for a work-up cruise in the Caribbean Sea. It was to be the final preparatory work-up prior to her scheduled deployment to the Mediterranean Sea that same year. While in the Mediterranean *Shangri-La* would be ridden by the battle group commander reporting to the Sixth Fleet commander. Fighter Squadron Sixty-Two, an element of Air Wing Ten, would take to sea aboard the "Shang" with twelve brand new F-8E Crusaders.

The airplanes were new in several respects; first, the airplanes had come fresh from the production line at the plant at Dallas. Second, and just as important, they were the first U.S. Navy carrier airplanes equipped with an automatic throttle, or approach power compensator (APC) as it was called. The APC ushered in a whole new concept in carrier aircraft operations because it represented the first, and one of the major, components of a system which would ultimately become known as the automatic carrier landing system (ACLS).

The third new feature on these airplanes is one which still raises eyebrows today, thirty years later. It was an infra-red search and track system (IRST). Today the research and development community is still struggling to master the technology necessary to field an operational capability in the 8 to 12 micron range of the infra-red spectrum. Thirty years ago, I keep reminding today's aircraft design engineers, the U.S. Navy fielded an operational IRST capability in the 3 to 5 micron range which actually worked.

Infra-red detection devices in the 3 to 5 micron range can see the radiation from hot metal tailpipes of jet engines. Devices under development now in the 8 to 12 micron range can see the much more subtle radiation resulting from skin heating of high speed air vehicles.

Unfortunately, in 1963 IRST was more of a novelty than anything else. This was before effective radar jamming became a reality. It was also before there was such a thing as stealth. So, therefore, IRST was a complementary detection device that was not as good as radar . . . therefore not as useful.

Today, IRST is enormously important. The designers of stealthy aircraft have mastered the challenge of reducing the radar cross-section of airplanes as well as the 3 to 5 micron radiation signatures associated with their engines. However, no one has figured out how to reduce the 8 to 12 micron signature of airplanes. So, IRST devices in development today represent a significant technological break-through which will essentially negate most of the advantages which presently accrue to stealth vehicles.

Our new F-8Es had an IRST which, as a passive detection device, had enormous potential. I immediately became a devotee of IRST and practiced with it as often as possible while on cruise. It was mounted on top of the fuselage just forward of the canopy center windscreen, was slaved to the radar dish and employed a four bar raster scan which enable the pilot to search with it in the same way he scanned with the radar. I found that I could detect an F-4 engine signature at high altitude at night at ranges out to thirty miles with ease.

Unfortunately, the system was not adequately supported with spare parts and had a high failure rate. Our electronics technicians had received no training in its repair. As a consequence, after the first month of the cruise not one set was working and the system fell into disuse from which it never recovered.

But, the automatic throttle was another story . . . a success story. It had been installed in our F-8Es as a conscious decision by the Naval Air Systems Command. As a principal component of the automatic carrier landing system, the APC needed to be evaluated in the fleet. Meanwhile, the other, more sophisticated, components of the system were being developed and tested at Patuxent River in a modified F-3D Skyknight.

The Crusader, with its notorious speed instability problems, was the obvious choice of fleet airplanes on which to conduct the evaluation. When it was engaged, the APC caused the throttle to be adjusted continually to maintain optimum angle-of-attack (speed) for a carrier arrested landing almost regardless of what the pilot did with the flight controls. The device was very sensitive and whenever the control stick was moved, however slightly, the throttle

The author's second crash landing in two days on U.S.S. Shangri La, 1962. Note the aircraft is exactly on centerline. (U.S. Navy photograph)

would move in immediate response to correct for the resultant change in aircraft angle-of-attack. If, for example, a pilot lowered the nose of his airplane to correct for being above glide slope, the throttle would automatically make a compensatory reduction in power. The system could only be engaged when the airplane was in the landing configuration (wheels down and wing up).

There was a general reluctance on the part of many of the squadron pilots to use the APC. Their concern was that if they ever became too dependent on it, what would they do when the system was inoperative? Would they gradually lose their basic carrier landing skills achieved with such effort in the "manual" mode? In fact, the system had enough maintenance problems at first, that the pilots got plenty of practice in the manual mode whenever a component malfunctioned or the system went out of adjustment. The APC came with a few built-in safety features.

If, for example, a pilot wanted to shut the system off, he could do so by simply turning off the engage switch. When in a hurry or in extremis, a pilot could physically disengage by exerting a pushing or pulling force on the throttle of over forty pounds. This would physically disengage the system by opening a clutching device.

Regardless of which disengage method was used, the system could be reengaged electrically with the engage switch. Personally, I discovered that the APC could be employed as a training device whenever it was engaged. It proved to be excellent in that regard. All a pilot had to do was rest his left hand lightly on the throttle and pay close attention to its movements during the carrier pass. The system would sense the need for a power change much earlier than even an experienced Crusader pilot could. Furthermore, it knew exactly how much of an adjustment to make. The secret of how much to move the throttle was something a pilot learned only from long experience, (and he never learned it as well as the APC).

Therefore, by following the throttle movements with his hand, an attentive pilot could improve his landing technique as well as his flying skills. As the squadron operations officer, the task fell to

CHAPTER 21: CRASH LANDINGS

me to get the pilots to use the system and to believe in it as a training device. Some of the older, more diehard pilots did not believe my thesis, but I proved it to be true at least for me. As a true believer in the system, the training aspect of it worked for me. My manual landing grades (assigned by the landing signal officer) improved steadily over the course of the Mediterranean deployment in spite of the fact that I never made a manual pass unless the APC was inoperative.

Unfortunately, the squadron was not scheduled to take possession of its new F-8Es until shortly before deployment. Consequently, we were forced to complete our last Caribbean work-up in the old F-8Bs which we had been flying for several years, now. There was one feature about our old F-8Bs (some of them) of which we were unaware. It caused the squadron no small amount of grief and ultimately caused me to make the only two carrier landing accidents of my career. This particular feature was far too subtle for anyone to notice in the due course of events. It took two carrier landing crashes to disclose the problem. Both of these crashes, unfortunately, occurred to me and on two successive landings.

The problem centered around the airplane's nose wheel trunnion. This was the point at which the F-8's nose wheel attached to the airplane. The trunnion was a rotating mechanism. Whenever the wheels were extended or retracted, the nose wheel strut rotated about the trunnion as it folded into, or extended from the wheel well.

The two attach points on the left and right bulkheads of the nose wheel well bore the weight of the forward part of the airplane as it sat parked on deck. However, during a normal arrested landing, the entire weight of the airplane as well as the dynamic arresting load was borne momentarily by those two attach points. This tremendous combined load occurred near the end of the arrestment roll-out when the airplane's mainmounts actually left the flight deck for a fraction of a second. When it became apparent that this phenomenon was occurring, not enough concern was expressed by the operators as to its long-term effect on the structural integrity of the airplane. Finally, when metal fatigue was revealed during routine structural inspections, the engineers at Dallas came up with a fix. A large metal stiffening patch was riveted to the inside walls of the nose wheel well just under the trunnion attach points.

A modification kit was fabricated at the plant and sent into the field for installation on all fleet F-8s. The installation was accomplished by "tiger teams" (skilled metalsmiths) from the plant who actually went aboard the carriers at sea to install the kits. A similar stiffening modification was introduced into the assembly line at Dallas for all new production Crusaders.

Of course, wise engineers knew that the fix also accomplished the side effect of shifting the load elsewhere in the load path structure of the airplane . . . location unknown. Eventually, we all knew, the effect of these enormous loads would reappear elsewhere in the form of some other symptom . . . maybe worse . . . maybe not.

By some unusual accounting oversight several old F-8Bs existed in the fleet without having had the stiffening patch installed. Unbeknownst to us, two of these airplanes ended up in Fighter Squadron Sixty-Two . . . bureau numbers 145498 and 145455.

On the night of 25 April 1963 I manned up F-8B, bureau number 145498 for a night air intercept training mission. My wingman and I flew the mission without incident, alternating as target and interceptor until the time came to return for recovery. The flight had been otherwise uneventful as I approached the ramp for my night arrestment. In the last few seconds of the approach the ball went just a hair high. Since I was also about two knots fast, I decided to make only a slight nose-down correction to stop further upward movement of the ball. To have done any more would have been to invite an immediate wave-off by the LSO. I did not want that. The correction effectively stopped the ball about half a ball high on the Fresnel lens. The Crusader crossed the ramp, touched down on the ship's centerline and engaged the number four cross-deck pendant. The LSO had already begun instructing the LSO trainee, who was taking notes, to give me an OK grade with the comment "nice catch." What ensued in the next second and one half devastated me!

As the airplane began to decelerate there was a bright flash underneath the forward part of the fuselage accompanied by a loud explosion. The forward fuselage slammed down onto the flight deck so violently that my left hand slid off the throttle and my left elbow struck the wing incidence handle. The impact was so great that my entire left arm immediately went numb. At the same time I was thrown forward in the straps much harder than was normal, to the extent that my upper shins slammed into the bottom edge of the instrument panel. In addition, I noticed that the airplane stopped rather abruptly accompanied by a loud, grinding and scraping sound. I also noticed that I was now sitting much closer to the flight deck than normal.

I immediately shut down the throttle and, as the engine wound down, the emergency flight deck flood lights came on, blinding me. The crash crew and the hot suit man appeared out of nowhere and swarmed over the airplane SWOOSHING their CO_2 fire extinguishers into the tailpipe. I opened the canopy, unstrapped quickly and stepped out onto the flight deck without needing the airplane's boarding steps. Reviewing the damage, I saw that my airplane sat awkwardly on its engine intake duct which was flattened to the almost closed condition. The mangled nose wheel assembly was scattered in small pieces all over the flight deck. I couldn't believe what had happened. Probably, I was in a state of semi-shock.

Later, down in sick bay, X-rays were taken of my badly bruised elbow. There were no broken bones but the elbow was black and blue. X-rays were also taken of both legs but, again, there were no broken bones. There were just black and blue bruises across the shins where the sharp lower edge of the instrument panel had made contact with them.

The wrecked airplane was taken below and spotted in the starboard aft corner of hangar bay Three, its nose supported by a hydraulic jack. There it sat, looking as dejected as I felt, awaiting the results of the inevitable accident investigation.

The sense of depression which I experienced was alleviated somewhat by the landing signal officer's written statement on the accident. In it he assessed my landing as an above average pass and not contributory, in any way, to the accident.

Quickly, an accident investigation was commissioned and the squadron executive officer (he had to be senior to me) was its president. The event was duly reported up the chain of command. All of this was accomplished within four hours of the crash. A flurry of high precedence message traffic was generated between the ship and various agencies, some in Washington, others in Norfolk . . . and thence to the manufacturer's plant in Dallas.

All of this, of course, took time, and there was the imperative of continuing training which needed to be conducted to get ready for the imminent Mediterranean deployment. As a consequence, I was scheduled for a routine training mission on the following day, black and blue bruises notwithstanding. For this flight I manned up F-8B bureau number 145455!

Again, the flight was uneventful until the landing. I recall paying extra special attention to my carrier landing pass. My F-8 crossed the ramp on centerline, with the ball dead center on the lens and my angle-of-attack indexer showing an on-speed amber donut. The tailhook engaged the number three wire (target wire) and rolled out on centerline. Near the end of the roll-out a series of events occurred identical to the ones of the previous evening!

The airplane's nose slammed down onto the flight deck crushing the engine intake duct. The entire nose wheel assembly exploded into hundreds of pieces. My poor left elbow took another wallop on the wing incidence handle as, simultaneously, my sore shins again made sharp contact with the bottom edge of the instrument panel exactly in the same painful place as the night before. The only difference was that, it being a day landing, my heart was not pounding quite as rapidly. But, the pain was intense. I'm sure I screamed into my oxygen mask. Numb with pain, I sat in the cockpit for a few seconds, but the swarm of emergency rescue personnel spurred me to evacuate the cockpit as quickly as the night before. One never knows what is going on in such cases. Their appearance in a swarm is always intimidating.

Again I stepped out of the cockpit onto the flight deck without the need of the boarding ladder. Viewing this new damage with total frustration, I felt like drawing my .38 service revolver and firing a round into the fuselage of my stricken airplane, the same way John Wayne might have shot his horse with a broken leg. As the crash crew swarmed over my airplane the thought occurred to me that I had single-handedly given them more live practice in the past twenty-four hours than they had gotten, thus far, in the entire deployment. As I dejectedly watched the crash crew SWOOSHING their fire extinguishers I felt a tug at my sleeve.

Turning my head, I found myself looking into the innocent blue eyes of a young marine orderly. "Sir", the young man said, "The Captain wants to see you on the bridge right away!" A few minutes later I stood on the carrier's bridge facing a very solemn-faced senior Naval officer. "Paul", he asked with genuine concern in his voice, "What in Christ's name are you trying to do to my airplanes?"

Later that same day a message came from Washington, D.C. directing us to inspect all of our F-8s prior to the next flight for the presence of the required nose wheel trunnion stiffening patches. We duly inspected them all and found proper patches in all but the two which were now wrecks in hangar bay Three.

My unease at being the wrecker of squadron airplanes was again assuaged by the fact that the LSO gave me a carrier landing grade of (OK)3 which is about as good as you can get. Again, the accident board concluded that I had contributed in no way to the accident.

No one adequately determined how the accounting mistake in incorporating engineering change proposals (ECPs) had occurred. But, it had, nonetheless. I flew F-8s for twelve years before moving on to F-4s. Never again did I crawl into the cockpit of a Crusader without first painfully crawling up into the nose wheel well and inspecting the trunnions!

22
MORTE TU CAPO

Every American over age forty can probably remember where he was, what he was doing, how he was dressed and a host of other details about the 22nd of November 1963. I know that I can! That was the day President John F. Kennedy was assassinated!

A great deal of flying occurred from the beach in the course of carrier operations during *Shangri-La*'s deployment to the Mediterranean Sea that year. This was because the carriers always used the airfield near Naples as a base of operations for a shore detachment whenever they were operating in the Tyrrhenian Sea west of the Italian peninsula. This shore detachment at Capodichino would conduct routine corrosion control work on airplanes which could be done much more easily ashore where there was plenty of space and a plentiful supply of fresh water. Neither of these conditions existed on any deployed U.S. aircraft carrier. As a consequence, flights to and from carriers, out of Capodichino were frequent and routine events.

It was exactly one hour before official sunrise when the Crusader went roaring down the catapult track and into the black closet with me hanging on for dear life! It was not an uncomfortable event

VF-62 F-8E (U.S.S. Shangri La) over the Tyrrhenian Sea, 1963. (Author photo)

117

for several reasons. First and foremost, the F-8 took a beautiful catapult shot. There was no ugly, ungainly rotation of the nose necessary as the bird left the flight deck. It just flew away as clean as a whistle! You could let go of the control stick if you wanted to be so foolish. It wouldn't matter. The Crusader would sail away no-handed.

However, that was the only thing that airplane did around the ship that was pleasant. Coming aboard was another story. It was the meanest, orneriest and most bad-mannered airplane ever built as it approached the ramp of an aircraft carrier.

The second reason I felt good about this particular launch was that I could just barely discern the hint of a horizon off to the east. That made the world of difference! It made the night catapult much less uncomfortable. And, finally, there were no clouds at all. It was a beautiful, clear night and the star-studded sky was breathtakingly exquisite.

There were two of us, and we were enroute to Naples, Italy from a point about three hundred miles west in the Tyrrhenian Sea. The military base at Capodichino was our destination. Even at military thrust (no afterburner) the Crusader climbed like a homesick angel and I realized that we were going to experience an unusual phenomenon in just a few minutes.

For two reasons the sun was going to come up faster than usual. One reason was that we were headed due east, toward the rising sun. Secondly, we were climbing to forty thousand feet; and that very fact would make the sun appear to rise earlier and faster than if we were standing back on the deck of the *Shangri-La*.

We were passing fifteen thousand feet and the horizon was very distinct with a powerful hint of daylight just beyond it. Our section of Crusaders was climbing at about six thousand feet per minute and I decided to change our cruising altitude from forty to fifty thousand feet. "All the better to enjoy the birth of a new day", I explained to my wingman, Stu Harrison.

Passing twenty thousand feet I could detect a faint pink glow along the rim of a black sea. There were some small clumps of cumulus clouds along the horizon to the northeast. Those fluffy little things were quickly changing from a deep pink on their lower halves to a pale gold on their tops.

Passing twenty-five thousand feet, I saw the same clouds begin turning from pink to a rich red at their bottoms while their upper halves turned to a rich white gold; and the ocean at the edge of the horizon was changing from a black to a deep purple before our very eyes.

It was almost as if an invisible painter, wielding an invisible brush, was busily plying it as we watched in awe.

"Beautiful, isn't it?" I recognized Stu's voice on the radio. He sounded as awe-struck as I felt. I clicked the microphone button twice in quick succession indicating my agreement. Somehow, I didn't want to spoil the moment by saying anything! There weren't words in my vocabulary to describe my feelings, anyway.

Just as we passed through thirty thousand feet, the upper edge of the sun appeared. Through a low haze layer, the sliver of sun was a deep reddish, gold color. But, it was rising rapidly. We could actually see it moving. It was a golden thunder!

Within a minute or two the entire sun disk cleared the horizon. As it emerged through the haze layer, it quickly turned to a burnished, bright golden color and so also did the cloud puffs ... like golden cotton balls. As we passed forty thousand feet we were forced to lower our visors because of the brightness of the sun directly in our forward vision.

Finally, we leveled off at fifty thousand feet and were clicking off a mile every six seconds. It took maximum continuous engine exhaust gas temperature (nearly full military power) to cruise our airplanes at this altitude, but the view was breath-taking, and the fuel flow was very low.

By this time the Tyrrhenian Sea had turned to its usual cobalt blue color and the edge of the Italian land mass appeared on the horizon. My thoughts turned lyrical. Mere mortals rarely ever get a chance to see a sunrise from a vantage point above ninety percent of the earth's atmosphere! What a privilege, I thought! It's nice to have secret, beautiful things like this that are hidden forever from the eyes of nearly everyone else in the world!

The approach and landing at Capodichino were uneventful and, after conducting some business at the base, Stu and I grabbed a taxicab for the short ride into Naples and the Hotel Mediterraneo. It was not a posh hotel, but it was comfortable and the rates were reasonable. Therefore, it was a choice of most of the Navy people who worked at the base. Stu and I shared a room. After a nice, long, hot shower we went down to a small restaurant on the Via Manzoni and enjoyed a sumptuous dinner.

I distinctly recall the strolling violinist coming over to our table and playing a particularly haunting tango. I liked it so much I asked the name. He answered "Caminito." No one I've ever asked since has heard of it or could play it.

We lingered over dinner. The evening was pleasant. The wine was good. We felt good. It was past ten o'clock when we returned to the hotel. I recall being tired and thinking of an early get up the next morning for a flight back to the ship. We would be bringing two corrosion-proofed airplanes back to *Shangri-La*.

We decided to stop at the roof-top bar at the hotel for a nightcap before turning in. On a clear evening the view of the harbor was excellent. What we found when we walked in was a cluster of people gathered in front of the television screen. The excited voice of the Italian announcer told me little until I edged up to view the screen. Then, the news was obvious!

President John F. Kennedy, the 35th President of the United States had been assassinated! I was stunned!

We sat at a table, after the commentator had finished, too devastated to speak. Finally, we tossed down the remains of our snifters of brandy and headed for our room.

CHAPTER 22: MORTE TU CAPO

I couldn't sleep for hours. I couldn't believe that dynamic young man was gone forever. The man who had fired up the aspirations of millions of Americans with his vitality and "vigah." The man who had written *Profiles In Courage* to inspire our youth to bigger and better achievements. The man who had stood before hundreds of thousands of cheering West Berliners and said, "Ich bein ein Berliner." The man who had admonished all Americans to, ". . . ask not what America can do for you. Ask what you can do for America," the tough-minded President who went to the brink of war with Nikita Khrushchev over the Cuban missile crisis. And, finally, the young President who charted a course for America to put the first man on the moon . . . was dead by an assassin's bullet. I felt sick and strangely empty. Sleep still would not come. The lines written by Herman Melville on the assassination of president Lincoln began creeping out of the deepest recesses of my memory. Finally, I fell asleep.

The telephone wake-up call roused me out of a deep sleep, but I woke still feeling tired. At the breakfast table the morning newspaper contained some details of the shooting and the theory that Lee Harvey Oswald had done the deed from an upper window of a book depository building along the route of the presidential motorcade. The newspaper account even contained a diagram of the ballistics trajectory of the Oswald theory.

As a member of my high school rifle team and a holder of the expert rifleman medal with the Marine Corps M-14, I thought I was pretty good with a rifle. I found it ridiculous that anyone who knew anything about shooting could imagine Oswald doing what was theorized. I knew that I couldn't do it. My opinion still hasn't changed.

We stepped out the front entrance of the hotel and paused at the top of the steps to look at the morning traffic and search for a taxi. Several young women were passing on the sidewalk directly in front of us. I guessed they were stenographers on their way to work. Dressed in our blue uniforms we must have stood out like sore thumbs. The three young ladies looked at us and one of them sang out. "Morte tu capo! Morte tu capo!" Your chief is dead. Your chief is dead! It was not said with malice; but rather a statement of sad acknowledgement . . . but its impact on us was chilling. I can still hear her voice.

The flight back to *Shangri-La* was uneventful. That night as I tossed in my bunk the rest of Herman Melville's poem came back to me. I got up and, sitting at my desk, wrote it down from memory:

*"Good Friday was the day
Of the prodigean crime,
When they killed him in his pity,
When they killed him in his prime.*

*They killed him in his kindness,
In their madness, in their blindness
And they killed him from behind.*

*He lieth in his blood,
The Father in his face.
They have killed him, the forgiver.
The avenger takes his place.*

*There is sobbing of the strong,
And a pall upon the land,
But, the people in their weeping
Bear the iron hand.
Beware the people weeping,
When they bear the iron hand."*

I couldn't help but note the strange similarity and applicability of Melville's words, written over one hundred years earlier, to Kennedy's death. Just before dropping off to sleep I reviewed the events of the last forty-eight hours during which the Good Lord had shown me the promise of an exquisite sunrise and the basest of man's malevolence!

developed during the flight. In order to attach the strap, the mechanic had to lie on his side with the top of his head just inches away from the whirling propeller blade. The thought of trying such a crazy stunt in a well lighted repair facility boggles my mind. Actually doing it on that work stand high in the skies over Mississippi seems beyond belief. But, those intrepid men did just that, and many more even greater death-defying deeds in that environment.

Thirty years later, in a letter to the Smithsonian Institute, Al Key describes one of the greatest problems they encountered during their record flight. It seems that he developed an abscessed tooth which he, ". . . successfully lanced with a surgical needle that had been wrapped in absorbent cotton and dipped in iodine . . ."

Many years were to elapse before aviators seriously applied themselves to the task of getting more fuel into high performance tactical airplanes; from other airplanes. Thus it was, that in the early 1950s the U.S. Navy made aerial refueling a tactically feasible operation by refueling F2H Phantoms from a tanker version of an AJ-1 nuclear bomber. It worked!

It wasn't until the Vietnam War, however, that aerial refueling became a real tactical evolution for aviators flying from aircraft carriers.

An external tanker store was developed and deployed to U.S. Navy carriers which gave the carrier group commander a great deal more operational flexibility than, heretofore, he had enjoyed. The tanker store contained a fuel storage tank, a hose, a fuel pump and a take-up reel. Features were gradually added to enhance this nascent capability. The stores station was plumbed so that fuel from the airplane could be pumped into the store, thereby adding measurably to the "give-away" capacity of the system. Lights were added to the refueling basket to enable night refueling. Then, lights were added to the receiving aircraft to illuminate the basket and make night refueling even more predictable and simple. No, it was never simple!

At first, the Douglas A-1H Skyraider was used as a tanker until its limitations became painfully apparent. Then, jet bombers were assigned the onerous mission of carrying the tanker stores. Then, came the day when the A-6 airplane was modified to carry the store system internally and the Navy had its first real dedicated tanker airplanes . . . though they could still drop bombs.

Other airplanes were also assigned the tanker role, notably the RA-5C Vigilante. Tanking was an ancillary role, but an important air wing capability because of the large load-carrying capability of the basic airframe. In the late-1970s the U.S.S. *Saratoga* air wing began using S-3 airplanes for a secondary tanking mission and, it turned out, the airplane was a wonderful tanker platform . . . and still is today.

About the same time frame, the idea of extending the range of helicopters with aerial refueling was born. What a great idea! Helicopters were notoriously slow and short-legged. Tanking couldn't help the speed but it certainly could add immeasurably to the operational utility of rotary wing aircraft. For special missions like support to Fleet Ballistic Missile Submarines and for Special Operations Forces, long range tanking missions added a great deal to the capabilities of rotary wing forces around the world. The most notable example, of course, was the Tehran Rescue attempt in 1980.

By the mid-1960s aerial refueling had become a fact of life for U.S. Navy carrier battle group operations. The amount of fuel that a battle group commander kept in the air and managed was determined largely by what mission he had in mind for his embarked air wing. The more demanding and complicated the mission, the larger the "fuel package" he kept airborne.

Joint use of aerial tanking capabilities achieved a hall mark in the Persian Gulf war when the war-fighting ability of the joint commander's forces hinged, to a large extent, on his aerial refueling assets.

In 1960, Air Group Ten, on *Shangri-La* had a night fighter squadron which operated twelve F3B Demons. The day fighter squadron operated twelve F-8B Crusaders. The night aerial refueling mission was not contained in the Crusader squadron's required operational capabilities (ROC). But, shortly, just before deployment, in fact, the Crusader squadron was to be assigned F-8Es which were purported to have a limited night fighter capability.

In anticipation of this additional capability, I asked for the Skipper's permission to try out some night aerial refueling. As the squadron operations officer I assigned myself to try it out. One of the light attack squadrons in the air wing volunteered to rig up an A-4 with a tanker store on which to practice.

It was not a good night for trying something new. The area off the coast of Jacksonville, Florida set aside for Navy flight operations had several storm clouds with scattered rain showers and lightning forecast. Nonetheless my A-4 tanker and I headed out into the operating area over the Atlantic ocean. When I noted how overcast and black it was I rationalized that it would be a better test of the idea than if it had been a bright and starry night.

The tanker picked a line of flight headed 090 degrees (due east) and set up at fifteen thousand feet . . . low enough to let tanking be comfortable, but high enough to minimize turbulence. The tanker reeled out his store and the first thing I learned was that beyond thirty feet the refueling basket was invisible, even with the little peanut lights around the periphery of the basket.

The second thing that I learned was that our F-8Bs, without the built-in refueling probe light (it was supposed to shine on the probe and also illuminate the basket when it was close in), made visual acquisition of the basket more difficult.

Nevertheless, we set up for tanking practice and I moved in. Using the tanker airplane as a reference I began my approach on the area behind and below the tanker's tailpipe where I expected to find the basket. After several attempts I managed to establish a pretty good starting point and got to where I could actually see the basket in time to make a reasonably good attempt at engaging it.

CHAPTER 23: AERIAL REFUELING

About this time I saw ahead of the tanker some lightning flashes and suggested to the tanker that he alter course to miss the "bumper." He did so but the turbulence in the vicinity of the clouds began to increase. With the tanker and the receiver bouncing around, the basket became an ever more difficult moving target. But, we pressed on.

On my final attempt, I had a fairly good start, the basket and the tanker had settled down and my own F-8 seemed to know that this was it!

At the very last second, the two airplanes hit the same bump and suddenly I heard the most horrible screeching sound I have ever heard in my life. I reduced power and began to gingerly back out looking hard for the basket. It was nowhere in sight.

The screeching turned to a deep rumbling sound and I could feel a heavy vibration in my ejection seat. I pulled some more power and kept on looking. Still, I could not find the basket.

"What's wrong?", asked the tanker pilot. "I'm slowing down." I couldn't think of an answer because I hadn't the foggiest idea what had happened.

Suddenly there was a bright flash as lightning struck nearby. In the brief instant of illumination I saw what the problem was. My F-8 had swallowed the in-flight refueling basket. The lightning showed the refueling hose leading straight down my intake. Furthermore, the hose was only one quarter its normal diameter. The tanker was literally towing me along and the refueling hose had an enormous amount of tension on it. Deep inside my Crusader, a powerful J-57 engine was trying mightily to eat the basket. It was suspended somewhere down that thirty feet of intake hung in the balance of an engine trying to eat it and a tanker trying to yank it out of my Crusader's belly. Suddenly, I knew I had to add power to ease the tension on the hose lest it snap off.

If that happened my engine would come apart while it was eating the basket and I would end up in the water. If, on the other hand, I added too much power, the added suction of my powerful engine compressor would cause the hose to part with the same effect. Again, my tanker pilot asked the question.

"What's happening back there?" My mouth was so dry I wasn't sure any sound would come out even if I knew what to say. All the while my poor engine had been protesting loudly over this thing that has been caught in its throat. The noises and vibrations were absolutely blood-curdling. Enormous gargling, screeching, rumbling sounds which made me feel ill. I decided to tell my tanker pilot . . . something. "Your basket has gone down my intake!"

Over Patuxent River, an F-8B and F-8E testing USAF hose and drogue used on the KC-135 tanker. The adaptation is shown here on a specially configured JD-1 in 1958. (Photo by R.M. Hill)

"My what?" It was a croak.

"Be very careful what you do with the throttle. I'm afraid the hose might snap. Keep it straight and level and let's see if we can get it out without breaking the hose."

"Okay." It was a weak response. "But, keep me informed", the voice added. Jesus, I thought, you can be damn sure I will keep you informed. What a stupid thing to say! He leveled his wings and seemed to steady out. By another lightning flash I saw that I needed to move laterally to the left to line the hose up with the centerline of the F-8. Once it was lined up I asked him to reduce power just a little. he did so and the hose seemed to grow in diameter just a hair. I began to pull off power slowly but steadily. The caterwauling under my seat increased and I could feel the rumbling and vibration seem to be moving forward. Suddenly out it popped right in my face, and I yanked back a whole handful of power. I was free! We decided to quit for the night.

A month or so later, we got our new F-8Es and the refueling probe light made an enormous difference. By start of the cruise, night tanking was a relatively normal evolution for everyone in the air wing.

Since that night I have tanked at night literally hundreds of times in an F-8 and dozens of times in Phantoms, A-7s and F-14s. Never have I even come close to eating another basket! That's called a steep learning curve!

24

HARMONICA

As a group, carrier pilots are probably the most resourceful professionals in the world . . . for one over riding reason . . . they have to be! Although it might be argued, Crusader pilots, among all carrier pilots, are the most resourceful of all. The harmonica flights are living proof of this fact.

Back in the 1960s, when the *Essex* class carriers still constituted the majority of what was then called the attack carriers, the F-8 represented a substantial portion of the Navy fighter community. To be sure, the F-4s on the larger deck carriers had a better all-aspect, all-weather, air-to-air weapons capability. But the Crusader community was not to be outdone, and bragged that they were the better clear air mass fighter because they had a gun. Certainly they fared better in the skies over North Vietnam. In that war the weapons combination with the greatest kill probability was the Sidewinder and the gun . . . and those were the F-8's weapons suite.

The most important competitive exercise in which F-8 pilots were tested was the live firing of a Sidewinder, heat-seeking air-to-air missile against a towed target which simulated the radar and infra-red signatures of a Soviet jet fighter. But the target system was a Rube Goldberg arrangement, fraught with failure modes, which often made the winning of a Battle Efficiency "E" by a Crusader pilot extremely difficult.

The towed target was a styrofoam shape about 6 feet long, looking much like a bomb, complete with 4 stubby tail fins. It was towed at the end of a steel cable (usually no less than 2,000 feet long). Inside the styrofoam target was a radar reflector which simulated the radar return of a Soviet fighter. Affixed to the after tips of each of the tail fins was a flare wired to a radio-controlled igniter inside the target. The tow plane, usually another F-8, had a pre-set ultra high frequency (UHF) radio channel which was designed to actuate the flare igniter.

The competitive exercise would be carried out under the watchful eyes of an airborne judge, flying his own F-8. The tow plane would head inbound toward the pilot being judged, under the radar control of the weapons range officer. The Crusader pilot would have to find the target with help from the controller, take over the intercept, lock on, get an infra-red tone from the flares, fire and hit the target. Naturally, the reliability of the flare ignition system was critical to the success of the exercise.

The pilot being evaluated would accept radar vectors from the ground controller until he was able to make radar contact with his own air intercept radar. Once he had good contact. the pilot would transmit, "JUDY", over the radio and take control of the remainder of the event. Once he had maneuvered his airplane into the heart of the firing envelope he would call for the flares.

"BUZZER, BUZZER, BUZZER." The airborne judge, of course, was also a safety observer and had veto power at all times. His silence would signal approval for the tow pilot to actuate the flares.

To do this, the tow pilot would switch to the pre-set radio channel and press the microphone switch. In theory, the tone generated by the radio would ignite the flares, and the shooter, if his Sidewinder sensed the flares, would then transmit, "HOTSHOT, HOTSHOT, HOTSHOT."

If the airborne judge determined that the range was clear, and that the shooter had the target (not the tow plane) locked up on his radar, the shooter was cleared to fire. All the shooter had to do then was turn on his armament switches and squeeze the trigger on his control stick. When the Sidewinder came off the rail he was required to transmit, "FOX TWO." That was the theory!

In actual fact, the failure rate of the flare ignition system was so high that many an "E" award was lost by virtue of target system failure. This weakness was generally considered by the F-8 community to be an outrage!

After countless unresolved complaints the F-8 community set about the business of fixing the flawed system.

An engineer from the contractor that built the flares blamed the manufacturer of the flare ignition generator. Another engineer from the builder of the generator blamed the manufacturer of the flares. Of course, the operators blamed the Navy system which designed and procured the system and equipment. It was vintage bu-

CHAPTER 24: HARMONICA

reaucracy... like a Three Stooges comedy, with the pilots from the F-8 community taking it in the shorts.

The first engineer carried the day with the argument that his flares fulfilled their design specifications... as test results proved. The problem, he claimed, was that the UHF radio ignition signal was broadcast over such a small bandwidth that often it did not produce the level of RF energy required at the particular frequency needed to ignite some of the flares. It was a specious argument... but we all suspected that he was right.

We also knew that the logistics system was not about to replace all of the hundreds of thousands of flares in stock. What was clearly needed was an adjustment to the radio signal to cover a slightly broader bandwidth. The Naval Ordnance Plant, Indianapolis was told to fix the problem. But, the lackluster way with which it moved to do so told all of us in the F-8 community that we would long be dead and buried before the fix ever made it to the fleet. If it were to be fixed in our lifetime, we would have to do it ourselves. A bright young defense industry electronics engineer, a friend of mine, came up with a simple interim solution... a harmonica... no kidding!

The engineer explained to us that if the tow pilot blew on a harmonica over the pre-set radio frequency in a warbling manner, running up and down the scale as he did so, the radio signal was bound to hit the right combination of frequency and signal strength to ignite the flare. It worked!

Unfortunately, we fleet pilots, having found a working interim solution to the problem, knew we could never sell the idea to the Washington bureaucracy. So, we decided to do it on our own. We found a specialty musical instrument retailer in Jacksonville who stocked a miniature harmonica, all of two inches long, which turned out to be perfect for the job!

Now, with the fleet fix, all the tow pilot had to do when told to ignite the flares, was disconnect the left side of his oxygen mask, insert the harmonica in his mouth, reconnect the mask, switch to the pre-set radio channel, push the microphone button on the throttle and blow on the harmonica in such a way (by twisting his mouth around the instrument) that a warbling musical sound was broadcast that would ignite the flares.

The system worked so well that we were sorely tempted to tell the logisticians about it. Fortunately, cooler heads prevailed.

But, no solution is without its pros and cons. One of the cons was the harmonica. The musical retailer had to order the instruments in quantity. The reason was hygiene! Since all squadron pilots had to fly their fair share of tow missions, all had to master the intricacies of tow target harmonica ignition. However, none of us wanted to share our harmonicas any more than our tooth brushes. So, we all acquired or own instrument. What better place to carry it than the small zippered pocket on the upper left arm of the standard issue flight suits? Of course, we all had to pay for them from our own pockets.

The only other disadvantage, other than the contortions required to get ignition, was in keeping it a secret. Naturally, the Navy hierarchy would never accept the knowledge that its fleet pilots were arcing around the oceans of the world with miniature harmonicas occasionally jammed into their mouths. Even though all such missile shoots were closely monitored by range safety officers, none of them ever asked me what that strange noise was that they kept hearing just before the shooter fired his missile. I have never understood why no one asked.

The acquisition process finally fixed the problem with an acceptable fix which improved the reliability of the flare ignition system. The harmonicas were no longer needed. Every once in a while when I was flying combat missions in the Tonkin Gulf I would hear a short blast on what I'm sure was one of those harmonicas. It would start me to wondering which of my old squadron buddies was out there somewhere reminding me of days gone by. There are probably those of us who still have one of those little babies squirreled away in a cruise box in the attic. The secret has been guarded pretty closely even to this day. I'm sure there are a few green-eye-shade bureaucrats from that era who would deny such a preposterous story.

My biggest concern was the conviction that, sooner or later, some pilot was going to cough during a critical part of the missile shoot and swallow his harmonica! I knew in my heart, that I would never be able to explain to the pathologist performing the autopsy, how one of my squadron pilots ended up on a slab with a harmonica lodged in his esophagus!

I recently came across my own harmonica in a box filled with memorabilia, while researching this book. When I tried it out I found that it still produced that same pitiful tone that it had thirty years ago when fleet F-8 pilots defied the system! My wife, hearing the strange noise, came into my study to determine its source. When I tried to explain it to her we both ended up doubled over with laughter!

25
BARRICADE

There is something about the very word, "barricade" that tends to make a carrier pilot's mouth go dry and his palms get sweaty. It is an intimidating device to begin with and the implications of engaging it in an airplane at one hundred fifty miles per hour are daunting.

The barricade is essentially a twenty foot high fence made up of clusters of enormous nylon straps oriented vertically. Its purpose is to capture an airplane flown into it as a last ditch effort to save the airplane and aircrew. Every U.S. aircraft carrier is equipped with a barricade. It is periodically erected as an emergency drill procedure. The carrier's air boss will routinely, without warning, direct his flight deck personnel to "RIG THE BARRICADE!" over the bullhorn. Then, he punches his stopwatch to see how long it takes. Dozens of selected members of the flight deck crew rush out onto the flight deck and connect the heavy nylon webbing to stanchions located on each side of the flight deck landing area. The stanchions (heavy steel arms) normally lie semi-submerged in the flight deck's surface. They are raised hydraulically on cue from the air boss, once the webbing is attached, and the fence stretches across the landing area looking like some enormous volleyball net.

If a ramp strike is the worst of a carrier pilot's nightmares, and a cold cat shot the next, the third worst of a carrier pilot's nightmares ought to be a barricade engagement. There are several reasons for this. As previously mentioned, barricade arrestments are rarely attempted and only as a last ditch effort. Serious problems must develop with an airplane before the decision is made to subject it to a barricade engagement. Then, the engagement, much like an arrestment attempt on the old straight deck carriers, is a landing attempt with no options. Because of it's height above the flight deck, there are no go-arounds, no bolters . . . no options. Once committed, the airplane is going to come to rest, somehow, somewhere.

Because of the bulk of its port side stanchion, the pilot's view of the optical landing system is blocked at the most critical part of the landing approach, as it crosses the ramp. Because of this, the LSO must talk the pilot down to the landing. This requires judgement, the most exquisite kind of judgement on the part of the LSO. He must read the weather, deck movement, the relative wind and the state of mind of the pilot as well as the condition of the damaged airplane.

Lt. Miottel was the first fleet pilot to carrier qualify in the Crusader. (All photographs in this chapter are courtesy of John Miottel)

The landing signal officer and his team – U.S.S. Hancock (CVA-19) off Hawaii, 1957.

126

CHAPTER 25: BARRICADE

Miottel launching from U.S.S. Hancock (CVA-19) off Hawaii, 1957.

The glide slope of the optical landing system during a barricade engagement must be shallowed which concomitantly reduces the hook-to-ramp clearance.

But, far and away the most terrifying aspect of a barricade engagement is its very nature. The stricken airplane is literally flying into a nylon net at 150 miles per hour. There is simply no way anyone can predict the outcome. Certainly, there will be damage to the airplane. Hopefully, there will be no injury to the aircrew. I recall watching an A-6 make a night barricade engagement with its starboard main mount missing. As the airplane touched down, the starboard main gear stub dug in to the flight deck causing the airplane to swerve sharply to the right and to roll into the starboard stub. As the heavy nylon webbing wrapped itself around the canopy (making ejection impossible, and fatal), the LSO screamed over the radio to the crew of the lurching A-6, "DON'T EJECT. DON'T EJECT, DON'T EJECT!" It was a terrifying scream. It is small wonder that no one likes to commit an aircrew to a barricade engagement. It is also understandable, then, why there are so few such events.

So, when Lieutenant John Miottel became the first person to ever make a barricade engagement in a Crusader it was not an inconspicuous event. However, when the same John Miottel became the only person to make two barricade engagements, again in a Crusader, that event took on all the aspects of a second lightning strike in the same place and also a Guinness book of records circumstance. What follows is an account in his own words:

"The mind is a strange and wonderful grab-bag of memories. Included among mine are a number of vignettes from my stint as an active duty naval aviator in the 1950s. Highlighted among these are a few savory incidents which occurred when I was with Fighter Squadron One Hundred Fifty-Four, the first Crusader squadron to deploy. Our air group, CAG-15, was assigned to the U.S.S. *Hancock* (CVA-19) in 1957. 'Hannah' had been a World War II *Essex*-class straight-deck carrier which was the first of the class to be modified in 1955 with an angled deck and steam catapults.

The particular episodes described in the following paragraphs involved my achievement of a somewhat dubious distinction as the holder of certain Crusader 'records.' The F-8U-1 was one of the

PART III: THE WORLD FAMOUS SEAGOING BOOMERANGS

NL 404 in the Groove, U.S.S. Hancock (CVA-19) off Hawaii, 1957.

first truly supersonic aircraft to become operational in the U.S. Navy. In early December 1957 I was lucky enough to become the first fleet pilot to carrier qualify in the F-8. A somewhat less enviable distinction derives from the fact that I subsequently became the first pilot to take the F-8 into the barricade; and then the first and only Crusader driver to be so snagged twice.

The first of these episodes found me prepared with just over 700 hours of total flight time; 150 of them in the Crusader. My preparation also included a grand total of 11 carrier landings in the Crusader plus 6 traps aboard the U.S.S. *Saipan* in the SNJ during flight training two years earlier. It was sort of like going to driver's school, commuting to work for a couple of years in a VW bug, and then being thrown into the Indy 500!

March 5, 1958 started early for me. I was scheduled for the first launch. 'Hannah' was cruising some 130 miles south of Oahu in moderate seas. The ceiling was broken from 7,000 to 10,000 feet with a visibility variable from 2 to 7 miles. The first thing out of my memory grab-bag is my attempt to have a nice quiet breakfast at 0500 after which I headed for the ready room for a flight briefing. I was assigned Crusader NL404, flying as a photo reconnaissance escort for an F9F-8P Cougar from VFP-61. After we launched at 0730 the weather turned sour over Oahu. This made things interesting because, if a return to 'Hannah' became problematical, our only alternative (or Bingo) airfield was Barber's Point on Oahu.

As we started our reconnaissance run over Maui headed for Lanai I noted my fuel gauge reading 8,000 pounds. At this end of the island chain it was a glorious day and we were looking forward to a great flight.

At 0900 we returned to *Hancock*. Holding above the ship in the dog pattern were four of my squadron mates, another photo Cougar and four AD-7 Skyraiders. At this point I had about 2,700 pounds of fuel remaining. My bingo fuel (the amount necessary to reach Barber's Point) was 2,000 pounds. Although the weather had deteriorated somewhat around the ship it was reported to be a lot worse at Barber's Point. Our bingo field was in the middle of the worst storm in 50 years – not a happy situation.

Meanwhile aboard *Hancock* an obscure electrician's mate had wandered up to the 0-2 level. He'd been directed to repair an electrical cable serving a lifeboat hoist. Of course, the first thing to do was deenergize the local circuit breaker panel. No big deal except that this coincidentally happened to disable the #5 and #6 arresting gear engines! This was of more than casual significance because these engines happened to control the tension on the #4 and #5 arresting cables. This in turn meant that the arresting gear engine settings, which must be changed rapidly for each different type of aircraft (depending on landing weight and approach speed), could now only be changed manually – a very slow process.

This situation degenerated quickly and by the time the word got to the Air Boss, it was garbled. Consequently he didn't grasp the fact that he couldn't change landing aircraft very quickly because his crew had to make the changes using a hand crank. He was only aware that a serious flap was developing on his flight deck because of some problem with the #4 and #5 arresting wires. His response was pragmatic: remove the offending wires and you remove the problem. He was obviously satisfied with this temporary

View of Hancock's flight deck and optical landing system as seen from the LSO platform.

CHAPTER 25: BARRICADE

"Rigging the barricade – a race against time!" U.S.S. Hancock (CVA-19) off Hawaii, 1957.

solution since it still left the #1, #2 and #3 wires available for arresting airplanes.

I attempted one landing and was waved off because of congestion problems on the flight deck. Those of us in the landing pattern were beginning to feel the fuel crunch and told the ship so in no uncertain terms. The tower responded by telling us all to shut up, clean up, climb up and dog it again while the offending wires were being removed. At this point one of the Cougars informed the tower that he had only 800 pounds of fuel remaining. As a consequence the ship ordered the two Cougars down first leaving the ADs and F-8s in the dog pattern.

At 0912 the first Cougar was trapped. One minute later his partner made it aboard and our five Crusaders were ordered down into the landing pattern. As luck would have it the four ADs somehow slipped in ahead of us. There were some harsh radio transmissions from the Crusaders and then the tower, before the Skyraiders were ordered back up into the dog pattern.

As the five F-8s entered the landing pattern there was a little uneasiness over the fact that there were only three wires remaining but the assumption was made that appropriate adjustments had been made to the optical landing system to ensure that the target touchdown point had been moved a little bit aft.

My room mate, Chuck Ramsey was leading the flight and I was number two in the echelon as we approached the 'break.' Chuck's approach looked good but he didn't catch a wire, 'boltered' and went around for another try. I also made what felt like a good pass – perhaps easing power a skosh because of the absence of #4 and #5 wires, but I too 'boltered.' In the process of my 'bolter' my tailhook actually struck one of the wires and was now canted to starboard about 60 degrees. In spite of repeated hook recycling, rudder stomping and verbal abuse the hook remained jammed out of position.

This was the 7th or 8th incident of this type. I didn't like it but knew that other pilots had arrested with canted hooks. My discomfort was aggravated by the fact that my fuel state was now down to bingo fuel – 2,000 pounds! The Crusader behind me caught a wire and was safely aboard. I 'boltered' on my next pass, as did the fourth and fifth member of our flight.

This was beginning to look very strange. The landing passes all looked good but resulted in only one arrestment. Things were becoming more pressing but they were not completely out of hand -yet! I found myself between the 180 and 90 degree position in the pattern behind Chuck Ramsey as he made another try. He 'boltered' again but this time, as he left the angled deck, he was trailing a stream of brownish white vapor from under his left wing. His port landing gear had broken off.

Landing gear failure was a frequent event in early Crusader carrier operations and resulted in loss of vital hydraulic systems depending on which gear failed. Although airborne and apparently under control Chuck's engine was running on fumes. He couldn't bingo and he couldn't make a normal arrestment aboard ship with a broken main mount.

He called the ship. 'Tower, this is 411 I have a utility hydraulic failure. I'm going up and eject.'

In the barricade, U.S.S. Hancock (CVA-19) off Hawaii, 1957.

This turn of events jolted me but Chuck seemed to have a handle

Above and across to next page: Barricade engagement – doing it right the second time. U.S.S. Hancock (CVA-19) June 1957.

on things and besides he didn't have much choice. I was waved off behind him and as I added power I heard the tower come up on the radio. '411, bail out, bail out, now!' I was stunned.

Apparently the assistant air boss, who had no jet experience himself, mistook 411's hydraulic fluid vapor for smoke. Not aware that the minimum safe altitude for bailing out was at least 1,500 feet he had given Chuck a death sentence. Before any of us could say anything to countermand the order Chuck bailed out.

As I passed abeam the ship's island I looked ahead and saw 411, now in a steep climb, suddenly pitch over from an altitude of about 800 feet and hit the water in a near vertical attitude. Chuck had ejected. His parachute streamed but didn't blossom before he hit the water. As I passed the crash site I saw nothing but churning white water. I realized that there wasn't anything I could do to help. The helicopter was on the way and I had no fuel to spare.

"Hannah" now had a real mess on her hands. Only three wires, three F-8s to recover – one with a broken hook, four ADs to follow and all aircraft at or below emergency fuel state levels. In addition, there was a crash to sort out and a search and rescue operation to conduct. In desperation the tower next ordered all aircraft to clean up and head for the beach. But they were not going to get off that easy. I got my wheels up and, as a token, put my hook up, preparing to dog it again but I told the tower that I was not going anywhere but HERE on 1,200 pounds of fuel. The tower ordered Paul Hamilton in NL410 with 2,100 pounds and Karl Koen in NL403 with 1,900 pounds to bingo back to Barber's Point – good luck, Chums, enjoy the maelstrom!

I was told to stay in the pattern and keep on trying to get aboard. I turned downwind, got my wheels back down and returned the hook handle to the down position. Of course, the hook didn't budge. As I approached the 180 degree position again the tower called the chopper requesting a status report on Chuck Ramsey. The chopper pilot's taciturn response still rings in my ears. 'The pilot in the water is dead!'

I try to relax when I fly. I even chew gum – this time it is a mistake. When I heard those words – my nice juicy wad of Clorets gum immediately turned to powder. I had no saliva to swallow it and no place, time or opportunity to get rid of it. I don't really know what happened to it. My only thought was, 'you'd better get it right or you're next.'

Chuck was the fifth squadron fatality in a little over a year's time. Three fatal accidents have now occurred in and around 'Hannah' in the last ninety days. Chuck was my second room mate to be killed. My original room mate had not survived an earlier ramp strike. Adding D.O.R.s and wash outs, we lost 50% of our pilots. The loss of 411 was one of 14 strikes or major accidents in our two year training and deployment cycle. This was the equivalent of losing an entire squadron of airplanes. That's one record I surely didn't want any part of.

In my ensuing pass, since I was now even more sensitive to the absence of the #4 and #5 wires, I eased gun a bit more at the ramp and touched down right in front of the #1 wire. I could actually feel the wheels running over the cables. According to the LSO this and the following pass also 'should have engaged a wire if the hook were centered.'

The LSO then came up on the radio with the laconic but truly nasty observation, 'Tower, we're never going to trap #404.'

At this point in the proceedings I was somewhat touchy about my situation, and any deprecation of my piloting skills was not appreciated. Calling the tower, I asked them to be more specific about any shortcomings in my technique – they assured me that my passes had all been excellent, but allowed as how unfortunate it was that that didn't appear to compensate for a broken hook. A brief but ominous pause ensued.

Then they uttered those fateful words, 'Stand by, we're going to rig the barricade.'

I was ordered to climb to 2,500 feet in order to permit a safe ejection if necessary while waiting for the barricade to be rigged. I

CHAPTER 25: BARRICADE

recall thinking that this would be the first live test of the barricade and the Crusader. Again I cleaned up and climbed. I even put the old hook handle up just for luck. As I leveled off I heard a plaintive call from one of the orbiting ADs. My God, they were still up there dogging it. This AD pilot complained because he had only 350 pounds of fuel remaining and, in a transparent bid for sympathy, he tossed in a report of a rough running engine. This caused the tower to also remember the existence of the ADs. They were forthwith condemned by the tower to join NL403 and NL410 in bingo hell.

I adjusted the throttle to maintain 250 knots maximum endurance speed and noted that my fuel state was now just under 700 pounds. Pretty soon I screwed up enough courage to call the tower to see how things were going – 'pretty soon', they reply. Within a four minute eternity they cleared the deck except for the A3Bs clustered around the starboard mirror – there was no time to re-spot them – and the barricade was rigged!

After grinding through two more orbits, the tower finally called and said to come on down and give it a try. I pushed over with less than 500 pounds of fuel remaining. It was now about 0933, but it seemed like midnight!

At first I throttled back leaving the gear up and the wing down. Stay clean and save fuel. I was about two to three miles astern, and able to see a tiny bright spark, the meatball shining at the center of a fleck of mirror.

Decelerating to 130-135 knots, I put the wing up and the wheels down; and even put the hook handle down (just for luck). I was carefully monitoring airspeed and alignment. Suddenly, about 300 to 500 yards from 'Hannah's' stern a large object on the port deck-edge loomed in front of the mirror. It looked as though someone had planted a telephone pole directly in front of the mirror and draped a tarpaulin over it. I couldn't believe it! What now? Now I couldn't see the mirror! I eased slightly over to the right to try and see around the god damned obstruction. When I finally was able to see the mirror again, there was no meatball! Whoa, look again – it's like the end of a beautiful sunset – there's just a faint orange glimmer at the very bottom edge 'horizon' of the reflecting surface. The LSO screamed. 'Power!, POWER!'

It didn't take long for me to realize my predicament. I was low, a bit slow, off alignment and beginning to get a very nice, close-up view of the spud locker. I deduced from this and the LSO's frantic call that I ought to pour the coal to the airplane, hold steady and fly straight in to reintercept the glide path at or near the barricade. With two or three feet hook-to-ramp clearance I hit the barricade slightly left of center. The airplane landing weight was just over 18,000 pounds and engaging speed was 96 knots – well within the prescribed tolerances.

I cut power on the LSO's command. As the barricade whipped past me all seemed well. But, the deceleration seemed milder than the wire engagements I was used to and I began to worry about dribbling off the angle. Suddenly, the airplane veered to the left as the lower horizontal load strap of the barricade ripped off my port main mount. The airplane teetered at the port side deck-edge, slid into the catwalk and began to roll over the side of the flight deck. I jettisoned the canopy (underwater ejection or other egress with the canopy in place is not recommended). The canopy separated with a BANG and was gone.

NL 404 rolled over and went inverted toward the water. I looked 'up' and saw the ship's wake rushing toward me – not a good escape attitude. I reached for the ditching handle as the airplane plunged into the water, and pulled it. I was immediately released, with my chute attached, from the seat and felt myself sucked from the cockpit at a water depth of about thirty feet.

My immediate concern was that NL404 was going to come down on top of me. I pulled the lanyard to activate the parachute bail-out oxygen bottle as I needed to breathe about now – but it didn't work! The ship's wake and screw turbulence was tossing me about violently. I believed, at this point that I had pretty much bought the farm and began to think about what that implied.

But – I'd almost drowned a couple of times before and had some experience with the phenomenon. Looking up I saw a circle bottle. My body temperature on the *Manley* was 86 degrees and I shivered for over two hours.

As it turned out my limp left arm was caused by root evulsions of my 5th and 6th cervical nerve which control the motor functions. A board of neurosurgeons recommended amputation. I declined on the advice of a wonderful nurse who I later introduced to one of my good friends. Thirty years later they are still married. For three years I had total loss of the use of my left arm. It has now come back about 80% and I carry a 14 handicap at golf. I was discharged, disabled, in 1963 and now am a successful inventor and chairman of a small medical device company. I have returned to flying status and have been active in competition aerobatics in my Great Lakes biplane. Additionally I am part owner of a Cessna 310R and for the Confederate Air Force I fly an SNJ and the F-8F-2."

Ron Luther is the only man I ever heard of who rode his airplane through the screw of an aircraft carrier and lived to tell the tale. As his airplane went through the screw it must have sliced through the fuselage just inches behind his ejection seat thereby freeing from his stricken airplane. The blade of the screw needed extensive repair back in the shipyard but obviously saved his life. His survival from such dire straits is nothing short of miraculous. His successful life after such an event is a testimonial to his guts, fortitude and perseverance.

27
Cold Cat Shot

Jim Foster enjoys an unenviable place in the history of carrier aviation. He is the only aviator, to my knowledge, who has ever ejected from an aircraft underwater. It all occurred during an air wing fly-off.

That state of extreme anxiety which afflicts carrier aviators in the last few hours of a deployment has been the cause of normally sensible men doing incredibly stupid things. After being separated from loved ones for six months, carrier aviators begin to get anxious as the cruise winds to an end. The return from a cruise always includes a transit either from the western Pacific or perhaps the Mediterranean which usually involves no flying for periods ranging from ten days to two or three weeks. Boredom sets in after the intensive flying tempo ceases and there is time for introspection, brooding and yearning for something... anything to happen. Tempers grow sharp. Tolerance grows thin. People lose their appetites. They find sleeping difficult. All of these are symptoms of Channel Fever.

As the last few hours of the deployment wind down and preparations begin for the flyoff all of the above symptoms worsen. The flyoff of all of the flyable air wing aircraft usually begins when the carrier reaches a point three or four hundred miles from the continental United States. For example, a typical flyoff from a carrier returning from a Mediterranean deployment might occur with the first launch three hundred miles east of Norfolk, of seventy-five or eighty airplanes. One squadron would land at Naval Air Station, Norfolk, Virginia, home base for the E-2 Hawkeyes. Three squadrons would land at Naval Air Station Oceana, Virginia, home base for two F-14 fighter squadrons, one A-6 medium attack squadron and an EA-6B Electronic Warfare detachment. The two F/A-18 strike fighter squadrons and the S-3 antisubmarine warfare squadrons would land at their home base at Naval Air Station, Cecil Field, Florida. The following morning, when the carrier is a few hundred miles closer to the east coast, a squadron of antisubmarine warfare helicopters might launch for their homeplate at Naval Air Station, Jacksonville, Florida.

Cdr. Jim Foster, C.O. of VF-13, manning up for the fly-off, August 1964. (Photo by R. Wright)

The arrival of these airplanes is always received with great fanfare. There are champagne, bands, banners and of course wives and families dressed to the nines to greet these anxious aviators as they step out of their airplanes. It is a joyous occasion fraught with high emotion and these reunions remain vivid in aviators memories forever.

Cdr Jim Foster ready for his catapult shot. (Photo courtesy of J. Foster)

It is painful to miss a flyoff. The most junior aviators usually are assigned to ride the ship into port. They, in turn, are greeted, along with all of the ship's company officers and enlisted personnel by an equally memorable pierside reception. For air wing aviators forced to ride the ship in, that last evening onboard is the longest of the cruise. A bunch of horny aviators lie in their bunks all night . . . eyes wide open . . . a group of over-stimulated insomniacs trying desperately to remember exactly what it felt like the last time each of them put his arms around his wife or girl friend and held her close . . . the longest night of the cruise.

And so it went on the evening of 10 August, 1964 on board U.S.S. *Shangri-La* (CVA-38) as preparations were winding down for the flyoff. Some of those same thoughts were running through the mind of Commander Jim Foster, Commanding Officer of Fighter Squadron Thirteen (VF-13). His was one of the two fighter squadrons in Air Wing Ten. The Airwing Commander, Commander Tom Hayward, had already told Jim Foster that he would be manning one of Jim's airplanes since VF-13 was the senior of the two fighter squadrons in the wing. It was a mixed blessing to have the CAG fly a VF-13 airplane in the fly-off. On the one hand it was a point of pride to be selected by the CAG as the squadron of choice (he could have selected any of nine other squadrons). On the other hand it meant that there would be one less airplane available for Jim's pilots to fly. One more junior officer would have to ride the ship into port . . . an ignominious way for a fleet fighter pilot to have to return from a carrier deployment.

Jim's maintenance department had worked hard and all twelve of his Crusaders were scheduled to fly ashore. Twelve out of twelve in the air at one time was always a banner achievement for any fleet Crusader squadron. According to tradition the CAG (code named OO) was assigned to airplane (side number) 100 and would be the first pilot to catapult from *Shangri-La* that warm morning in the Atlantic ocean three hundred miles east of Jacksonville, Florida.

The air wing pre-flyoff briefing had been uneventful. Extra time had been scheduled for pilots manning their aircraft to load their belongings into whatever compartments were available. The rest of their personal effects would be picked up after the ship docked at Naval Station, Mayport, Florida. In the Crusader, the only storage space was in the two ammunition gun bays on top of the fuselage and just behind the cockpit. The ammunition cans had been removed and Jim Foster loaded some necessary clothes, uniforms, a shave kit and some special gifts for his wife and children.

The signal to start engines was given by the Air Boss over the bullhorn and the adrenalin began flowing in the veins of all the pilots and aircrews scheduled for the launch.

"Now, check chocks, tiedowns and loose gear about the deck," the booming voice said, "Stand by to start engines." After a pregnant pause the voice again boomed, "Start the jets!" Jim noticed,

CHAPTER 27: COLD CAT SHOT

with no small degree of satisfaction, that neither he nor the CAG were having any airplane problems. The yellow jerseyed flight deck director spotted Jim onto the port catapult in side number 101 while the CAG was spotted on the starboard catapult in side number 100.

"Letter perfect," he remembered thinking to himself, "just exactly by the book."

The ship was in a starboard turn, looking to put the wind directly down the ships axial deck. The sweep second hand on the Crusader's clock had already been synchronized. The ship was timing its turn into the wind so that exactly as the sweep second hit straight up and the clock read 10:00 hours, the starboard catapult would fire and flight operations would be underway. On a truly professional attack carrier personnel throughout the ship would set their watches when they felt the audible "thump" of the catapult bridle assembly crashing into the arrestor at one hundred thirty miles an hour.

As the sweep second hand hit twelve, the CAG's Crusader roared into the air quickly tucking away its ugly assemblage of landing gear, flaps and slats. Jim had already saluted the catapult officer and sat with his head crammed against the headrest, his left hand jamming the throttle full forward, his right hand firmly gripping the control stick, his eyes straight ahead and his heart pounding like a trip hammer. "What a joy it is to be a carrier pilot!", he remembered thinking as the initial shock of the catapult piston broke the holdback and started his Crusader down the track . . . all thirty-two thousand pounds of aluminum, steel, fuel and Jim Foster . . . headed for home and the warm loving embrace of his wife.

The Crusader hadn't moved twenty feet down the track before Jim knew that something was very seriously wrong! The catapult deck edge operator had, upon seeing the catapult officer's signal, reached down and pressed the launch button on his control panel. The signal went electrically to the launch valve which opened letting high pressure steam into the catapult shuttle piston chamber. The proper steam setting to get a 32,000 pound Crusader into the air under the existing wind conditions was 300 pounds per square inch. Later investigation discovered that the valve opened only partially letting only 30 pounds per square inch into the chamber. There is only one inescapable and utterly inevitable outcome from this set of circumstances: Jim Foster was going into the water.

Jim knew this in an instant and did several things almost simultaneously. (Carrier aviators mentally practice these procedures their entire careers). He shut his roaring engine down by stop-cocking the throttle. He also locked his toe brakes holding them down with all the strength in his legs. But, unfortunately, there is no way either of these two actions could have the slightest effect on the terrible and inexorable rush of that frail airplane and its terrified occupant towards the bow of the ship. Jim even tried to steer the Crusader off of the catapult track and into another parked airplane, a tractor . . . anything to keep from falling off the bow and dropping sixty-five feet into the water directly in front of a fifty thousand ton ship steaming at twenty-five knots. As might be expected the powerful arm of the steam catapult, ignoring all of his puny efforts to the contrary, hurled his Crusader off the bow well below flying speed and into the water a hundred feet or so in front of *Shangri-La*'s bow.

Ejection at the moment he crossed the forward part of the flight deck was an option, but Jim, calculating that he was at the ragged edge of the ejection seat's escape envelope, made an instant decision to stay with his airplane and ditch it. However, ditching demands jettisoning the airplane's canopy. Jim failed to do and, sitting in his airplane, hit the water in a slightly nose-down attitude with bone-jarring violence.

On *Shangri-La*'s bridge a horrified but extremely alert ship's Captain barked out orders to the helmsman for an immediate "right standard rudder." This was intended to swing the ship's bow away from the stricken airplane which had already disappeared from his view under the bow. Immediately afterward, the same alert Captain barked out another order to his engine order telegraph man to "emergency stop both port engines." Before either of these orders were responded to he gave a third order, this time to his helmsman to "reverse his rudder." This last order was intended to swing the ship's still churning screws away from the wreckage which he knew was passing down the port side of the ship, close aboard, and still out of his view.

The Air Boss simultaneously called the rescue helicopter hovering astern. "Angel, plane in the water. Port side." These quick actions by the Skipper and the Air Boss saved Jim Foster's life.

Meanwhile, back in the sinking Crusader, Jim Foster disconnected his two shoulder harness connections and his two leg restraint lanyards; and unlocked the canopy preparatory to swimming clear of the wreckage. The canopy refused to open because it was being held shut by the increasing force of water pressure on the rapidly sinking Crusader. Jim was now standing crouched forward pushing upward frantically trying with all of his strength to lift the canopy open with his shoulders. Meanwhile he heard two very ominous sounds. One was the approaching crescendo of the four huge screws which were propelling *Shangri-La* through the water. To be sure, the ship's chief engineer was trying to stop the two port side screws closest to Jim's airplane but they were just beginning to slow down. The terrible sound of the approaching CHUNK, CHUNK, CHUNK of the screws was awesome. Jim could imagine them making minced meat of him and his Crusader as they ran over him. The second sound was even more sinister. It was the "crump, crump" sound of the metal skin of the Crusader's fuselage as it "oil canned" inward under the increasing water pressure . . . a sound similar to a beer can being crushed. Jim noted, as the seconds ticked by, that it was getting darker and the sea water was beginning to fill up the cockpit around his ankles.

Cdr. Jim Foster, soaking wet, being "escorted" to sick bay. (Photo courtesy of J. Foster)

Realizing that he was "in extremis", Jim decided his only recourse was to eject underwater. So he sat back in his seat, in the rapidly darkening cockpit, and began frantically to hook up the two shoulder and two lap fittings which connected him to the seat. No one had ever ejected from an airplane underwater, he knew. But, he also knew that he had no other choice. Finally, after what seemed like an eternity he was all hooked up again. Saying a silent prayer, he reached over his head, with water now up to his waist and grabbed the ejection seat face curtain firing handle. With a strange feeling of terrible finality Jim took a deep breath, closed his eyes and pulled down hard on the face curtain.

The observers in the port side catwalk were looking almost straight down at the stream of bubbles left by the sinking wreckage as it passed under the port side deck edge elevator. To their amazement the ejection seat emerged from the water like a Poseidon missile launch, upside down, and rose majestically upwards almost striking the underside of the elevator. At the apex of its trajectory the ejection seat and its occupant parted company and the two started back toward the surface of the water. As the pilot started back downwards his parachute was observed to deploy but it never fully blossomed before he hit the surface with a large splash. The force of the impact from such a height could easily have killed Jim Foster. It stunned him. He was still connected to the oxygen in his bail-out bottle and was breathing from it through his oxygen mask when he became aware of the barnacle encrusted hull racing past him almost within arm's reach. Jim estimates he was about twenty feet below the surface and sinking. He also noticed the deployed parachute canopy being sucked inward and downward into the vortex of the outboard port screw which, still turning, was now terrifyingly close.

As quickly as he could, Jim unsnapped the two shoulder harness fittings which connected him to the parachute just as the monstrous screw churned by, sweeping it into the maelstrom roller coaster of the ship's wake. Jim pulled the two toggles of his flotation vest and, feeling it inflate under his armpits, watched the dim surface of the water far above him begin to brighten.

After what seemed like another eternity Jim reached the surface of the water astern of the ship and almost immediately was blinded by the pelting spray being whipped to a frenzy by the beating rotor blades of the rescue helicopter hovering overhead. Jim signalled he was OK with a thumbs up signal and slipped his arms through the waiting horse collar rescue sling. Once properly positioned in the sling, Jim gave another thumbs up signal and the hoist began slowly to lift him clear of the water and upward toward the outstretched arms of the helicopter rescue crewman.

Within minutes the helicopter deposited Jim back on the flight deck not far from where he had begun the shortest flight of his aviation career. Jim climbed out of the helicopter and began walking aft down the flight deck towards the VF-13 Crusader waiting behind the jet blast deflector, next in line for the starboard catapult. Jim pounded on the side of the fuselage just below the cockpit to get the pilot's attention above the engine's roar. When the pilot of the airplane looked his way Jim gave him the signal to get out. Implicit in his body language was the intention to man the airplane and fly it to the beach. The look on the pilot's face was incredulous. There was his commanding officer standing there soaking wet with his hard hat still on and sea water trickling out of his "G" suit pockets. The pilot just stared in disbelief. Jim turned and ran back to the next Crusader in line and repeated the comic opera performance.

By now the Captain of the ship had had enough of this and so informed the Air Boss on the squawk box. The booming voice of the Air Boss on the bullhorn could be heard above the roar of eighty jet engines saying, "SOMEBODY DOWN THERE PUT THAT GUY IN A STRETCHER AND TAKE HIM TO SICKBAY," adding on a wry postscipt, "AND TIE HIM IN IT IF YOU HAVE TO!"

28

BINGO

One of the greatest hazards of naval aviation has been "bingoing" to an unknown field, especially if it happens when one is alone and at night. For openers, a "bingo" is, by definition, an emergency. It is a flight from the ship to the nearest suitable shore base because, for whatever reason, one could not land aboard the ship. This usually involves some sort of mechanical failure of an airplane system . . . but, not always. I can remember at least twice, having to "bingo" because of the failure of the ship. One of them required at least ten airplanes going to the U.S. Air Base at Danang, South Vietnam. The other required about half a dozen airplanes going to another carrier, called the "ready deck" carrier. But, those are two other stories.

The hazards of fumbling around in unfamiliar airspace with a malfunctioning airplane always low on fuel seem to compound themselves. Their effect can often be seen etched on carrier pilots' faces and can be the cause for premature gray hairs. The classic example occurred at night (as always) in the Western Mediterranean in 1962 from U.S.S. *Shangri-La*. I observed this event from the office known as Air Operations when I was the squadron representative. For some reason one of the air wing A-4 pilots couldn't get aboard. He was bingoed to a small Spanish Air Force Base on the northeastern coast of Spain. The name, San Sebastian sticks in my head. The duty A-4 tanker was sent to join up on the A-4 and escort him to the base. Meanwhile, the ship got on the radio and tried to raise the necessary agencies to clear the way for the pilot to go to San Sebastian. They were unable to contact the base so several operators on the ship were simultaneously trying to make contact by radio with the embassy in Madrid to facilitate the bingo flight which was already underway.

By some mix-up the air base had just been shut down. Normally, only bases were designated as bingo fields when prior arrangement had been made by the ship through proper state department channels to keep them open during the time flight operations were going on. Nonetheless, someone had turned out the lights at our designated bingo field and gone home for the night. The two A-4s miraculously found the airfield in the dark and were circling, waiting for the ship to get the place reopened and the lights turned back on.

The two planes circled . . . and they circled, able to make out the runways, taxiways and other airfield features by the light of a full moon. The poor bingo pilot was getting critical on fuel and his calls were becoming more frantic. Still, the ship, now on the single sideband radio with the military mission in Madrid, was unable to get the lights turned back on. *Shangri-La*'s operations officer, thinking it might help to have airplanes on hand with more loiter time than the A-4 tanker, sent an entire flight of four A-1 Skyraiders, who happened to be airborne, to circle San Sebastian. Now, instead of one airplane circling San Sebastian, there were six of them.

Listening on the radio in AIROPS I heard the bingo pilot finally say he was so low on fuel that he was going to try to land at San Sebastian in the dark without permission. He did it, and he made it. The only problem was that he couldn't see well enough to taxi in the dark so he elected to shutdown his airplane on the runway informing his airborne compatriots of his attentions before doing so. By now, the tanker pilot was also too low on fuel to attempt to return to *Shangri-La* and continued to circle San Sebastian also waiting for someone to throw the electrical switch that would illuminate the duty runway.

So he circled and circled. Finally, he also announced over the radio to the ship that he was going to land, also in the dark. This occurred just about the time the flight of four Skyraiders showed up "to help." Sure enough, the "Spads" ran low on fuel also and elect to land on the same runway. Of course, with each successive landing attempt in the dark, the cluster of parked airplanes on the far end of the runway increased in size. The pilots all ran clear, anticipating a crash on the runway when the landing airplane plowed into the others in the dark. But, no one did, thank God.

When daylight came to San Sebastian, and Spanish Air Force personnel arrived to man the tower and open up the base, they were astonished to see that there were six U.S. Navy airplanes parked on the duty runway in disarray. By some miracle, all six managed to land on a darkened runway and not hit each other in the process. It was a minor miracle!

So, when a young Lieutenant nicknamed "Duke" determined that his F-8 had experienced a major electrical failure and elected to bingo from his carrier in the Tyrrhenian Sea to the Italian Air Force Base at Capodichino near Naples, Italy, his squadron commanding officer properly began to worry. But, he reassured himself, "Duke" was one of his best pilots and, if anyone could handle the situation, it was this young man in whom he had such confidence.

"Duke" was above a solid overcast and his stricken F-8 had no navigation equipment (by which to find his way), nor any communications equipment (by which to talk to anyone). The communications, navigation and identification, friend or foe equipment (CNI) was located in a pressurized compartment known as the biscuit located just behind the cockpit. It had a notoriously high failure rate; and when it failed all of it was gone.

But, F-8 pilots are, by definition, survivors. "Duke", realizing that "first things first" was the applicable rule, ran the throttle to full military power, turned toward the last known position of Capodichino, and went to the maximum range cruise (bingo) profile of forty thousand feet. Then, using his airspeed indicator and the sweep second hand of his eight day clock, began what is known as "dead reckoning" navigation. The magnetic compass on his radio magnetic indicator was inoperative so he used the standby (wet) compass for heading information.

Using the last known distance to Capodichino (which he had gotten before take-off), his estimated ground speed from his airspeed and the clock he headed for the Italian land mass somewhere to the east.

The investigative body was never able to reconstruct what the actual upper winds were in the vicinity of the carrier so the mystery of "Duke's" bingo flight will probably never be solved. When his dead reckoning calculations dictated starting a descent into the Naples area, "Duke", always careful about his low fuel status, pulled the throttle to idle and descended into the cloud bank. He was clearly worried about his navigation and his inability to communicate which, of course, was compounded by his low fuel state.

He breathed a sigh of relief when he broke out of the clouds at an altitude of almost 5,000 feet. He quickly scanned the horizon to the east expecting to see the Italian land mass, and hopefully identify Mount Vesuvius by its smoking crest. To his dismay he saw nothing but water as far as the eye could see. Damn, his calculations must have been in error or there must have been some kind of headwind. So, our intrepid aviator did the right thing, or so he thought. He leveled off just beneath the cloud bottoms for maximum cruise efficiency and continued east, scanning the horizon frantically to catch the first glimpse of the Italian land mass. Still there was nothing but blue water . . . everywhere!

So, he pressed on, heading 090 degrees magnetic and now watching the needle on his fuel gauge with mounting anxiety. Still, there was nothing. The seconds ticked by and the fuel needle continued its inexorable movement toward the empty reading. His low level fuel warning light came on, as advertised, at 1,100 pounds . . . and still no sight of land.

"Duke" grimly pressed on. Finally, with the fuel needle approaching empty he caught sight of the thin, purple line of land on the eastern horizon. "Duke's" heart fairly leaped for joy! He was saved at last, or so he thought. Slowly, he approached land, searching now almost frantically for Mount Vesuvius with its telltale swirl of smoke coming off the summit. But, there was no Vesuvius. He must have been off heading, he decided. The wet compass was notoriously inaccurate. What should he do? Should he continue inland hoping to find any airfield or should he turn and follow the coastline? If he did so, which way should he turn, north or south?

He opted to turn south hoping he chose the right direction, and began searching for a familiar landmark now even more frantically because he didn't even have the heart to look at the fuel gauge. Nothing seemed familiar. As many times as he had flown into Naples from the west, he couldn't recognize anything familiar in the way of landmarks.

Finally, when he was about to give up all hope he spotted an airfield. He didn't recognize it; but, it was an airfield. He was unable to determine how long it was but he almost didn't even care. It was an airfield. What more could he ask at this point? Since there was no traffic in sight, he headed for the runway that was closest in hopes of getting the airplane on the ground before the engine quit running.

With no radio, he entered what he thought would be the correct "break" altitude (1,000 feet) and rocked his wings in the international signal for an airplane with no radio wanting to land. There was no responding green flashing Aldis lamp so he turned onto the downwind leg and dirtied up, praying that the engine would somehow keep running until he was on the deck. He touched down with the welcome squeak of rubber on concrete and almost was overcome with gratitude to the dear Lord for having saved him just one more time.

Since he was unsure of the length of the runway, "Duke" raised the nose of the Crusader and began the standard aerodynamic braking technique he always used. The Crusader flashed by a parking ramp and a row of airplanes parked on its edge parallel to the runway. Glancing quickly at them he received the shock of his life. Later describing it as the equivalent of a glass of ice water poured in his face, he recognized the familiar red star of Warsaw Pact military aircraft painted on the tailfin of each. They were MiGs!

"Jesus Christ," "Duke" muttered out loud. Where in the world had he landed? The priority of slowing his airplane down took over from the near total paralysis which had seized him. As he turned off

CHAPTER 28: BINGO

at the end of the runway and onto the parallel taxiway, he saw a bevy of vehicles racing down the taxiway towards him. There were armed soldiers in most of them.

The sight of the guns reminded him that he was behind the dreaded iron curtain. His kneeboard contained classified radio frequency designations, his missile station on the right side of the fuselage carried the best air-to-air missile in the world, the famous Sidewinder. How the Russians would love to get their hands on one. He, the "Duke" had just delivered one to them free of charge. He didn't know whether to commit suicide or what. Resignedly he followed the first vehicle which signalled for him to follow. The other vehicles followed, their armed riders had their weapons aimed at him.

When signalled to stop, he did so and resignedly shut down his engine. The last thing he did before climbing down the boarding ladder was to pull a long black, Puerto Rican hair from his head and lay it across the canopy rail. Then, as he locked the canopy he considered the thought that at least (if he ever returned to his airplane) he would know if the canopy had been opened in his absence. The thought represented small solace to an enormously chagrined U.S. Navy fighter pilot.

Almost as he was doing this a series of coded military messages of the highest priority flashed from the base communications facility to Belgrade, and thence to Moscow, thence to Washington, D.C., thence to NATO headquarters at Brussels, thence to U.S. CINCEUR headquarters in Vahingen, Germany, thence to CINCUSNAVEUR in London, thence to U.S. Sixth Fleet headquarters in Gaeta, Italy, thence to the flagship of the 6th Fleet, thence to the flagship of Commander Task Force Sixty and finally to the battle group commander riding the carrier.

There followed several minutes of shocked inactivity then the sequence of flash message traffic reversed itself and a messenger delivered a piece of paper to the base commander who ultimately showed it to "Duke."

It contained instructions for him to return to the ship, from Dubrovnik, Yugoslavia (so this was where he had landed!) with his airplane if at all possible, at the last PIM (position of intended movement). Recovery time was 0900 the next morning. There was one admonition, obviously added at the insistence of some intelligence weenie. It advised him to exercise extreme caution with all of his "equipment." It was the worst possible gratuitous advice that could have been put in the message.

What ensued is the stuff of which legends are made. "Duke" was wined and dined like some visiting dignitary. The food was soso but the vodka was wonderful and, as it turned out, much too powerful and bounteous. "Duke" learned the lesson many others have learned . . . that fighter pilots are the same the world over.

"Duke" was something of an anti-hero, having inadvertently overflown the entire Italian Peninsula and the Adriatic Sea to come to rest in the one Warsaw Pact nation from which he could have been extricated without disappearing into a salt mine in Siberia forever!

Because of the quantity of vodka which he had consumed the night before, and because of the fact that his CNI package had not been repaired, "Duke" opted to make the far less demanding flight from Dubrovnik (escorted by Yugoslavian MiGs to the edge of Yugoslavian airspace in the Adriatic Sea) to Capodichino. The ultimate irony is that a much more experienced squadron maintenance officer, Lieutenant Jack Barnes was sent to get the airplane fixed and fly it back to the ship. When he took off from Capodichino the next day IN BROAD DAYLIGHT, Jack entered the history books by lifting off with his wings folded!

But, that's another story!

Fighter Squadron Fifty-Three, the Iron Angels, had a history which went all the way back to World War II and the Pacific Campaign. It was one of the most distinguished fighter squadrons in the United States Navy. Of course, every fighter squadron worth its salt will, and ought to, make that sort of claim! After all, that is why they are out there on the cutting edge of United States' foreign policy.

Often misunderstood by the general public (the tax payer) and the Congress is why we have armed forces in the first place. We have them, quite bluntly to kill people and to destroy things. If they can't do that well, then nothing else that they do is really very important.

The Vatican has this group of bumpkins who go about caparisoned in medieval garments carrying medieval weapons. Their original purpose was to protect the Pope. Now, they couldn't even enforce parking regulations in Saint Peter's Square. They are what is called, in current military parlance, a symbolic force!

The British Royal Navy, once feared around the globe for its power and competence, is now also a symbolic force.

The current administration, lured by the temptations of "collective security", is in the process of turning the United States Navy into another "symbolic force." That, if it happens, will truly be the saddest legacy of the 1990s!

This part of the story contains a few nuggets showing what it was like to be a part of a squadron full of professionals who understood that they were supposed to be good at punishing the enemy!

The "Iron Angels", Tonkin Gulf 1967, U.S.S. Hancock (author, Squadron C.O., is fourth from right, rear row). (Author photo)

146

CHAPTER 32: FOLDED WINGS FLIGHT

Early carrier suitability test Crusader with wings folded. (U.S. Navy photo)

Nonetheless, the first F-8 wings folded flight occurred in broad daylight. The place was an Italian air base named Capodichino; the pilot was a feisty young Navy Lieutenant named Jack Barnes. All of the events resulted in the safe recovery of the airplane. Jack Barnes' feat was a famous first in aviation because it was the first such jet powered folded wings flight . . . a record for which he would just as soon not be remembered.

Three F-8 folded wing flight events all occurred at the same place, doing the same thing and by the same squadron! It was what is known in naval aviation as a "GOTCHA." The pilot takes off and makes a series of night field carrier practice landings. But, at the end of his last landing, he is sent to the refueling pits for a hot refueling then a return to the "night bounce" pattern. The wings are carefully checked by the plane captain before the airplane leaves the parking line the first time. But, when he folds them to enter the refueling pits, he must remember to spread them by himself because he goes directly from the gas pits to the end of the runway. During a four year period, three pilots all from the same squadron at Miramar did a repeat of the above sequence. All three safely recovered and there was practically no damage to any of the airplanes. The damage to the pilot's psyche was, however, quite traumatic. In each of these three cases there is, of course, a bottom line. That is the one which the incident investigation board would ultimately settle upon; failure to go over the take-off check-off list was the primary cause factor . . . that is irrefutable.

Some months ago, I was approached at a Naval Aviation convention by a nice young man whose face I recognized. It was one of those "You don't remember me but . . ." conversations. He reminded me that as an air wing commander I had occasion to meet him immediately after he had taken off from Naval Air Station Cecil Field, Florida one night in an A-7 with his wings folded! This nice young man reminded me that my reaction to his error was one substantially affected by the fact that, through superb airmanship, he had gotten the airplane back on the runway unscathed. To my knowledge, he is only one of two pilots I ever heard of who actually got that terribly underpowered airplane into the air with his wings folded. That, in itself, was no small achievement!

But, the winner of the prize for folded wings flight still goes to that absolutely marvelous naval aviator who took his Crusader off at night with the wings folded . . . AND SPREAD THE WINGS IN FLIGHT. If Naval Aviation had a Nobel Prize for aviation "chutzpah", it should go to this unnamed pilot.

His air wing was conducting night carrier qualification landings off of San Diego. Weather conditions dictated that the carrier operate further west than is normally done. As a result, the air wing operated a carrier qualification detachment on the island of San

Clemente about sixty miles off the coast. This detachment did routine maintenance and airplane refueling and made San Clemente the bingo base rather than Miramar. The use of San Clemente allowed the ship to keep the bingo distances short and made the carrier qualifications a more efficient operation. However, the field possessed only minimal support . . . and that became a key factor in the harrowing events which ensued.

Our intrepid aviator was experiencing a little difficulty on his first series of night qualification landings and was "bingoed" to San Clemente for refueling. His instructions were to return to the U.S.S. *Ticonderoga* after refueling to continue landing practice. When he taxied into the transient line at San Clemente the airplane captain signalled for him to fold his wings because of cramped parking facilities on the ramp.

A gas truck refueled his machine and, in about twenty minutes he climbed back in to start up and return to the ship. He taxied out of the line and, because of the paucity of servicing personnel, a final checker was not available to check him over prior to departure. It was left to the pilot to complete the take-off check-off list by himself. Printed clearly on the instrument panel placard were the words: "wing fold."

The break in the routine of standard operating procedures, the pilot's concern over his landing performance, the darkness of the night, the inexperience of the tower personnel and perhaps several other factors all compounded themselves into the final oversight by the pilot. He taxied into position for take-off on the duty runway with the outer ten feet of each wing unlocked, folded and pointed straight up!

Had the pilot not been so concerned with his own landing performance and been rushing to make a "charlie time" back at the ship, had the tower personnel been more alerted to recognize the peculiar position of the wingtip lights and had the plane captain not been too busy to watch the wingtip lights to ensure the wings had been spread while taxiing out; none of what ensued might have happened!

Our aviator was fairly senior, a Commander aspiring for command of a fleet fighter squadron with plenty of experience in carrier aviation. He made an afterburner take-off and noticed a much longer than normal take-off run. He noticed further that the airplane felt sluggish when responding to control stick flight control inputs as the airplane lifted off. He raised the wheels and followed shortly thereafter by moving the wing incidence handle to the down position. But, the wing wouldn't come down!

The investigation board later concluded that at that particular light fuel loading, the center of pressure of the wing was too far forward to permit the wing incidence hydraulic actuator to overcome its lifting force. That may or may not have been the case. Regardless, it was at this moment, with several unexplainable clues giving him severe discomfort, that the pilot looked out, and back, and up and saw, much to his horror, the wingtip lights . . . not where they ought to be!

Visions of crashing, dying or ejecting washed over the pilot like a tidal wave until he remembered happily, that Jack Barnes had done the same thing a few years earlier . . . and had gotten away with it! Jack had left the wings exactly where they were, and had returned to a relatively uneventful landing on the long runway from which he had just departed.

Moments later our aviator began to reflect on his own circumstances. For him things were a little different! It was at night, for one thing. But, much more importantly, our aviator had already screened for executive officer (XO) of a fleet squadron. This meant that after a year as XO he would, in the normal course of events, "fleet up" to command of the squadron.

Unfortunately for our intrepid aviator, flying around in the dark with one's wings folded might not be considered "normal" by the command screening board. They would have to evaluate his performance as XO before recommending that he be advanced to command the squadron. Normally that was a perfunctory process, but it wouldn't be perfunctory this time! All of these thoughts flashed through his mind in the few seconds after realization of his plight.

But, all was not lost, our pilot thought, for he had guts, a fair understanding of aerodynamics and an appreciation of the hydraulic engineering aspects of his problem. After a few moments of reflection he decided that zero "g" was the answer to his problem.

In straight and level flight, even if he slowed to 140 knots, he was certain that the puny wing fold hydraulic actuator would never overcome the aerodynamic forces on the outer wing panels and move them to the spread position, holding them there long enough for the locking pins to go home . . . no way! It was hard enough to accomplish that with the airplane parked on the aircraft carrier flight deck with a little bit of wind blowing. However, if he flew a zero "g" maneuver the wing fold actuator would have no aerodynamic loads to overcome.

The question was; could he hold that zero "g" condition long enough for the actuators to move the wing panels through ninety degrees of throw and then hold them there for the several more seconds it took to run the locking pins home? Our intrepid aviator estimated that the entire sequence would take a full ten seconds. He had flown countless zero "g" maneuvers during his flying career, but never for that long!

To add to the problem he could only start the maneuver at the limit maximum speed for the wing up condition; which was only 200 knots. But, he thought, if he did the first half of the maneuver in afterburner, he could start at a much steeper climb angle and therefore hold the condition longer! Then, of course, there was the recovery from the maneuver. If his stunt didn't work there would be no way to pull out from the resulting steep dive angle which the prolonged zero "g" maneuver would produce. It was all or nothing!

CHAPTER 32: FOLDED WINGS FLIGHT

Thinking, "no guts, no glory", our intrepid stunt man decided to "go for it!"

Leveling off at twenty thousand feet our hero accelerated to the limit speed of the F-8 wing up and smoothly pulled the nose up all the way to the forty-five degree mark on the attitude gyro indicator. As the airspeed decelerated through 180 knots he banged the throttle into full afterburner. Then he quickly shifted his left hand to the control stick and his right hand grabbed the wing spread handle. While programming the control stick forward into the zero "g" profile, he squeezed the interlock detent on the wing spread handle and pushed the handle to the spread position. He was now floating off his seat and up against the seat belt restraint in the zero "g" condition. He offered a silent prayer and started counting, ". . . one potato, two potato . . ."

Watching the wingtip lights in his rear view mirrors, he exulted to see the starboard wing panel begin to move downward and outward. Unfortunately, it began to move a fraction of a second before the port side began to move. This imbalance created a right yawing motion which inhibited movement of the port panel . . . consequently it moved slower. He was still counting, ". . . three potato, four potato . . ."

The pilot shifted his gaze to the attitude gyro to see the Crusader had passed through the apex of the zero "g" maneuver and started back down into the black hole. Our pilot let go the stick with his left hand long enough to snatch the throttle from full afterburner to idle and simultaneously thumbed out the speed brakes. He quickly returned his left hand to the stick.

Airspeed was now his enemy and he knew it. "Five potato, six potato . . ." He thought he felt the slight "clunk" of the starboard outer wing panel hitting the stops. Without waiting he squeezed the wing lock handle and pushed it down, saying another prayer as he did so. Miraculously, he felt it go home! "Seven potato, eight potato . . ."

A quick glance to the left told him that the port outer wing panel had not spread. It was stopped about halfway to the horizontal. The gyro told him that he was pointed almost straight down, and speed was building rapidly. "Nine potato, ten potato . . ." Not daring to look at the airspeed indicator, our hero started a gentle pull back to level out. In its new wing configuration the airplane yawed and "dutch rolled" wildly but leveled out at about twelve thousand feet. The pilot thumbed in the speed brakes and advanced the throttle to mid range.

He knew he was going to have to repeat that harrowing maneuver. Now, however, at 200 knots, with the wing still up, the right outer wing panel spread and locked and the port panel at the forty-five degree up position, he was holding almost full right rudder and left aileron to keep the airplane flying straight and level. This was his worst nightmare . . . contemplating a repeat of the first maneuver. He thought about all of this as he laboriously climbed his airplane back to twenty thousand feet.

As if things weren't bad enough, *Ticonderoga* Strike was calling him . . . wondering where he was and what he was doing. "If they only knew!" our hero muttered to himself. He called them and said he was having a minor problem with his airplane and would call them right back as soon as he had done some trouble shooting. What an understatement! Our hero was ready to try again! He repeated the complicated maneuver and, Allah be praised! It worked. The locking pins went home and he was able to lower the wing. Finally, at long last, our hero now had his Crusader in the configuration it was meant to be in when airborne. The ship called him again for an estimate of his arrival time at Marshall.

But, our stunt man had experienced enough high drama for one night. He couldn't face the thought of returning to the ship and resuming night carrier qualification landings! Besides, he correctly reasoned, he may have done some damage to both the wing incidence and the wing fold mechanisms. It was the prudent thing to cancel the CQ mission and take his airplane to Miramar to be scrutinized by the maintenance experts. He called the ship and told them of his decision to bingo to Miramar for maintenance. It was not the whole truth . . . but it was the truth, nevertheless! The rest is history!

By my count, there have been ten successful attempts by Navy or Marine pilots (we must not omit the Marine who took off from Danang one night with a load of bombs!) to take-off with their wings folded; once in a Skyraider, six times in Crusaders, twice in a Corsair II and once in a Phantom. Incidentally, the Marine pilot successfully jettisoned his bomb load and returned to land at Danang with his wings still folded. Unfortunately, he snatched defeat from the jaws of victory by landing with his wheels up! In addition, one Navy Crusader pilot had his wing fold on him after a catapult shot. I wouldn't be one bit surprised if there were others. The system can be remarkably close-mouthed about reporting such events depending upon the seniority of the perpetrator. But, in all honesty, it is a stunt that is very difficult to keep quiet. To this day I cannot understand how the A-7 pilots did it; but, they did and survived to tell the tale.

The practice of aviation safety is sometimes thought of as a black art. One never really knows, for example, why some airplanes are more prone to some kinds of accidents. In 1993, for example an airliner barely escaped what could have been a major aircraft tragedy when he nearly landed wheels up at a major U.S. airport in broad daylight. Except for a timely radio call from another airline pilot taxiing out to take-off, it could have been a disaster.

In this chapter on folded wings flight, there have been, curiously, no reported examples of this phenomenon by F-4 Phantom II aircrews . . . that is, until the near disastrous event on U.S.S. *Franklin D. Roosevelt* on 10 May, 1966.

An F-4B, side number AB-112, was taxiing forward for take-off during daytime flight operations in the Caribbean Sea. The pilot, Lieutenant (junior grade) Greg Schwalber stopped behind the jet blast deflector on the starboard catapult. The airplane on the catapult seemed to be having difficulties. It was the aircraft handling officer's nightmare. The airplane finally went "down" (for mechanical problems) on the catapult and had to be taxied clear, turned around and taxied back through the crush of airplanes proceeding forward for the launch. Greg and his radar intercept officer (RIO) Lieutenant (junior grade) Bill Wood saw that their flight deck director was signalling for them to fold their wings (which they had just finished spreading and locking) to help make room for the "downed" airplane to taxi past. Greg unlocked the wing lock mechanism and then moved the wing fold control to the fold position. Nothing happened! This was not an unusual malfunction in the old F-4Bs but the director seemed more than usually vexed by this frustrating system failure. The flight deck crew finally managed to get the downed airplane past them and thread its way toward the stern.

Then the launch sequence continued for Greg and Bill with one serious exception. Their wings were now unlocked but no one seemed to have remembered the fact, including the aircrew.

Here begins the same old story of so many other aviation tragedies. Notwithstanding the mechanical wing lock indicators on the upper surface of the Phantom's wing, as well as the wing unlock indicator light in the cockpit the launching process ground on. The warning indications were ignored by the catapult officer, the flight deck director, by the final aircraft checker, by the airplane's radar intercept officer and, finally, by the pilot. Everybody missed it.

When Greg Schwalber gave the catapult officer the usual snappy salute indicating that they were ready for flight, both he and Bill really believed it to be true!

The catapult fired, the F-4 roared down the catapult track and off the front end of the mighty *FDR*. As soon as the airplane became airborne, and the wings began to generate lift, both outer wing panels folded upwards. The airplane began a sickening settling trajectory and the pilot, sensing a problem, slammed both throttles into full afterburner . . . thereby saving the first of their nine lives. Commander Joe Elmer, the air boss, saw the wing panels start up and immediately called the plane "off the starboard cat", advising them not to turn because their wings were folded! It was the first time the aircrew knew what the problem was. Their alert air boss saved the second of their nine lives. The Phantom did what that marvelous airplane does so well . . . it accelerated to flying speed . . . even with the wings folded.

Meanwhile, the aircrew headed for the nearest "bingo" field at Guantanamo Bay, Cuba, several hundred miles away. Fortunately, they had plenty of fuel. Unfortunately, no one had the slightest idea how far a Phantom could fly with a full bag of internal fuel, its wings folded and its wheels and flaps down. As it turns out it wasn't far enough.

A photo Crusader which happened to be in the air flew alongside to assist and took a rather dramatic photograph of the Phantom in flight with both wing panels folded and over center. It was decided, after a quick check of fuel flow and rate of progress over the water, that they would never reach "Gitmo" in that configuration. So, the pilot raised the wheels. There was no apparent effect on the flying qualities of the airplane and its speed over the water increased substantially.

Unfortunately, it was still not enough. Another fuel flow versus airspeed check revealed that they still would not be able to make it to "Gitmo." So, with bated breath, the pilot raised the flaps to one half, and the wonderful Phantom continued to fly like a trooper. Another check revealed they still wouldn't make it to "Gitmo" so they raised the flaps to the full up position and began to believe they may have spent the third of their nine lives. The remaining internal fuel, when compared to their new ground speed, indicated that they could now make it to Cuba. Of course, there was still the problem of landing their airplane. But, that was ahead of them. For the moment, at least, they were saved again. The fourth of their nine lives had now been expended.

After an anxious approach to the long runway at Leeward Point airfield at Guantanamo Bay, Cuba, and with barely enough fuel left to keep the engines running, our intrepid aircrew crossed the fence at about 200 knots, touched down and, shutting down both engines, whistled into the long field arresting gear grinding to a halt on the over run . . . safe and sound. There were still several of their nine lives still left and a harrowing story to tell their grandchildren . . . including a few lessons learned about take-off check-off lists!

The last F-8 wings folded take-off attempt occurred in 1969 and set the aviation safety experts to wringing their hands. What to do? What kind of warning signal could be devised to prevent this terribly dangerous lapse of attention from recurring?

About this time Naval Air Stations adopted the Air Force practice called "Last Chance." It was simply, a group of talented enlisted men stationed at the take-off end of the duty runway, night and day. They inspect each airplane just before it taxies onto the duty runway for take-off. It is a safety as well as a configuration check. It is a good practice. Since its institution, there have been no more folded wing flight attempts. Amen!

Author's Note: By sheer coincidence, the author was describing this chapter with one of naval aviation's legends, now a retired captain. When I told him about the five F-8 wing folded events he corrected me with the cryptic reword, "Paul, there were six." He had to repeat it to me before its significance sank in! He admitted to taking off from NAS Dallas after "hot refueling" on a night cross-country flight from NAS Norfolk to NAS Point Mugu, California. In his case, he told me, after realizing what he had done, he held a nose-high altitude, climbing in afterburner at 140 knots and spread and locked his wings. He then continued on to his destination. This conversation was his first admission that he had joined an exclusive club, and as a reward for his candor I have left him unnamed.

33
SOUTH CHINA SEA GRAN PRIX

My wingman for the flight from U.S.S. *Bon Homme Richard* to NAS Cubi Point was Lieutenant (junior grade) Dexter Manlove. He was one of our nuggets and a fine young man . . . with a name right out of central casting! It was going to be a large fly off to loosen up the deck for "Bonnie Dick's" eight hundred mile voyage from Yankee Station to the Philippines.

The fly off was to consist of six F-8 Crusaders, eight A-4 Skyhawks, four A-1 Skyraiders, three A-3 Skywarriors and two E-1 Trackers. The entire group had been given a general intelligence briefing then repaired to their individual ready rooms for the specific briefing for the fly off. At the general brief I learned that our air wing commander, Bruce Miller was going to lead a flight of four F-8 Crusaders from our sister squadron Fighter Squadron Fifty-one. We were only sending two. Since Bruce normally flew the Skyhawk, I judged that he chose to fly a Crusader to get to Cubi Point first. In fact, I overheard one of his wingmen, a VF-51, pilot comment about having no difficulty beating "brand x" to Cubi Point. The two fighter squadrons referred to each other as "brand x." I resented what I had heard. Up until that point I hadn't really cared who got there first. Now, it had become an ISSUE! During my briefing to Dexter Manlove I made it very clear. We were going to get to Cubi point first, AT ALL COSTS! He nodded his understanding . . . but I sensed he didn't fully appreciate just how serious I was!

In our own ready room I briefed Dexter in nauseating detail how we were going to win the South China Sea Gran Prix! Since CAG was, by convention, always catapulted first, he already had a head start. However, he had to rendezvous four Crusaders before heading for Cubi Point and I had only two airplanes in my flight. I intended to capitalize on this advantage.

Dexter and I taxied onto the two catapults watching the four VF-51 Crusaders, already airborne, proceed upwind in a running rendezvous. That gave them a two minute head start! Dexter was catapulted about fifteen seconds behind me. I was watching in my rear view mirror as I raised the wheels and lowered the wing to the clean configuration. I left my throttle at full military thrust remaining at the prescribed five hundred feet VFR departure altitude. About two miles ahead of the ship I executed a starboard turn to a heading (opposite to the ship's course), of one hundred thirty-five degrees magnetic. Dexter, as briefed, cut across my circle joining me in a rendezvous which was a clear violation of the standard VFR rendezvous procedures. But, I was certain no one would complain. They didn't. Meanwhile, CAG's four Crusaders, following proper course rules were rendezvousing while heading off in a direction opposite to that necessary to get to Cubi Point.

I was exultant! We had already stolen a march on the opposition and they didn't even know it! As soon as Dexter was joined up I began a military power climb, but on a schedule that was on the fast side. I knew that Bruce Miller, being relatively unfamiliar with flying the Crusader, would fly the climb profile printed on his kneeboard. It might get him to altitude with a little more fuel than I, but I would be substantially closer to Cubi Point, and we would get to cruising altitude at roughly the same time. We had plenty of fuel for the trip. I could afford to give some away to gain a distance advantage. Furthermore, Bruce's kneeboard card told him to cruise his flight of Crusaders at forty-two thousand feet. I was planning on cruising my flight at forty-six thousand feet. Since we were both limited to maximum continuous turbine inlet temperature power settings, my cruising altitude advantage would net me a higher ground speed.

I knew that when we passed through the contrail level (thirty-five to forty thousand feet), Bruce would see my markings ahead of him and realize he was being "euchered" by "brand x"! As soon as Dexter and I saw each other's contrails we watched carefully to see where they stopped. It was at forty-three thousand feet. Dexter had been enjoined to maintain absolute silence for the whole flight, UNDER PAIN OF DEATH!

Now, I knew, it was CAG's move. He would have no other recourse but to use afterburner. Unfortunately for him, he couldn't use very much afterburner . . . and it was an unfamiliar device for a Skyhawk pilot. The eight hundred mile flight would have to be made mostly in basic engine (without afterburner). Bruce knew that

Three VF-53 F-8Es on BARCAP station off of Haiphong, North Vietnam, 1969 – U.S.S. Bon Homme Richard (CVA-31). There is a little bit of showmanship in every fighter pilot. (U.S. Navy photo)

fuel flow at high supersonic speeds can be as much as three times the fuel flow used at subsonic speeds at military power. I was planning on using afterburner only when necessary.

We cruised along, Dexter and I, in majestic silence. I heard no chatter on the radio from the "brand x" flight. They had either switched channels or also elected to remain silent. At forty-six thousand feet the sky was a very deep blue. It was beautiful! There was a scattered layer of small, white puffy cloud formations far below . . . snowy white against the deep, cobalt blue of the South China Sea!

We made the turn to the east after passing the small cluster of tiny Paracel Islands which marked the southern boundary of Mainland Chinese airspace. "What a beautiful day!" I said out loud in my quiet cockpit. "What a great way to end my combat flying career!" This was my last flight in the combat zone. In all likelihood it was the end of a very important chapter in my life!

It was in this reflective mood that a radio transmission jolted me back to reality and nearly made me wet my pants!

"This is Badman. Gates out!" It was CAG's voice and it sounded very near. What he had just told his flight was to come out of afterburner! He had made his bid. I was certain he had not passed me. I believe we would have seen them. Dexter was known to have the sharpest eyes in the air wing.

What this meant was that the CAG's flight was probably very close behind with a very high overtake speed . . . probably about to go right on by us if we did nothing. They were probably below us by several thousand feet and did not have us in sight. CAG never would have made that transmission if he could see us . . . or knew we were near. He probably didn't know where we were and had come out of afterburner out of sheer necessity . . . fuel! This meant he couldn't go back into afterburner!

CHAPTER 33: SOUTH CHINA SEA GRAN PRIX

I looked at the TACAN DME indicator and saw that we were one hundred thirty miles from Cubi Point. My fuel gauge read 2,800 pounds remaining. A few minutes earlier I had bobbled the nose of my Crusader up and down once or twice . . . the silent signal for a join up. When Dexter was close enough we exchanged fuel states by hand signal. He was within two hundred pounds of me.

"I've got him!", I shouted into my mask at the top of my lungs. I pushed the throttle into full afterburner and watched Dexter to see if he would catch the action. Within a few seconds I saw his burner light. That was when I knew we had it made! I put the leading edge droop into the supersonic position and watched the mach needle climb. When it reached 1.3, I began slowly modulating the throttle backwards in the afterburner detent to the fully aft position. The airplane decelerated slowly to mach 1.25 and stopped there. It was urgent that I watch the fuel gauge very carefully, but I was sure I could make it almost all the way to Cubi at this speed.

The TACAN DME indicator was clicking off a mile every five seconds. When the distance to Cubi dropped to ninety miles I eased the nose over and began a gentle, supersonic descent. Dexter's airplane began to drift closer. He was getting ready for a quick join up.

We had a great view of Subic Bay as we descended though thirty-five thousand feet. We briefly made contrails as we descended and I knew that Bruce Miller would see them and do one of two things. He would either give up the race or, depending how close he was, light burner and try to catch us. I knew he couldn't do the latter because I didn't intend to go subsonic for another fifty miles. There was no God damned way he was going to beat me to Cubi! Not today!

The rule book prohibits supersonic flight below thirty thousand feet within thirty miles of land. It also prohibits all supersonic flight over land, regardless of altitude. I was going to cut a few corners in the rule book today!

At thirty miles out my fuel gauge read eleven hundred pounds, Dexter signalled one thousand pounds. I broke our self imposed radio silence and called Cubi Tower for landing instructions. I knew the CAG was somewhere close behind me . . . perhaps in hot pursuit.

The tower cleared us for a VFR break and landing to the west. That meant that we had to make a one hundred eighty degree turn into the break and then a right turn to a right hand pattern for the western runway. I signalled Dexter to cross over to my left wing just as we descended through five thousand feet and indicating seven hundred twenty-five knots! Things were going to happen quickly here. I hoped Dexter could handle it.

As we crossed the beach line we leveled at fifteen hundred feet and came out of afterburner; making a level six and one half "g" turn into the 'break." The sonic boom we laid on Subic bay and the

VF-211 Crusader "at the ramp", Tonkin Gulf, 1966 – U.S.S. Hancock (CVA-19). (U.S. Navy photo)

town of Olongapo must have been something to hear! Our two F-8s decelerated to subsonic speed during the turn and at the "break" I decided to continue down the runway a mile or so to give Dexter some room to slow down. Above all, I knew that neither of us must have to make a go around on this landing approach. That would be a disaster!

We were down to 550 knots when I broke and as I got my wheels down and wing up on the down wind leg, I saw the CAG's flight of four chagrined Crusader drivers approach from the west. All I could think of saying was a heart-felt "GOTCHA!"

When I touched down on the long runway at Cubi, I realized that I was "off the line" for the last time. The Bureau of Naval Personnel had some silly rule early in the war about restricting people to three tours of duty in the Tonkin Gulf. They later abolished the policy out of necessity. But, for me, I was finished with combat flying.

Dexter and I walked to the makeshift ready room in the fleet hangar showered and changed into khaki uniforms. My wingman then went to the ready room blackboard and picked up a piece of chalk. In large letters, twelve inches high, he wrote:

THE WINNERS OF THE 1968 SOUTH CHINA SEA
GRAND PRIX ARE:
PAUL GILLCRIST - "FIREFIGHTER ONE"
AND HIS WINGMAN
DEXTER MANLOVE!!!

With that, we gathered up our flight gear and headed for the Cubi Point Officer's Club and a cold beer!

One of the first things historians of the southeast Asian conflict may observe is that, although the statistical sampling was small, the F-8 enjoyed the highest exchange ratio (number of enemy killed divided by number lost in enemy aerial engagements) of all U.S. tactical aircraft: 6:1! As will be pointed out in a later chapter (where I try to analyze why this is so), the reasons for this are manifold. But, setting aside the whys, for a moment, it is certainly true that the U.S. Air Force was quick to note this fact. Their exchange ratio (F-4s versus all MiGs) was, in the early stages of the conflict, a horrible 1:1.

At the time I had just returned from three combat tours in southeast Asia and was working in the Pentagon as a southeast Asian air-to-air warfare analyst. I worked daily with the Air Force and deeply sensed their embarrassment.

That is the reason why, at the operational commander level, the Air Force invited an F-8 squadron commanding officer, a veteran of 300 combat missions and also a MiG killer, to come to Thailand with four airplanes and show them what they were doing wrong! I leave that to "Pirate" to tell the story (see Chapter 45).

The format for this part of the book is verbatim quotes from the MiG killers. This was done four ways: 1.) notes from personal interviews (transcribed by me and committed to paper in my words, then edited by the individual); 2.) similar notes taken from tapes sent to me by the individuals (also edited by the individuals); 3.) written accounts by the individuals (unedited) and, finally; 4.) notes taken from telephone interviews with those who chose neither to write, nor to talk into a tape recorder, nor to meet with me in person. Fortunately, there was only one in the last category!

So, this is bona fide oral history! The reader will notice, that in many cases the MiG killers chose to editorialize. This was neither encouraged nor discouraged. I find the phenomenon fascinating, and attribute this trend to two things; the passage of over twenty-three intervening years and the sad fact that, as participants in an unpopular war, these aviators still feel the need to make a statement! So be it. I feel the same way!

Five VF-194 F-8Es over southern California, 1966 – U.S.S. Hancock (CVA-19). (U.S. Navy photo)

34

Commander Harold "Hal" L. Marr VF-211

Commander Harold L. Marr, USN, as Commanding Officer of Fighter Squadron Two Hundred Eleven has the distinction of being the first U.S. Navy pilot to down a MiG in the Viet Nam conflict. The incident occurred on 12 June 1966. His squadron was the senior fighter squadron in Air Wing Twenty-One aboard the U.S.S. *Hancock* (CVA-19).

The flight on which the MiG encounter occurred was a midday flight and it was a strike mission composed of eight A-4s and four F-8 escorts. The strike was led by the CAG himself, Commander Jack Monger and consisted of four A-4Cs from VA-216 and four A-4Es from VA-212. Two F-8Es from VF-211 and two F-8Cs from VF-24 made up the fighter escort group.

Commander Hal Marr was flying in F-8E side number 103, bureau number 150924. His call sign was CHECKMATE ONE ZERO THREE and his wingman was Lieutenant (Junior Grade) Phil Vampatella. The section of F-8Cs from VF-24 was led by Lieutenant Commander Richardson who was one of the few black naval aviators at the time. In Hal's words, "He was a good stick (good pilot)."

Commander Marr observed that the F-8Cs had some limitations which did not encumber the F-8Es. For one thing, the C had no air-to-ground capability. But, another, perhaps more serious limitation was the fact that it carried only two AIM-9 Sidewinders, and therefore was not normally employed in the fighter escort mission over the beach.

The strike group rendezvoused and proceeded north from a relatively southern Yankee Station position. They proceeded almost due north, passing Hon Gai to the east and turned in at an abandoned airfield called Ti Hien where they could approach the target area at low altitude and be hidden behind the karst ridge which lay just north of Haiphong.

The target was an army barracks and supply storage area which was just west of Dong Trai and lay about 30 miles east northeast of Hanoi. The leg northwards over the Gulf was flown in a fairly tight formation at altitude. At the turn in point they descended under high power settings (in the A-4Cs) to low altitude taking refuge behind the karst ridge for their overland approach to the target.

The weather in the target area was a heavy broken cloud layer at about 6,000 feet above the terrain. When they reached the valley at Do Bien, the A-4s turned southwest toward the target. Meanwhile, the four F-8s established a combat air patrol about halfway between the target and the major North Vietnamese airfield at Kep. The strategy was to let the A-4s approach the target, traversing the short distance from the mountains to the target in such a way as to entice the MiGs at Kep to intercept them.

The A-4s were in and off the target very quickly and egressed back over the same route toward the mountains where the F-8 escort picked them up in a loose escort position. The egress route was east northeast to get behind the karst ridge then turn east to go feet wet north of Haiphong and Hon Gai. The A-4s were at low altitude and the F-8s had stepped up on the strike formation to about 5,000 feet just below the bottom of the broken cloud deck. Hal Marr and Phil Vampatella were on the left (north) flank of the strike group. Richardson and his wingman were on the right (south) perch.

About this time Lieutenant Vampatella called tally ho on a flight of four MiGs at a position seven o'clock to the strike formation. The MiGs were about 10 miles astern and right down on the deck. Commander Marr turned his section of fighters into the MiGs and closed rapidly in roughly a head-on geometry. The ensuing engagement lasted about four minutes. The four MiG-17s were in what Commander Marr termed a "mini-loose deuce." The two sections of MiGs were spread no more than 1500 feet apart, flying abreast with the wingmen tucked in pretty tight to their leaders.

Commander Marr's section attempted to gain position on the MiGs by executing a high yo yo maneuver. To his surprise the MiGs commenced a left turn and executed two left three hundred sixty degree turns before departing the engagement area. Commander Marr recalls being astonished by this tactic which seemed to run counter to the slash-and-run tactics which the intelligence officers had led them all to expect in their briefings.

In the foreground: Lt. Vampatella, Cdr. Hal Marr, Cdr. Cole Black. Marr is congratulated on his MiG kill. (Photo courtesy of H. Marr)

Hal Marr describing his MiG engagement to the "Intelligence Weenies." (Photo courtesy of H. Marr)

On the second or third yo yo maneuver Hal Marr "squeezed off" two Sidewinders but, despite the fact that they were the brand new AIM-9D missiles, the combination of angle-off (forty-five degrees) and the "g" which the MiGs were pulling caused the missiles to track but miss their targets. Also, during that process, Commander Marr got off two or three bursts of 20 millimeter cannon fire at high "g" and high deflection.

As often happens in the heat of battle, division integrity broke down for both the F-8s and MiGs as well; and the engagement degenerated into two separate two-versus-two dogfights. The two MiGs which Commander Marr was engaging completed their second 360 left turn and headed back in a westerly direction toward Kep. At that point the airplanes were over the small valley that runs from Dao Binh to Luc Namh.

Commander Marr squeezed off his remaining two Sidewinders and one of them tracked the MiG impacting it in the empennage blowing the tail off the airplane. The MiG crashed near the river that runs just north and east of Luc Namh. He also managed to ease in behind the second MiG and, being out of missiles, pressed a gun attack. There were some hits, and Commander Marr saw pieces coming off the MiG but he either ran out of ammunition or his guns jammed (he doesn't remember which, now).

At this point, he noticed that he was "horribly alone." There were no other airplanes in the vicinity so he headed due west at high speed, with his head on a swivel until he went feet wet just north of Hon Gai. From that point he proceeded back to the carrier. The strike group had followed the same egress route as he. However, the other three fighters had gotten fairly low on fuel so they egressed to the southeast passing over the water between Haiphong and Hon Gai.

Commander Marr, flush with the exuberance of his first combat victory asked the ship for permission to make a victory fly-by.

Of course, this was granted and Hal Marr came by the ship's port side at flight deck level and high speed. He executed a slow roll and followed with a break into the downwind leg. Lowering his wheels and raising the wing he proceeded to make a hook up landing attempt. After ricocheting back into the air, very embarrassed, he made an uneventful carrier arrestment. The flight deck directors parked his airplane on the bow of the *Hancock* and the Captain of the ship, Big Jim Donaldson called down on the bull horn inviting him to come to the bridge. There he congratulated Commander Marr for his victory and shook his hand telling him, "Don't worry, Hal, I'll pay the five dollar fine for the hook up pass."

Some background information was also provided by Commander Marr. He pointed out in his tape interview that VF-211 and 24 accounted for well over half (12) of all of the MiGs shot down by Crusaders (18) during the Southeast Asian conflict. The three combat deployments during which the command of VF-211 shifted from Commander Phil White to Commander Hal Marr and finally to Commander Paul Speer marked the period during which most of these successful MiG engagements occurred. Commander Marr makes the point that there was a powerful show of interest by the carrier group commander, Rear Admiral Eddie Outlaw, to ensure that his fighter squadron skippers knew that one of his primary objectives was to punish the enemy.

At a time when many senior U.S. Naval officers were essentially gun shy of behaving like the warriors they were supposed to be for fear that they would inadvertently tread on the toes of their civilian masters in Washington, Eddie Outlaw gave some pretty clear guidance. Because of this, Air Wing Twenty-One scheduled plenty of MiG sweeps and MiG CAPS which were clearly more effective in making contact with enemy aircraft than were fighter escort missions.

CHAPTER 34: COMMANDER HAL MARR

Like many of his contemporary fighter squadron commanders, Commander Marr emphasized the inclusion of a few minutes of dog fight training at the end of every training mission during their stateside training respites. The importance of stressing the combat bottom line in every aspect of their training was understood by Commanders Marr, White and Speer. Regardless of the training limitations of only six missiles per squadron during their turn-around cycle, every pilot in VF-211 got to fire a training missile during Commander Marr's twelve month tenure as commanding officer. The concept of training the way they wanted to fight had many positive aspects when their air crews actually encountered MiGs for the first time.

As always happens under these circumstances the question of operational safety versus realistic training reared its ugly head. For example, for all air combat maneuvering training in the states, a certain altitude was established as "the ground." Flying below that altitude was prohibited in the interests of keeping "gung ho" flight crews from over-extending themselves in the heat of simulated battle. Ten thousand feet was often a nominal "ground" altitude for combat training flights for Navy flight crews.

It turns out that a good many of the combat engagements over North Vietnam occurred totally below that altitude. The basic aerodynamic performance of the Crusader above 10,000 feet was markedly different than that below that altitude. Below 10,000 feet the Crusader accelerated better, could sustain a higher "g" load, had a higher turn rate and a better turn radius than it could above that altitude. That can be said, by the way, for almost all tactical aircraft.

Commander Marr summed up the dilemma with a comment shared by most of his peers at the time. "You don't create safety by designing safer equipment or by legislating safer rules. You do it by pounding safe operating procedures into the heads of your aircrews."

Author's Note: Hal Marr retired as a Navy Captain and lives in Florida where he devotes a goodly amount of his time to learning the finer points of golf!

35

Lieutenant Eugene "Gene" J. Chancy VF-211

Gene Chancy's MiG kill was a particularly troubling event for several reasons. As the "fighter guy" in OP-05W I had the responsibility for keeping the box score for all of the air-to-air engagements in the Southeast Asian air war. There were security aspects about Gene Chancy's MiG kill which made the Navy refuse to confirm the kill. As of January 1994, almost thirty years later, some of those restrictions still are in effect. However, the system finally relented and announced that Chancy was indeed a MiG killer and deserving of the Silver Star medal. Contained in the memorabilia which Gene Chancy sent me as a part of my research was an item which I had long ago forgotten. It was a short congratulatory note from me to him, written twenty-five years ago from my office in the Pentagon. For months I had been trying to convince the "powers that be" to break the code of silence on Gene Chancy's MiG kill and award him his just desserts. They finally agreed and I was, I believe, the one to break the news to him. As of this writing, we still haven't met!

Gene's story begins back in a little town called Dothan in southern Alabama. Gene was a high school football star and toyed seriously with the idea of accepting one of several football scholarships, before rejecting them all in favor of an appointment to the United States Naval Academy. An accomplished athlete, he lettered in football, boxing and gymnastics. After a year as a surface Navy officer aboard the destroyer U.S.S. *Fiske*, Gene went into flight training and ended up assigned to Fighter Squadron Two Hundred Eleven as his first fleet tour of duty.

The squadron deployed to combat operations in Southeast Asia in 1965 as an element of Air Wing Twenty-One on U.S.S. *Hancock* for a seven month cruise and later in 1966 for a nine month tour. It was during the latter deployment that Gene Chancy got his chance and downed a North Vietnamese MiG-17 in aerial combat. Shortly after his return from that cruise he resigned from the Navy.

The second deployment was not without its ups and downs. Gene Chancy remembered one mission in particular over South Vietnam where they did some strafing underneath a relatively low overcast. Gene made the mistake of getting a little slow during one of his runs and bottomed out of his run so low that he brought back some foliage and mud on the underside of his airplane. The skipper, "Hal" Marr was so upset over the incident that he grounded Gene for a short period. On another occasion, Gene had some difficulty getting aboard at night, experiencing a few hook skip bolters and then was sent to an aerial tanker who couldn't give him enough fuel so the decision was made to put him "in the net" (the barricade). The barricade engagement inflicted minimal damage to the airplane because his airplane engaged one of the arresting wires and brought him to a halt before serious damage could be done by the nylon webbing of the barricade. The final incident was an attack on some gun emplacements at Hon Me Island off the coast of Vietnam during which Chancy's airplane was hit by anti-aircraft fire forcing him to bail out. After being picked up by a helicopter and spending a few hours on the radar picket destroyer, he was returned to the carrier and back into the air the next day flying combat operations.

On 21 June 1966 Gene launched in a flight of four F-8Es with Cole Black, Phil Vampatella and Dick Smith on a fighter strike escort mission against a target in North Vietnam. Cole Black was the fighter division leader. Chancy was his wingman. Dick Smith was the leader of the second section and Phil Vampatella was his wingman.

The mission was relatively uneventful and the A-4 strike aircraft had just egressed the target area when the four fighters were reassigned as rescue combat air patrol (RESCAP) to cover a downed F-8 photo reconnaissance pilot who had been shot down over the North.

The division arrived over the area of the downed pilot and Commander Black, after assessing the situation, dispatched his second section back to the ship to minimize the exposure of four airplanes on a mission which only required two. Cole Black and Gene Chancy were circling overhead the downed aviator at an altitude of two to three thousand feet while awaiting the arrival of a rescue helicopter which was enroute from the northern search and rescue (SAR) destroyer on station off the coast. Gene was crossing over

CHAPTER 35: LIEUTENANT GENE CHANCY

Left to right: Cdr. Tex Birdwell (C.O. VA-216), Lt. Chancy (VF-211), Rear Admiral Reedy (CARDIV Commander), Lt. Vampatella (VF-211). (Photo courtesy of E. Chancy)

Lt. Chancy in the cockpit of his F-8E after his MiG kill mission. (Photo courtesy of E. Chancy)

the top of his section leader and had momentarily lost sight of him when he heard somebody yell, "MiGs" over the radio.

Chancy leveled his wings and, looking up at two o'clock, saw two MiG-17s. Chancy describes his reaction as more instinctive than calculated. It was not even a classic guns tracking fire control solution. The MiGs, in his words, "... apparently screwed up their run", because his climbing turn into them put him almost head on. He instinctively fired his guns and then the three airplanes passed at a high closure rate so close that he could almost see the lead pilot's face. The second MiG passed even closer and Chancy thought they might actually collide.

However his burst of gunfire hit the second MiG and Chancy saw pieces of airplane and a liquid vapor streaming from him as the passed almost touching. He banged the throttle into afterburner and brought his airplane back to the left, after the disappearing MiGs, in a climbing turn which almost stalled the Crusader.

At that point, Chancy saw one of the two MiG-17s low over the ground pulling out of a dive. He unloaded his Crusader and accelerated toward him, still in afterburner, as he watched the MiG pitch back up in a climbing left turn to engage him. At this point Gene Chancy fired a Sidewinder missile which did not guide. Then, looking back over his left shoulder, he saw two MiG-17s at his seven o'clock position, turning inside him and "tracking" him (at this point he realized that there had been at least three and probably four MiGs involved in the attack). He also saw flashes from the lead MiG and believes they were flashes from the single 37 mm gun which the MiG carries.

Sometime during this melee Dick Smith and Phil Vampatella reappeared on the scene. While enroute to the ship they heard Cole Black's frantic MiG call and immediately reversed course; heading back toward the fray. When they arrived on scene, only a few minutes later, they found four MiG-17s attacking Cole Black and Gene Chancy who had gotten separated at the very outset of the engagement.

Vampatella saw a MiG close behind a Crusader and called, "MiG on your tail. Break right!" Chancy never heard that transmission but, looking around the sky, did see a MiG so close behind a Crusader that it seem to be "... joining up on him." That, he assumed was Black.

Shortly afterward Chancy heard Black's voice announcing, "I'm hit. I'm going to have to get out," (meaning eject). Chancy still didn't know that the other two Crusaders were in the area and decided that his circumstances were fast becoming critical.

Chancy realized that he was losing this particular fight and his only chance of survival was to immediately "bug out" at high speed. He unloaded his airplane and headed for the tall grass. Because of his low fuel state, Chancy elected to climb to an economical cruise altitude, disregarding the SAM and AAA threat and head back to the ship. Unbeknownst to him Vampatella and Smith were also inbound to the ship. Chancy asked for a tanker which was dispatched toward them. Eventually Vampatella and Chancy joined up and Phil's airplane, having been hit by enemy fire, needed to be escorted. They rendezvoused with the tanker and ultimately recovered aboard U.S.S. Hancock.

Chancy's decision to resign his commission in the Navy is partly because of the frustrations of many aviators of the day in the futility of a policy of limited war as it was executed in Vietnam by seemingly incompetent bureaucrats in Washington, D.C. rather than the professionals in the field.

Author's Note: Gene Chancy flies for Continental Airlines and is comfortably ensconced in the Atlanta, Georgia area.

36

Lieutenant (JG) Phillip "The Skinny Guinea" Vampatella VF-211

If anyone in the fighter requirements business ever questions the need for a large volume of internal fuel, he should reread the account of how Lieutenant (junior grade) Phillip Vampatella shot down his MiG! More than any other aerial engagement in the Southeast Asian war, his points out the importance of a fighter characteristic which, in a later chapter I call its durability in combat.

Durability, the third characteristic in priority in my list of fighter requirements means many things . . . but, it must be able to stay in the fight once the battle has been joined. So, durability includes a large internal fuel capacity. The most dangerous thing a pilot can do in an aerial engagement is try to "cut and run." Durability also means that an airplane can take punishment and not go to pieces (literally). It also means that when it is hit, its ability to continued to fight is not seriously impaired. "Graceful degradation" (a nice phrase) of its weapons systems after sustaining damage is part of the durability equation. So, vulnerability and susceptibility (the elements which make up survivability) are part of the durability description. If it was nothing else, Phil Vampatella's Crusader was durable on the 21st of June 1966.

As a pilot in VF-211, an element of Air Wing Twenty-One in U.S.S. *Hancock* Phil was assigned as a spare on a fighter escort mission to strike a target in North Vietnam about 30 miles south of the major North Vietnamese airfield at Kep, north of Hanoi. Three fighters from VF-211 were assigned to the fighter escort mission. One of the fighters went down after engine start and Phil was launched to fill in. The side number of his airplane was #104. The bureau number was 150300. One other squadron F-8 was launched on that event as a photo escort. Phil's room mate, Dick Smith was the photo escort pilot and the photo pilot was a young man named Len Eastman.

The strike was relatively uneventful and the A-4 bombers had just completed their runs and were being escorted to the beach when Dick Smith called the fighter leader, Lieutenant Commander Cole Black, and informed him that the photo pilot had been shot down in the target area to which they had been assigned in the general vicinity of Kep about thirty miles north of the strike target. Dick Smith was "marking the downed aircraft" (circling it) while Cole Black and his three Crusaders completed escorting the A-4s to "feet wet." Then Commander Black took his flight to the site of the downed airplane and assumed what is called rescue combat air patrol (RESCAP) responsibilities. The flight, now consisted of four fighters with Cole Black leading. On his wing was Gene Chancy. Leading the second section of fighters was Dick Smith with Phil Vampatella as his wingman.

It was at this point that Phil Vampatella's airplane was hit by "something." The airplane behaved badly but there was no cockpit indication of system damage. Aerodynamically the airplane didn't fly like a normal F-8, but there was no indication of imminent catastrophe.

The entire flight was low on fuel (almost at "bingo" fuel), so Black directed Smith to take his section of airplanes out to sea to find a tanker and refuel. Then, they were to return and assume RESCAP responsibilities while Black and Chancy went to the tanker for aerial refueling. Smith and Vampatella were no more than fifteen miles away headed outbound for the tanker when Cole Black was heard to call, "Tally Ho, MiGs" on the radio. Smith immediately reversed course and took his section back into the fray.

It was at this point that Vampatella got his first inkling of how badly he had been hit because he was unable to stay with Smith during the turn which was executed as almost a maximum performance maneuver. Vampatella fell well behind his section leader when the flight rolled out of its turn, now headed inbound to the fight. The airplane buffeted heavily during the turn, didn't turn nearly as well as it should have and felt generally sluggish.

When they arrived on the scene, the first impression Vampatella recalls was seeing a bunch of airplanes, "like a bee hive" whirling around through the air. As they grew closer to the "fur ball" Vampatella, who was now trailing his section leader by a considerable distance, recognized a MiG-17 close on the tail of a Crusader.

Vampatella describes the action which followed as three separate engagements. The first was with the MiG-17 which he saw on

174

the Crusader's tail. Not being certain who the F-8 was, Phil broadcast a general warning for all F-8s to "break right" because that was "what this guy needed." Vampatella dove after the MiG-17 with the intent of saving the pursued F-8 pilot. At that moment the tail of the Crusader blew up. It turns out it was Cole Black who ejected shortly thereafter and spent seven years as a POW of the North Vietnamese.

Vampatella was now closing on the MiG for an attack when he saw two other MiG-17s crossing directly beneath him from right to left. Realizing that these two MiGs represented an easier target at the moment he began what he called his second engagement. He rolled in after the two MiG-17s below him. During the early part of this attack, Vampatella became aware of a MiG at his seven o'clock position, guns blazing away.

This began Vampatella's third engagement. Now, in a defensive mode, Phil broke hard left into the attacking MiG-17 in a descending left turn. He describes his attacker as "not a very good pilot" since, even with a damaged F-8 he was able to defeat the firing solution of the MiG and stay out of his cone of fire. Unfortunately, he was unable to shake the MiG from his tail so Vampatella decided to try to "scrape him off on the trees."

Rolling into as steep a dive as he thought he could safely recover from, he headed for the trees. The Crusader is notorious for its poor aft visibility. No matter how hard he twists himself around in his seat a Crusader pilot physically cannot see directly behind him. Therefore, he must do a sharp turn in either direction ten degrees or so to clear his own tail. That's what Phil had to do at this point.

After he leveled out just above the tops of the trees, Phil "threw up a wing to look behind him." He saw the MiG still at his six o'clock and concluded that his only recourse was to outrun his adversary. The F-8, he knew, could easily outrun a MiG-17.

But Phil Vampatella was in what he called "an ugly situation." He was shot up, low on fuel, had a MiG on his tail and was now faced with the only solution of lighting the afterburner (an enormous fuel consumer) to escape. In combat a pilot does what he has to do. Lieutenant (junior grade) Phil Vampatella lit his afterburner and headed south. After what seemed like an eternity (but was probably no more than 15 to 30 seconds) his Crusader was thundering along at a speed far in excess of what the MiG-17 could do. Vampatella came out of afterburner and again "threw up a wing" to check on the MiG.

What he saw warmed his heart and flooded him with relief. The MiG pilot had apparently given up the hopeless chase and (probably low on fuel himself) was turning back to the north. This was the moment when Vampatella said he got "pissed off." He had a decision to make. Prudence dictated taking his badly shot up (and dangerously low on fuel) airplane back to the ship to fight another day. He could, he knew, get courts-martialed for going back into the fray under the circumstances. But, Phil also reflected on the fact that those bastards had put his executive officer on the ground as well as his friend the photo pilot. This had not been a good day for the Navy. It was time, Phil decided, to exact a little revenge. Throwing caution to the winds, he ran the throttle to the firewall and turned back after the departing MiG. At a range of about a mile, Phil fired a SIDEWINDER which guided "like a trooper" and exploded just at the right side of the MiG's tail. The pilot probably never knew what hit him.

The MiG began burning and, trailing a cloud of black smoke, began a steep nose-down right hand spiral. Vampatella never saw the airplane crash. He turned toward the ship and began "screaming for a tanker."

The response was almost immediate. But, it contained good news and bad news! Yes, there was a tanker in the area. It was an A-4 tanker carrying a "buddy store." That was the good news. The bad news was that it had already given away all of its fuel. But, the two airplanes rendezvoused anyway. The tanker pilot, named Art Culver had barely enough fuel to make it back to the ship himself. Neither did Vampatella. So, after discussing the situation, the tanker pilot decided to transfer his own internal fuel into the buddy store and give the Crusader exactly half of his fuel. Both airplanes finally made it back to *Hancock*. Each airplane arrived at the ship with enough fuel for only one landing attempt. They both made it.

Gestures of selflessness like this became commonplace in all of the air wings who served in the Tonkin Gulf. Phil Vampatella feels, to this day, an eternal debt of gratitude to Art Culver.

But, there is one last wrinkle to the story. Hancock, for some reason had only two of its four arresting gear wires strung . . . numbers one and three (the normal target wire). Not having enough fuel to make a second attempt (in case his arresting hook missed the number one wire), Phil Vampatella committed what would normally be considered a capital sin by the landing signal officer . . . he cut power right at the ramp (and went for the one wire. Under normal operating circumstances an LSO will never give a good landing grade to a pass which catches the number one wire. On many ships (Essex Class carriers) there is a standing fine for a one wire. But, on this day, approaching the ramp in a shot up airplane, with the engine running on fumes and a MiG kill in his game bag, Phil Vampatella was given a landing grade of "OK.#1."

In his account Phil was quick to acknowledge that, had he lost the airplane that day, because of his decision he might very well have been given a courts-martial. As it was, for re-entering a battle with a damaged airplane, critically low on fuel and bagging a MiG, Lieutenant (junior grade) Phil Vampatella was awarded the Navy Cross. In addition to the huge hole in his horizontal stabilator there were over eighty shrapnel holes in his airplane. Phil was lucky in that no important systems were damaged.

Author's Note: Phil Vampatella resides in Maine with his family, is an airline pilot and is living the good life.

37

COMMANDER RICHARD "DICK" M. BELLINGER VF-162

There are several aspects of Dick Bellinger's MiG kill which are worth noting at the outset. First, Dick was shot down by a North Vietnamese MiG-17 on 14 July 1966. He was recovered uninjured and continued on in his chosen profession. On 9 October of that same year, Dick got his second chance and shot down a MiG-21, becoming the Navy's first MiG-21 killer.

Of the nineteen Navy Crusader pilots who shot down MiGs over Vietnam, Dick Bellinger is the only one no longer living. He died in 1978 in the Veteran's Administration hospital in Northhampton, Massachusetts from Alzheimer's disease after he was retired. His story comes from several pilots who knew him and flew with him. Especially helpful was Lieutenant Dick Wyman, the wingman who was with him when he was shot down. Lieutenant Lee Prost, his wingman when he shot down the MiG-21 is also dead. Most of the material in this chapter came from the interview with Wyman.

Dick Bellinger ended up in Washington, D.C. sometime after his southeast Asian experience, and the author had several occasions to discuss his MiG engagements with him while we were both in the Pentagon. Whenever Dick began to wax eloquent on his victorious MiG engagement, we his audience were often wont to point out to him that he had achieved a 1:1 exchange ratio . . . which, in our view, was no great testimonial to his aeronautical ability! It was a cruel thing to say but then, Crusader pilots have never been known for diplomacy and restraint. Besides, we were envious of his combat experience.

In that regard, he was also quick to remind us that both he and Harry Blake, Skipper of VF-53 were fighting their third war; having accumulated combat experience in World War II, Korea and now Vietnam. Harry was three years older than Dick was when he took command of his squadron. Naturally, we were all duly impressed.

On 14 July 1966, Dick Bellinger was leading a fighter escort division of F-8s on an Alpha strike in the Hanoi area of North Vietnam. Bellinger's section leader, Lieutenant Commander Chuck Tinker, was attacked by a MiG-17 and found his opponent persistent and talented. Tinker ended up in a high speed, very low altitude pursuit over roof tops of buildings on the outskirts of the capital city. Despite his own acute circumstances, Tinker observed another MiG-17 sliding into a rear quarter position on his flight leader, Bellinger. Unfortunately, Tinker's radio was inoperative and he watched despairingly while the attacking MiG-17 began to pepper his Skipper's airplane.

Bellinger went into what I have always called the "survival mode." He ducked into the nearest cloud, slowed down to best cruise speed and headed back towards where he hoped the ship was. When it became obvious that he hadn't enough fuel to do that, he altered course for the South Vietnamese air base at Danang. Enroute to Danang he managed to rendezvous with an aerial tanker, but, because of the damage to his airplane, was unable to extend his inflight refueling probe. Continuing on toward Danang, Bellinger ultimately ran out of fuel and was forced to eject from his airplane about 40 miles off the coast of South Vietnam.

Bellinger was rescued and returned to Oriskany chastened but undaunted. At this point, I reminded him one time, his exchange ratio was zero divided by one which, mathematicians tell me, is an absurd number.

On 9 October 1966, Commander Bellinger led a fighter escort mission for an Alpha strike pod of A-4s from U.S.S. *Intrepid*. It should be remembered that Intrepid had an experimental air wing consisting of almost all A-4s with only a detachment of fighters led by Lieutenant Commander "Tooter" Teague. On this particular occasion, *Intrepid*'s fighter detachment must have been otherwise occupied because the strike planners called for Oriskany's fighters to provide the escort.

As the strike approached the "coast-in point", the E-1B observed unidentified radar contacts closing in on an intercept course. Bellinger's four F-8s were vectored to cut them off. MiG-21s were sighted and a dogfight ensued, this time with Dick Bellinger in a better position to even the score. Ending up in a rear quarter position on one of the MiGs, Bellinger closed in for the kill. The MiG-

CHAPTER 37: COMMANDER DICK BELLINGER

21 pilot must have realized his extreme circumstances because he entered a low altitude Split-S maneuver calculated to fly his pursuer into the ground.

I have always been convinced that the North Vietnamese pilots knew the combat performance parameters of U.S. tactical airplanes. It would be silly not to so assume. The data were too easy to get from unclassified sources. Someone must have told the North Vietnamese pilots that the F-8 had a 6.5 "g" limit. This was perfectly true. What the North Vietnamese pilots were obviously not told was that any Crusader pilot worth his salt would quite willingly pull 8 "gs" if the prize were a MiG kill. The odds were, we all knew, that the wings would probably not come off the airplane so long as not too many Crusader pilots had pulled similar 8 "g" maneuvers with this particular airplane.

In keeping with this tradition, Bellinger elected to follow the MiG-21 through this last ditch effort. The MiG pilot knew he could make it . . . and was obviously hoping his pursuer could not. He was wrong!

With his airplane inverted and in about a 20 degree nose down attitude, Dick Bellinger fired two Sidewinder missiles in rapid succession before attempting to roll out of the modified Split-S maneuver. He made it!

One of Dick Bellinger's Sidewinders detonated close enough to the MiG to inflict fatal damage. The MiG crashed into the ground almost immediately. This particular kill was important to all of us because it was the first of its kind to be killed by a Navy pilot.

Quite by coincidence, the Secretary of Defense, who was on a Southeast Asia tour, was in the area and showed up to hang the Silver Star medal on Dick Bellinger's left breast, on top of all those faded World War II and Korean combat decorations. I recall feeling good about the news when I heard it. Dick Bellinger, in spite of all his colorful behavior on the ground, was a true warrior in the air. I have always felt that he richly deserved this distinction.

38

COMMANDER MARSHALL "MO" O. WRIGHT VF-211

"Mo" Wright's aerial engagement with two MiG-17s on 1 May, 1967 proved to be one of the more interesting aerial encounters between the MiG-17 and the Crusader of the whole Southeast Asian air war.

It was the occasion for the first strike against North Vietnamese airfields. It was also a multi-carrier evolution and the *Bon Homme Richard* air wing's target was the major North Vietnamese air field at Kep, 30 miles north of Hanoi. The strike consisted of twelve A-4s and eight F-8s. The weapons selected for the strike were iron bombs, Mark 82 500#, Mark 83 1,000# and 5" Zuni rockets. The principal target was MiGs on the ground; in revetments (Mk. 83), on the ramp (Mk. 82) and taxiing (Zuni).

Commander Paul Speer, commanding officer of VF-211 was the senior fighter pilot, and therefore, the fighter leader. Of the eight fighters, four from VF-211 and four from VF-24, only six got airborne. Both Paul Speer and his wingmen had to abort the launch due to mechanical difficulties. So, the only two airplanes from VF-211 who participated in the strike were Lieutenant Commander "Mo" Wright and his wingman, U.S. Air Force Captain Ron Lord (an exchange pilot).

The strike leader was Lieutenant Commander Paul Hollandsworth (VA-76). The strike group rendezvoused and headed north up the Gulf of Tonkin, tanking enroute, to go in "the back door." This was one of the standard ingress routes which meant going north of the port of Haiphong, then turning west, dropping down to low altitude and proceeding inland behind the cover of the karst ridge which lay to the north of the major cities of Haiphong (on the coast) and Hanoi (forty miles inland. At any point, the strike group could pop up over the karst ridge and, turning south, strike any target in the general area.

While it is true that the ground control intercept (GCI) radars couldn't see the strike groups while they were behind the karst ridge, it is silly to imagine they didn't know that the strike group was ingressing along that route. There were simply too many other indicators of the approach of strike groups, not the least of which were small fishing vessels equipped with radio transmitters which were all over the Tonkin gulf.

So, the element of surprise really wasn't there. The only thing the North Vietnamese air defenses didn't know was precisely when and where the pop-up would occur. So, whenever the strike groups would pop-up, they were instantly met with a withering array of anti-aircraft artillery (AAA) and surface-to-air missile (SAM) fire.

The particular assignment for the four NICKEL F-8s was to establish two combat air patrol (CAP) stations, one to the north and the other to the northeast of Kep as a barrier to incursions of fight-

"Mo" Wright describes his MiG kill. (Photo courtesy of M. Wright)

178

CHAPTER 38: COMMANDER "MO" WRIGHT

ers from across the Chinese border. I never understood exactly why the strike planners feared such incursions. But, after all, this was the first time U.S. tactical aircraft were striking North Vietnamese airfields. It represented a distinct, "upping of the ante", and the thought which obviously worried some of the planners in the chain of command was a Chinese reaction to the escalation of the air war.

The other four fighters, the PAGE BOYS from VF-24 were to set up two CAP stations roughly between Kep and the other major airfield closer in to Hanoi known as Phuc Yen. This supposedly would protect the strike group from incursions of supporting North Vietnamese air defense fighters from Phuc Yen. Since the NICKEL force was reduced to only two fighters, "Mo" Wright decided to set up a single two plane CAP station directly north of Kep.

While the strike group was ingressing north of the karst ridge, the NICKELS were stationed on the left (south) flank of the strike group formation and the PAGE BOYS were on the right (north). When the strike group was about 5 to 10 miles from pop-up point, the PAGE BOYS called out MiGs at 1:00 o'clock high crossing from right to left. Moments earlier, "Mo" Wright had detached his section of fighters, accelerated to 350 knots, climbed to about 12,000 feet and begun his first CAP orbit.

He was halfway through his first orbit turn when the MiG call came. "Mo" immediately spotted two silver colored MiG-17s in tight formation down low, underneath the strike formation headed east northeast. Lieutenant Commander Wright is convinced the MiGs never saw the strike group and believes they were headed on a vector out to the northeast to establish a low orbit from which they could pick off straggling A-4s as the egressed the target area. The MiGs were at "Mo's" 2:00 o'clock position when he saw them and, before he could even roll into an attack, a third MiG-17 (presumably a straggler from the section) passed directly underneath them. All three MiGs were at extremely low level and seemed to be going at least 500 knots.

"Mo" lead his section of F-8s into a descending 270 degree right turn rolling out directly behind the third MiG at a range of about 1 1/2 miles. He got a good Sidewinder tone and fired an AIM-9D missile which tracked perfectly and went right up the tailpipe of the MiG. There was a large ball of fire as the tail section separated from the MiG. It's nose then pitched violently nose down from the loss of the empennage and almost immediately both wings came off the fuselage (probably from the enormous negative "g" load). Seconds after missile impact the MiG struck the ground and a second ball of fire was observed at the crash site.

"Mo" and his wingman pulled up and into a right turn headed back to their orbit to resume their CAP responsibilities. The first two MiG-17s were not in sight. By this time the strike group had completed its attack and individual elements were egressing at low altitude and high speed over the pre-planned egress route. It was essentially retracing their ingress track. Lieutenant Commander Wright hear one of the strike group, call sign BUFFALO, calling that he had, "... a MiG on my tail."

"Mo" opined that it was probably the fourth MiG-17 in a flight of four which had been scrambled rather hastily from Kep. They had seen the first two go by in formation. Numbers three and four were stragglers and, as usually happens with stragglers in aerial combat, one of them, number three had been shot down. "Mo" looked down and saw BUFFALO passing directly beneath them at high speed with a MiG-17 in a one half mile trail. "Mo" was high in a climbing attitude at the moment, and seeing that his wingman Ron Lord was in a better position, directed him to attack the MiG.

Captain Lord closed to gun range and fired a burst, hitting the MiG in the left wing root area. The MiG broke so hard to his left that Ron couldn't stay with his turn and yo yo'd high. In classic "loose deuce" style "Mo" rolled in on the MiG. What happened next often happens in combat. The complicated "switchology" of modern aerial combat may have caused "Mo" Wright not to get a second MiG.

VF-211 had been in the habit of carrying a mixed bag of missiles at this point in the war. They had some AIM-9C SARAH Sidewinders (an experimental radar guided version of the Sidewinder) and had been loading them on the first one or two missile stations in sequence of operation. In this instance, "Mo" Wright carried a SARAH missile on his left upper missile station, station #2. The squadron policy was to cross the beach with a SARAH missile selected (because of its forward quarter capability). Once an aerial engagement was joined however, the policy was to step the station selector around manually (if the SARAH hadn't been fired) to the standard infra-red missile (IR) on the next sequential station (#3) to take advantage of the IR missile's rear quarter capability. If, on the other hand, the SARAH had been fired, the weapons system would automatically move to the next sequential station.

In "Mo" Wright's case, however, he had manually switched from station #2 (SARAH) to what he thought was station #3 (IR) in the heat of battle. What he actually must have done, keeping his eyes on the enemy and groping with his left hand, was accidentally shift from station #2 to #4, skipping over #3. This meant that after he fired his first missile (station #4), the system sequentially shifted back to station #1, found it empty and then stepped automatically to #2, the SARAH missile.

Now "Mo" Wright was closing behind his second MiG and couldn't seem to find that magic spot on the enemy's tailpipe which should have given him a signal tone indicating the missile seeker head was "seeing" the heat signature of the MiG's hot tailpipe. This, of course, was because the SARAH missile did not have an IR seeker head.

Still in hot pursuit of the fleeing MiG and still "searching" for the elusive missile tone, "Mo" Wright found himself being sucked

PART V: AERIAL COMBAT

Left to right: Capt. Ron Lord (USAF), "Mo" Wright, Rear Admiral De Poix CARDIV Seven. (U.S. Navy photo)

Capt. Ron Lord (USAF), and "Mo" Wright describing MiG engagement. (U.S. Navy photo)

out of the area by the MiG as it headed across the northeast railroad line. Meanwhile, his wingman, Ron Lord, had momentarily lost sight of the two of them (his section leader and the MiG). Remembering his responsibility to protect the strike group, "Mo" made the tough decision to break off the engagement and return to his station.

Shortly thereafter, his wingman regained visual contact and the two airplanes joined up on the last several strike airplanes egressing from the target. Several egressing A-4s ahead of them reported two MiG-17s flew through their formation headed back in a southwesterly direction but kept on going. Again, "Mo" assumed that they were the same two who had gone by in the other direction just a few minutes earlier.

This was the same strike mission in which one of the A-4 pilots, one Lieutenant Commander T.R. Schwartz (a former Crusader pilot), shot down a MiG-17 as it lifted off the runway at Kep. He did it with a Zuni rocket. Since there is no known aiming method for shooting a Zuni (strictly an air-to-ground weapon) at an airplane, T.R. can be credited for an uncanny seat-of-the-pants sense of where to point his airplane (and using extremely short range) for his MiG kill. In total, the strike group killed seven MiGs (five on the ground and two in the air). The strike was considered to have been a success.

The switching confusion which occurred preventing "Mo" Wright from killing a second MiG is a classic example of what historians are fond of referring to as the "fog of war." It happens!

Author's Note: "Mo" Wright retired and is living in San Diego's north county area, working full time for IMED, a company which manufactures sophisticated intravenous mixing equipment for hospitals. "Mo" remains one of the most respected members of naval aviation.

"Mo" Wright landing after his MiG kill. (Photo courtesy of M. Wright)

39

COMMANDER PAUL SPEER VF-211

On 19 May 1967 Commander Paul Speer launched from U.S.S. *Bon Homme Richard* as fighter leader on the first U.S. Navy strike in Hanoi. At the time he was Commanding Officer of Fighter Squadron Two Hundred Eleven, an element of Air Wing Twenty-One on board U.S.S. *Bon Homme Richard*.

In the spring and early summer of that year U.S. Navy carriers began what came to be called "Doctor Pepper strikes." That was because, as the soft drink advertisement said, they occurred regular as clockwork every day at 10:00, 12:00 and 2:00 o'clock; the time for a soft drink work break. The schedule wasn't quite that rigid but it couldn't vary a whole lot. They would launch a maximum effort for the strike then maintain a ready deck. Many times the strike elements would straggle back, especially if they were engaged by MiGs or had to tank on the way. After all aircraft had been recovered, the respot and rearm evolution would begin in preparation for the next target time.

In "Bonnie Dick", however, something special was being cooked up for the North Vietnamese to think about. It was highly classified and was called Walleye, an electro-optically guided glide bomb. It was a standard Mark 80 series bomb with a special kit including wings and a special guidance attachment. All a pilot had to do was get close enough for the bomb to glide to the target, get its seeker head to lock on a contrasting part of the target (a window, or a door or a building's corner) and let it go. Then the pilot could ignore the weapon and head for ". . . the tall grass", to enhance his own survival. The Walleye would do the rest all by itself, and fly right through the window or the door to its destination. Attack Squadron Two Hundred Twelve was selected as the squadron to conduct the operational evaluation of the weapon in actual combat conditions. After several successful strike missions against softer targets to the south, it was decided to take Walleye DOWNTOWN!

This was the genesis for the first Navy strike against a target within the confines of downtown Hanoi, the Hanoi Thermal Power Plant! The specially selected target was situated on a lake near the center of town. The strike composition was unusual. It comprised only two A-4s, each carrying a Walleye, and twelve F-8s, six for flak suppression and six for fighter escort. The idea was to put the Walleyes into the large, west facing windows of the plant thereby avoiding any collateral damage and putting the city out of power.

Cdr. Paul Speer, C.O. VF-211, after his MiG kill mission. (Photo courtesy of P. Speer)

Left to right: Lt. (JG) Joe Shea, Cdr. Paul Speer, "Mo" Wright, and Capt. Ron Lord (USAF). (Photo courtesy of P. Speer)

181

43

LIEUTENANT COMMANDER MARION "RED" H. ISAACKS
VF-24

"Red" Isaacks is one of the most respected gentlemen in the history of carrier aviation. His account of the events of his successful MiG engagement on 21 July 1967 and of another one later on in the next deployment are surprisingly candid and, consequently, carry the powerful ring of credibility despite the passage of twenty-seven years.

Red had just joined the squadron (VF-24) on Yankee Station on U.S.S. *Bon Homme Richard*. Although, in a short time he had gotten his share of Barrier Combat Air Patrols (BARCAPS), he had only been over the beach over North Vietnam two or three times. He considered himself "fresh caught" and not heavy on experience over "Indian Country."

The target for that day was a petroleum storage facility at Ta Xa about 20 miles north of the harbor of Haiphong. Red was the division leader of four F-8Cs from VF-24 which were serving as MiGCAP. Since there hadn't been any MiG activity in over a month there was a general sense in the strike group that this mission was probably going to be a milk run. Red's division of four F-8s was assigned the strike call sign of Pageboy.

Everything had gone as planned and it was fairly quiet until the strike group reached the target area and roll-in point. Suddenly, Red spotted three or four MiG-17s at his 10:30 position and a little below them, at perhaps 8 or 9 thousand feet. It took just a minor turn to the left to put himself in a firing position on one of them. Red got a good "growl" with his Sidewinder and fired it. It didn't guide so Red tried to fire the second missile but it didn't leave the rail.

In retrospect Red assumed that he had not waited enough time for the armament selector to step over to the number three station. On his second attempt to fire the second Sidewinder, the missile came off the rail, seemed to be guiding, then flew right up the tail pipe of the MiG. There was a tremendous ball of fire and the MiG simply disappeared. Red felt fairly certain that the pilot could never have survived the explosion and was genuinely surprised to learn later that the pilot had, in fact, ejected and that a parachute had been seen on the ground.

This was when Red admits to violating the principal commandment of every fighter pilot. It had been a relatively easy kill and Red, lulled somewhat by his success, was still watching the fireball on the ground where the MiG had crashed when he noticed a stream of tracers whip by his airplane and seemed to disappear into the space at one o'clock. Glancing quickly over his right shoulder as he simultaneously heard the sound of cannon fire, Red was horrified to find himself looking into the intake of a silver MiG-17 with a red painted nose and its cannons flashing menacingly. What shocked him most was how close the MiG was. Red said he could probably have "counted turbine blades" in the MiG's intake duct.

As an instinctive reaction Red "reefed it into him" and saw the MiG roll inverted just as it passed under his Crusader. It was so close that Red felt the "WHUMP" as the MiG passed. He estimates it was a "matter of a few feet, or maybe even inches." Commander Isaacks never saw that MiG again. But, he immediately took stock of his situation and saw that the MiG had indeed hit him in the starboard wing. (Afterwards the maintenance people found four holes from cannon fire in the starboard wing fold area and around the trailing edge of the starboard flaps).

Cdr. Red Isaacks, C.O. VF-24, after his MiG kill mission.

39

COMMANDER PAUL SPEER VF-211

On 19 May 1967 Commander Paul Speer launched from U.S.S. *Bon Homme Richard* as fighter leader on the first U.S. Navy strike in Hanoi. At the time he was Commanding Officer of Fighter Squadron Two Hundred Eleven, an element of Air Wing Twenty-One on board U.S.S. *Bon Homme Richard*.

In the spring and early summer of that year U.S. Navy carriers began what came to be called "Doctor Pepper strikes." That was because, as the soft drink advertisement said, they occurred regular as clockwork every day at 10:00, 12:00 and 2:00 o'clock; the time for a soft drink work break. The schedule wasn't quite that rigid but it couldn't vary a whole lot. They would launch a maximum effort for the strike then maintain a ready deck. Many times the strike elements would straggle back, especially if they were engaged by MiGs or had to tank on the way. After all aircraft had been recovered, the respot and rearm evolution would begin in preparation for the next target time.

In "Bonnie Dick", however, something special was being cooked up for the North Vietnamese to think about. It was highly classified and was called Walleye, an electro-optically guided glide bomb. It was a standard Mark 80 series bomb with a special kit including wings and a special guidance attachment. All a pilot had to do was get close enough for the bomb to glide to the target, get its seeker head to lock on a contrasting part of the target (a window, or a door or a building's corner) and let it go. Then the pilot could ignore the weapon and head for ". . . the tall grass", to enhance his own survival. The Walleye would do the rest all by itself, and fly right through the window or the door to its destination. Attack Squadron Two Hundred Twelve was selected as the squadron to conduct the operational evaluation of the weapon in actual combat conditions. After several successful strike missions against softer targets to the south, it was decided to take Walleye DOWNTOWN!

This was the genesis for the first Navy strike against a target within the confines of downtown Hanoi, the Hanoi Thermal Power Plant! The specially selected target was situated on a lake near the center of town. The strike composition was unusual. It comprised only two A-4s, each carrying a Walleye, and twelve F-8s, six for flak suppression and six for fighter escort. The idea was to put the Walleyes into the large, west facing windows of the plant thereby avoiding any collateral damage and putting the city out of power.

Cdr. Paul Speer, C.O. VF-211, after his MiG kill mission. (Photo courtesy of P. Speer)

Left to right: Lt. (JG) Joe Shea, Cdr. Paul Speer, "Mo" Wright, and Capt. Ron Lord (USAF). (Photo courtesy of P. Speer)

Flight of VF-211 F-8Es preparing to land, Tonkin Gulf, 1967 – U.S.S. Bon Homme Richard (CVA-31). (Photo courtesy of P. Speer)

The target time was in the afternoon because they wanted the sun behind them as they dove in from the west. This not only helped against the flak sites but also gave the Walleye better contrast on the sunny side of the power plant.

The launch from the carrier went off without a hitch. They were getting pretty good at launching, rendezvousing and setting up for the approach to the target. In all, there were 12 Crusaders, 2 A-4 Walleye airplanes, 4 A-4 Iron Hand airplanes, 4 A-4 "wet wing" tankers and 2 A-3 tankers. The strike group climbed to 12,000 feet after rendezvous and headed for the coast. As the strike group neared the coast they commenced a descent building up airspeed to cross the beach at 3,000 feet with plenty of energy for jinking.

The strike group went feet dry south of the Red River delta and in a westward direction to the mountainous region marking the Laotian border southwest of Hanoi. As they crossed the beach and began jinking they were immediately taken under fire by AAA. This circuitous route had several benefits. First and foremost, it provided mountains to help mask the approach of a low-flying strike group. Secondarily, it added some confusion to the air defense problem by coming along what had been traditionally an Air Force strike group route . . . over "Thud Ridge."

The egress plan was directly southwest to the water, on the deck as fast as they could go. Needless to say, their plans were interrupted as they encountered about ten MiGs.

The strike group crossed North Vietnam at low level in a relatively tight formation. In the lead were Commander Homer Smith, Commanding Officer of VA-212 with two A-4s. On Smith's left flank were the fighter leader, Commander Paul Speer, Commanding Officer of VF-211 and his wingman, Lieutenant (junior grade) Joe Shea. On his right flank was another section of VF-211 F-8s led by Lieutenant Commander Kay Russell and his wingman Billy Foster. Directly behind Smith's section of A-4s was the third section of escort F-8s. The section of fighters positioned to the rear of the strike formation was developed specifically to require that MiGs attacking from the rear would literally have to fly through the Crusaders to get to the bombers.

As the strike group approached the Red River delta, the mountains began to flatten out and the strike group's first indication that they had been detected were the SA-2 SAM launchings. At about the same time a flight of MiGs was observed passing overhead in the opposite direction. Speer kept his fighters together, opting to let the MiGs go in the interest of getting the Walleyes to their assigned targets.

However, either the MiG pilots spotted them or the radar controllers gave them a vector, because the MiGs were observed to reverse course after they had gone by. They then came back and approached from the rear quarter.

Simultaneously, the two A-4s pitched up to an altitude of about 4,000 feet for their Walleye delivery. Paul Speer and his six F-8s lit afterburner and pitched up at the same time to about 6,000 feet. As soon as the Walleyes had been released the A-4s began their planned egress which was the most direct route to the waters of the Tonkin

CHAPTER 39: COMMANDER PAUL SPEER

Gulf. Racing right across the Hanoi suburbs and out the waterways that made up the port of Haiphong seemed like a short egress route and one that might surprise the city's air defenses.

As mentioned in the previous chapter, VF-211 had been in the habit of carrying one or two radar guided Sidewinder (SARAH) missiles on the port dual missile pylon (stations #1 and #2). The conventional infra-red Sidewinders were carried on the starboard dual missile pylons (stations #3 and #4). After the initial penetration of the target area (where the forward quarter capability of the SARAH might come in handy), it was squadron policy to shift from station #1 or #2 (depending on whether one or two SARAHs were being carried). On this mission, Paul was carrying two SARAHs and two IRAHS.

It was about this time that Paul Speer spotted the MiG to his left and immediately turned into him setting up a series of descending scissors maneuvers. It was a full afterburner, maximum "g", heart-pounding evolution during which Paul felt uncertain whether he had actually shifted his weapons control wafer switch from station #2 to #3. After the MiG had executed its 4th or 5th scissors maneuver it must have temporarily lost sight of him because the MiG pilot committed the fatal error of reversing his turn. Commander Speer felt he was close to minimum range for a Sidewinder shot but decided to try it anyway. Probably because of the short range and the heavy "g" loads the missile dropped away out of sight and Paul never saw it again. The MiG, now attempting to escape pulled fewer "gs" and made his second missile shot easier.

Commander Speer launched his second Sidewinder, at fewer "gs" and with a solid missile tone. The missile guided perfectly to a hit somewhere in the aft section of the airplane. There was a large fireball and the MiG appeared to fly out of the debris unscathed. Paul continued to close for a gun kill when the airplane suddenly began spewing flame from his tailpipe, rolled wings level momentarily, then rolled over and went into out of controlled flight and impacted the ground inverted with another equally large ball of fire.

As Commander Speer pulled off to the left his wingman called out another MiG at 7:00 to 8:00 o'clock on roughly a parallel course. The MiG didn't appear to see them or he wasn't interested because he continued on his straight and level course. Paul didn't see the MiG at first and, after a quick and anxious glance in the direction of Shea's call, told his wingman that he had the lead. This is standard practice. In the seconds immediately following a sighting, if the leader can't acquire the target immediately, he is duty bound to pass the lead to the member in his flight who has the visual contact. Shea, following Paul's apparent two shot policy, fired two missiles also at the MiG. Both tracked and both hit. The first missile was probably a contact hit. The second may have homed and fuzed on the fireball from the first impact. Joe's MiG crashed into the ground and the two Crusaders went feet wet following the same egress route taken by the bombers.

They were dangerously low on fuel when they got over the Tonkin Gulf and were very gratified to see the beautiful A-3 tanker show up with its standard wing up maneuver. The "Whale" is a huge airplane. In a rendezvous maneuver, when it briefly throws its wing up in a bank, it can be visually acquired twenty miles away. The wing up maneuver became a standard air wing procedure and made for much more expeditious tanking evolutions. Another air wing procedure was to take two shots of fuel when there were lots of airplanes and the fuel states were critical. Paul plugged the tanker and got enough fuel to stay in the air for a few minutes. Then he backed out and made the tanker basket available to the next thirsty fighter and so on down the line. After all six fighters had been "saved", they proceeded, at a more leisurely pace enroute back to the carrier, to plug in a second time for a more substantial load of fuel . . . enough to make it back to the ship and effect a recovery.

In all, the fighters scored four MiG kills during that engagement. But the strike group also lost airplanes to AAA and SAMs. In fact two F-8s went down and the pilots were captured. Commander Speer doesn't believe a single airplane did not suffer some battle damage during that raid. It was Paul's 150th Vietnam combat mission.

There were five major MiG engagements experienced by VF-211 pilots. They had lost an A-4 initially in one of them. Thereafter, the fighters had a "never again" rule of placing that extra escort section of fighters astern of the bomber group. The air wing never lost a bomber to enemy fighter action after that. Through all of his combat deployments Commander Speer stressed training the way they expected to fight.

The question of the value of the gun in the Crusader came up several times in the interview. It is the author's opinion, shared by now, Rear Admiral (Retired) Speer that although the gun never killed many MiGs in the southeast Asian air war, having it was critical. The most deadly tactics were always to maneuver aggressively for a gun kill. While so doing, the Crusader pilot usually passed through the heart of the envelope for a Sidewinder shot and took it. He was never required, as were many Phantom pilots, to pop speed brakes or do a lag maneuver to open the range for a Sidewinder kill. Having the gun enabled the Crusader pilot to stay aggressive throughout the engagements . . . and to stay alive!

Commander Speer was notified shortly after his return from the deployment that he had been selected to command an air wing. However, the Bureau of Naval Personnel had some peculiar rules about combat exposure. The rule was three combat deployments and no more. Paul's two deployments to Vietnam and one to Korea made him a candidate for command of an air wing that wasn't going to the combat zone in Southeast Asia. However, the untimely deaths of several air wing commanders left the Navy with no choice but to send him to Air Wing Fourteen which deployed to Southeast Asia on U.S.S. *Constellation*. Commander Speer joined his air wing at sea in the Tonkin Gulf in January 1970. He was now flying the F-

4, A-7 and A-6 on combat missions. And then it happened again in March of that year. He had an encounter with MiG-21s while flying an F-4 Phantom II.

During that at-sea period the flying hours were 6:00 PM to 6:00 AM. They were assigned the night hours because they were the "big deck" carrier on Yankee Station. However, after flying hours the fighter squadron still manned a five minute alert combat air patrol (CAP). It was while standing one of these alerts that Commander Speer and his wingman were scrambled on an actual intercept. They proceeded into the northern end of the Gulf of Tonkin under the radar control of the Northern Search and Rescue (SAR) destroyer, U.S.S. *Horne* where they were given a hot vector with clearance to fire. Commander Speer's wingman immediately moved out into the loose deuce wing position as they went into minimum afterburner. After a few minutes they met two MiG-21s head on at 15,000 feet. The MiGs initially had the altitude advantage so Commander Speer and his wingman went into a high yo-yo maneuver, jettisoning their external fuel tanks at the top of the maneuver. The MiGs did a turn over the top of them and it became obvious that the MiG wingman was getting "sucked" (trailing his section leader). There followed a series of high "g" maneuvers mostly in the vertical plane after which the two F-4s got into a position of advantage over the MiGs. Commander Speer's wingman was able to "bag" the MiG wingman with a Sidewinder missile while Speer was devoting his attention to the section leader. Speer ordered his wingman to go "feet wet" to refuel from a tanker which had been vectored to their support.

Commander Speer was now behind the other MiG and found that the MiG was very hard to keep visual sight of. At some point in the one versus one engagement the MiG dropped his nose, engaged his afterburner and ran for home. Commander Speer was a little late in detecting the maneuver and followed at a trail. He now found himself alone, deep over hostile territory, on top of an overcast and,

Cdr. Paul Speer and Lt. (jg) Foster. (Photo courtesy of P. Speer)

with SAM warnings going off, decided that discretion was the better part of valor. He departed the area and returned to the ship. This was his 199th combat mission.

In retrospect, Commander Speer concluded that, not having many hours of experience in the F-4, he was perhaps not as aggressive as he might have been had he been in a Crusader. The F-4 felt much bigger and certainly more powerful. But, the pilot sits up much more erect in the seat and he felt that he was driving the airplane more than flying it. By comparison the F-8 felt more like he was "wearing" the airplane rather than driving it!

Author's Note: Admiral Speer resides in Coronado, California where, after a successful second career as an executive in the private sector, he now serves only as a member of the San Diego Port Commission and master golfer.

40

Lieutenant (JG) Joseph "Joe" M. Shea VF-211

Joe Shea was assigned to the fighter escort mission during the first strike on a target within the 10 mile radius exclusion zone around Hanoi. Four F-8Es from Fighter Squadron Two Hundred Eleven, an element of Air Wing Twenty-One based on U.S.S. *Bon Homme Richard* launched on 19 May 1967 as the strike escort element of an Alfa Strike on the Hanoi Thermal Power Plant. The escort division of F-8s was lead by the squadron Commanding Officer, Commander Paul Speer. Joe was Paul's wingman. The leader of the second section of F-8s was Lieutenant Commander Kay Russell. Kay's wingman was Lieutenant Billy Foster. Russell, incidentally was shot down by enemy fire on this mission and became a prisoner of war.

As has already been mentioned in other accounts of this same raid, it was the first use of a "smart bomb" called the Walleye. The special munition was reserved for this first strike inside Hanoi. Only two A-4 bombers made up the bomber portion of the strike group. Each carried a Walleye. There were four other F-8s in the strike group which were configured with Zuni rockets and designated flak supressors. Also, as previously mentioned, the ingress route was carefully picked to minimize exposure of the strike group, enhance the element of surprise and take advantage of the sun angle on the target for the best chance of an optical lock-on by the Walleye seeker head.

Joe's airplane load-out was four AIM-9D Sidewinders and a full load of ammunition. The launch, rendezvous, tanking evolution and ingress to the target area were all relatively uneventful. However, when the strike group approached the target area there was a substantial amount of AAA and SAMs.

The escort fighters had positioned themselves above and behind the bomber group in a combat spread formation with Paul Speer's section on the left flank of the group. Kay Russell's section occupied the right flank. Joe remembers that the bomber group were just pulling off the target when a SAM was called at 9 o'clock. Speer dropped the nose of his airplane to build up speed preparatory for a SAM evasion maneuver. When he pulled up to a wings-level attitude, Joe observed the SAM pass directly beneath him at about 500 feet range. It was close enough for Joe to note that it was "as big as a telephone pole and painted yellow." The missile passed across the entire formation and hit Kay Russell's airplane on the other side of the strike group.

The SAM maneuver had driven everyone down to 400 knots and 400 feet above the terrain. About then, there was a MiG call. At the moment, Joe Shea was on Speer's right side in a "loose deuce" formation, and didn't see the MiG at first. He was concentrating on flying a proper wingman's position and still avoid his section leader. Speer saw the MiG and abruptly turned right directly into Shea. It was not until Speer fired a Sidewinder that Shea first saw the MiG.

That was just before the Sidewinder fuzed and detonated. Shea had the feeling that the missile had detonated just a little short of its target. However, it immediately pitched nose down and impacted the ground with a large explosion and fire ball. Paul Speer immediately turned back toward the strike group.

Joe's position on his leader was approximately 5 o'clock and 1,000 feet. As the section of F-8s was rejoining the strike group Joe Shea saw a MiG-17 fly directly between him and Speer in hot pursuit of an A-4. Joe recognized that the A-4 was in extremis and made the now famous radio transmission, "There's a MiG among us!" With that he turned hard right as hard as the Crusader could turn and drew enough lead on the MiG to fire a burst of cannon fire in front of and over the MiG's canopy. It was not a serious effort to hit the MiG as it was to cause him to break off his attack. It worked.

The MiG not only broke off the attack on the A-4; it also made the fatal tactical error of reversing his turn to the left. Joe slowly closed the range firing his guns at an estimated range of 500 to 700 feet. At this moment, Joe was beginning to fear that he might run out of ammunition before he could kill the MiG. So, with finger still depressing the guns trigger, he also mashed the missile button on the stick.

The Sidewinder came off the rails, flew through his own "hail of bullets" and again appeared to detonate a little short of the tail of the MiG. However, the MiG immediately pitched nose down and impacted the ground with another fireball. Shea recalls flying

185

through the fireball which puts him no more than 100 to 200 feet over the terrain. Joe pulled off and turned back toward the strike group. Paul Speer was not in sight and was, in fact behind him.

Joe recalls looking down at the ground right at that moment and the area directly under them seemed like it was all lit up by the sparkling flashes of anti-aircraft artillery gunfire. Right at that moment, Joe heard someone call on the radio, asking if anyone in the area had any Sidewinders left. They were in need of some missile fire power. The call came from someone in the other flight of F-8s from Fighter Squadron Twenty-Four. There were obviously more MiGs in the area. Although he didn't know it at the time, Joe's airplane had been hit by a round of small arms fire. The bullet went through the after fuselage skin, penetrated an after fuel tank which, happily was empty. Then the bullet proceeded into the afterburner section of the engine.

A post flight investigation of his battle damage showed that if he had attempted to refuel enroute back to the ship, his airplane would probably have caught fire.

Lieutenant Joe Shea was awarded the Silver Star medal for his performance on this flight.

Author's Note: Joe Shea resigned from the Navy and began flying for Eastern Airlines. He now flies for EVA, a Taiwanese Airline and lives with his family in Coral Gables, Florida.

Lt. (JG) Joe Shea after his MiG kill mission. (Photo courtesy of J. Shea)

41

Commander Bobby C. Lee VF-24

My first of three combat cruises was with VF-24 in Air Group Twenty-One embarked in *Bon Homme Richard*. Our squadron had some of the last F-8 Charlies. Brand X, VF-211 led by Paul Speer, was flying F-8Es. The Charlie had very good points; we couldn't carry bombs and we were cleaner than the F-8E so we burned less gas. Our jet attack squadrons were VA-76 flying A-4 Charlies and VA-212 flying A-4Es. We also had the Barnowls on their last Spad deployment before transitioning to A-7s.

We arrived on Yankee Station for our first line period in January and the weather was miserable. We would launch in sections to do road recce (reconnaissance) and usually find the weather over the beach 200 to 600 feet solid overcast. We would then patrol off the coastline looking for targets at which to shoot our Zunis (5" rockets). The bombers would usually have to dump their bombs because the low overcast precluded any kind of ordnance delivery other than Snakeeye retarded, and it was too low to go over the beach.

The monsoon season lasted until late March or early April and then (weather) was seldom good enough to mount a strike of any consequence. We would try to get strikes in every time the weather looked halfway decent and then we would have secondary and tertiary targets. We would usually break up the strike group and do road recce in the southern route packages or just dump the ordnance in the gulf.

Late April and May the weather started getting pretty good. We were enjoying some decent target assignments but nothing overly exciting because the weather in the toughest areas was still not good enough for Alfa Strikes. It was about this time we learned the big secret of the cruise. VA-212 had a new weapon they had trained with prior to cruise specifically designed to go up against point targets. We fighter jocks had heard nothing about this new weapon and had no idea what was in store for us and the delivery pilots of 212. Walleye had arrived in North Vietnam.

Homer Smith, the Commanding Officer of VA-212 and some of his pilots began to train with the weapon on coastal targets to get a feel for tracking and delivery tactics in a combat environment. We even had a training strike against a barracks area to ensure we had everything right prior to taking on a high value target that would draw attention to the new secret weapon. The training went well and we were waiting for weather and tasking for a major strike into the Hanoi area. We were going against the Hanoi Thermal Power Plant. It would be the first time the Navy had been downtown with an Alfa Strike.

The strike had been scheduled a time or two and the weather had caused an airborne abort each time. We began to almost think of the strike as usual because we had scheduled and briefed it so many times.

The 19th of May we were scheduled again and the weather was questionable. The briefing went as usual with everyone thinking it would be just another airborne abort. The route to the target took us feet dry in the Vinh area, crossing into the mountains just short of Laos and then north to Hanoi. The last seven miles was the only open, flat area where we would be exposed to everything. We thought it would be too risky to go directly up the Red River to Hanoi. (Very soon we would be going that way two or three times a day). Strike composition was 12 F-8s and two A-4s from VA-212 with VA-76 (acting as) Iron Hand with their A-4Cs. Six F-8s were TARCAP (target combat air patrol) and six were flak suppressors. The two A-4Es carried Walleyes. My wingman was Lt (jg) Kit Smith, we called him nugget and indeed he was. He and I were flak suppressors. We had loaded a Sidewinder on station one of each of the flak suppressor airplanes because of the high MiG threat in the Hanoi area. We could select the Zuni tubes on stations two, three and four and not shoot the 'Winder unless we went back to station one. This turned out to be a very fortuitous decision.

Paul Speer led the TARCAP and Joe Ellison led the flak suppressor group. We were assigned individual flak sites so we briefed to become independent sections in the target area. Homer Smith led the Walleye section and I believe Steve Briggs was his wingman. Again, they had different attack headings in the target area so they were to split up before we got to town.

As we progressed along the track it began to look as though we might get the strike in. The weather was good (enough) for routine strikes but there was a broken layer about 5000 feet causing shadows. (Not good enough for Walleye). We pressed on and as we were approaching the flat plain prior to the city we met a MiG head on. I told Phil Wood (who was part of the TARCAP) to go get him. I didn't want a MiG milling around in the area when we had work to do. Phil split off, with his wingman Bill Metsger, and went after him. (That is a story in itself).

When we left the safety of the hilly terrain the whole world exploded in our faces. There were three distinct levels of explosions with SAMs going off all over the place. I thought the best place in this situation was low. The RHAW (radar homing and warning) gear was so loud and constant it was annoying to say the least. I took Kit and myself low enough to get the missile warning system at least intermittent. We were so low all I could see in front of us was green, but it felt good. As we approached the lake just west of the power plant we pulled up to about 3000 feet, rolled on our back and pulled to point the Zunis at the flak site that was our target. I can still remember the size of the thing, it was huge and smoke and flashes were coming from it. As we were in our run this little lone A-4 crossed in front of us on his glide toward the target. It was Steve Briggs, and about that time he came on the air with, "Busy hands are happy hands."

Kit and I were now pulling out of our run to the west and heading for the grass again. Suddenly a MiG crossed our flight path. He looked as if he was turning toward the major part of the strike group. I watched him in disbelief. I leveled off at his altitude, about 700 feet, and pulled hard toward him. We were almost co-speed at the time, about 450 knots. I reached down between my legs for the wafer switch and selected station one. The Sidewinder. I kept pulling to match his turn and wound up very close to him. I didn't realize how close until I fired the missile. It completely disappeared outside my turn. I said to myself, "Oh shit, I'm too close", and at that time the missile crossed back in front of me and tracked into the MiG between his wingline and tail. The tail of the airplane rotated to the outside of the turn as the Sidewinder cut his airplane in half.

Kit and I then turned toward the hills and started our exit. As we completed the turn, off to my left I saw another MiG almost paralleling our track. He had obviously been hit and was descending about a half mile alongside us. He continued down and hit in a green grassy area and exploded in a big orange fireball. I don't know who had hit him, but he was shiny silver. No markings were visible. The one I shot was a darker shade, not shiny.

We couldn't have been out of the mountainous terrain more than four to five minutes but it seemed a long time. Kit had been hit just behind the canopy in the area of his ammo cans, but he never got out of position on the wing.

The strike group was starting to partially join up as we progressed toward the relative safety of the terrain along the border of Laos.

The trip back to the ship was uneventful except that the excitement of the previous few minutes was still very sharp in my mind.

I didn't know until we got back on board that we had lost Bill Metsger from VF-24 and Kay Russell from VF-211. Both would return when our POWs were released from Hanoi. I was one of the lucky ones of the day not to get a scratch. I believe we had a total of ten airplanes hit out of our strike. Also, ours was one of the three big strikes on Hanoi that day. Total Navy and Air Force losses were 11 airplanes. There were 32 confirmed SA-2 firings at our strike group and untold 23, 37, 57 and 85mm firing. It literally formed broken layers at three levels.

I had three combat cruises. The first in 1967 flying F-8s and two flying the A-7 airplane, one in 1969 and one in 1972. There was no comparison to the 1967 war at any other time except possibly the Christmas bombing just prior to the war's end. It was not even possible to explain the difference to those who were there earlier or later, the intensity of the day to day 10, 2 and 4 Alfa Strike schedule and the amount of stuff the North Vietnamese could put up around the major target areas. By the late summer of 1967 we could see the effect of our strikes. Very little was moving south. I believe if we could have continued to strike the lush targets of 1967 through 1968 we could have made air power make a difference. Alas, another strategic mistake by our civilian masters.

Author's Note: Rear Admiral Bobby Lee retired during the 1980s and resides in Lemoore, California with his family where he dabbles in real estate. This account is in his own words, and unedited.

42

Lieutenant Phillip R. Wood
VF-24

The Vietnam War was heating up in 1966 when I left the advanced training command as an instructor with some 1,500 jet hours. Being a fully-qualified LSO helped me get the type of cockpit and coast I wanted. Though the F-4 Phantom II had been in fleet service for six years, the idea of flying a fighter with a GIB (Guy in Back) didn't appeal to me. The trusty old F-8 Crusader was a macho fighter pilot's dream: single-seat, single-engine with a weapons system that made it a true fighter rather than an interceptor.

Each of the Navy's two fighter communities had completely different mentalities about how to engage and shoot down an enemy aircraft. The F-8 pilots were still "turnin' and burnin'," bringing guns to bear along with visual-range heat-seeking missiles as well.

The coast you went to was also important. East coast carriers were still meeting their Mediterranean commitments, whereas all the west coast boats were headed for the Tonkin Gulf. I wanted to get there as soon as possible.

NAS Miramar was called Fightertown USA. It was home of all the NavAirPac fighters including VF-124, the west coast replacement training squadron. Called the Gunfighters because F-8s had 20mm cannons, VF-124 was led by Commander Merle Gorder who had been on *Bon Homme Richard* (CVA-31) with me. Normally, replacement pilots were trained in their aircraft type for five months, then assigned to a fleet squadron just returning from cruise. This allowed the aircrews to train with the squadron throughout the workup schedule for the next deployment. I couldn't wait that long to get to Vietnam. I asked the Skipper if he would assign me to the first squadron headed west. Merle obliged me and I joined VF-24 just one week before it sailed in January 1967.

As luck had it, VF-24's home for the next seven months was "Bonnie Dick." It was like an old baseball glove for me. I was right at home with over 300 carrier landings during my two previous cruises. And my roommate was Lieutenant B.C. "Bobbie" Lee. We were the new guys on the block, having just joined the squadron.

Though we were senior to all the first tour "nuggets", we were assigned one of the smallest staterooms on the ship. It was not important – a stateroom was only used for sleeping and writing letters.

Having B.C. as my roomie worked out great. He was a tiger. We kept track of which pilots were getting the good missions, and insisted that we got our share. We were standing in line, so to speak, for the missions that might allow us to find some MiGs. Unfortunately, my section leader was afraid of his shadow and found any excuse to abort a mission scheduled "over the beach." Rules did not allow me to proceed independently, so we orbited over the water a lot.

As the air war intensified and the North Vietnamese arsenal increased in sophistication and numbers, we started losing more aircraft to MiGs, surface-to-air missiles (SAMs) and "triple A" (anti-aircraft artillery). Many pilots were killed or captured and the volunteers for deep missions became fewer and fewer. My section leader's fear had made him a basket case, and he started drinking a lot in his stateroom. Halfway through the cruise he was shipped off to a staff job in Saigon.

B.C. now became a section leader and we could almost pick and choose our flights over the beach. We already had lost Ken Wood and Stretch Tucker, with no replacement pilots, so most of us now flew two combat missions a day. Every seventh day we stood down so the carrier could replenish fuel, food and ammunition and the air wing could conduct memorial services – lots of them. From March to July we lost twenty-three aircraft and eighteen aviators.

One of the eight squadrons in the wing was VA-212, a light attack unit flying A-4Es. Their Skyhawks were specially configured to carry a new weapon called the Walleye, a glide bomb with a TV camera in the nose. It provided its own guidance to the target – what is called a "launch-and-leave" weapon. After the pilot released the bomb he was free to turn away and exit the target area, thus reducing his exposure to groundfire. Though the Walleye only had a 100-pound shaped-charge warhead, it was deadly accurate. Its seeker head locked on to a shadow contrast – a building window, for instance – and entered the building through that window, exploding inside. The building often collapsed from the extreme overpressure of the blast.

Air Wing 21 systematically started proving Walleye at the start of the deployment so testing actually was conducted in combat. It worked great. We started releasing Walleyes on buildings and bridges in Route Packages I and II in the southern part of North Vietnam. Though we were shot at occasionally, it was a fairly benign environment compared to the nonpermissive atmosphere up around Hanoi and Haiphong in Route Pack VI. We worked our way north, dropping the new weapon on progressively more important targets. The "smart bomb" had proven itself.

The "Bonnie Dick" berthed at Leyte Pier of Cubi Point Naval Air Station in early May 1967 following a forty-day line period. We had only been in port two days, and the air wing was widely dispersed all over the Philippines, recovering from the rigors of war. I had rented a cottage at Baguio, a mountain resort on Luzon, where there was plenty of golf, booze and good food. Five days of R and R usually was enough to rejuvenate body and soul for another month or so in the Gulf.

The knock on the front door came early – too early to be a squadron mate. I opened the door and there stood the biggest, blackest, meanest-looking air policeman you can imagine. All I could think was – what did I do wrong? The AP asked, Are you Lieutenant Wood?"

"Yessir, ah, that's correct, sir," I heard myself say. The "sir" wasn't necessary since he was an enlisted man, but I wanted to be on his good side.

"Sir, you are to report immediately to the airfield for a helo ride back to Cubi. I'll wait until you're packed and ready to go." I didn't ask any questions.

As we lifted off from Baguio the helo aircrewman told me the ship was getting underway two days early. Rumor had it that downtown Hanoi no longer was off limits to our bombing. A funny war, I thought to myself. Strategic targets all over the industrial north were selectively eliminated from the "frag list", the targets assigned by civilians working for Robert Strange McNamara back at the Pentagon.

The frustrations soon became humorous. Some of the targets were a joke. Our favorite was the Cam Loa boatyard southwest of Haiphong. The attack pilots were assigned to thread their A-4s through a maze of AA fire and SAMs to drop their bombs on a derelict, vacant lot that hadn't seen a boat in at least a year. But Naval Aviators are not known for slavish obedience to orders from civilians who don't know their ass from a hole in the ground. We made room for improvisation. During prestrike planning we would find another target in the vicinity worthy of the energy and risk of dropping bombs. If the battle damage assessment (BDA) photos showed no bombs on the assigned target, the aviators just said it was bad aim – they'd do better next time. Thus, we hit a lot of lucrative targets that were not "fragged" even some owned by U.S. companies and therefore "off limits."

We got our sea legs back and started planning in earnest the surgical elimination of military targets in and around Hanoi. Our planning put a lot of emphasis on avoiding civilian casualties adjacent to our targets. But the game was not always played the same by both sides. The North Vietnamese on several occasions placed their AA guns in the yards or on the roofs of hospitals. Even without the big red cross – and it always was – we knew where the hospitals were. We couldn't shoot back to defend ourselves because we knew we wouldn't intentionally bomb a hospital. Late in the war Bach Mai Hospital was damaged and the commies howled their outrage to the world. Nobody seemed to mind that the hospital was adjacent to an enemy fighter field.

Once back on the line, we used the Walleye on a couple of significant, well-defended targets south of Hanoi just to make sure our tactics were sound and the hardware functional. The word came down that our primary target was to be the Hanoi thermal power plant – downtown! This would be the first strike ever conducted inside the city.

It made sense for a lot of reasons, but more importantly it was a clear signal to the North Vietnamese Government that we would no longer place their strategic assets off limits. At our pay grade level it was a real boost to see some reason for even being there in the first place.

There were plenty of volunteers for this mission. Lots of risk, but it was for something meaningful. VA-212 had the lead on this one. Commander Homer Smith was the skipper and would lead the strike. Because it would be the first time downtown, we knew it would stir up a bees' nest. MiGs would be launched against us – we hoped. On normal Alfa strikes there would be a bunch of A-4s carrying iron "dumb" bombs and fighter escorts. Support aircraft would include suppressors to hit the SAM and flak sites immediately prior to the bombers' rolling in on the target. Presumably the suppressors would hit and render useless the gun or missile sites. If not, at least the defenders would have to duck and be unable to get a steady tracking solution on the bombers.

Then there were the jammers to zap electrons at enemy fire-control radar, and Iron Hands to shoot their Shrike missiles at the SAM guidance radars. The inevitable photo bird and his escort would take BDA photography of our hits, and there were various other players. Altogether, usually thirty aircraft comprised this coordinated effort.

However, this strike would be different with the Walleye "smart" bomb. We only needed two Skyhawks as opposed to twelve in a normal raid. But because of the anticipated MiG threat we needed to stack the deck with fighters. Twelve were assigned – six escorts and six flak suppressors. F-8s were terrible in the latter role, but we wanted fighters around while keeping strike composition to a minimum. It was a tradeoff. Crusader pilots weren't highly trained

CHAPTER 42: LIEUTENANT PHILLIP WOOD

in air-to-ground delivery, and the F-8 lacked an effective bomb sight. Furthermore, it carried five-inch Zuni rockets instead of the preferred weapon – Rockeye cluster bombs with hundreds of bomblets that obliterated a flak site. Actually, sending an F-8 on a flak suppressor mission over Hanoi was akin to opening your canopy and trying to pressurize the world. But that deep in "indian country" you wanted plenty of gunfighters around, and we filled the bill.

Our route to Hanoi would be circuitous, requiring more fuel than normal. As the crow flies, the distance from our launch point in the Gulf would be approximately 150 nautical miles, or 300 round-trip. But a straight-in route would lose the element of surprise. Hanoi lies on the western edge of the Red River Valley delta near the mountain ranges leading to Laos. Because of the mountainous terrain masking the radars, and minimal exposure to SAMs inbound, it was decided to head west from the ship, passing south of Than Hoa, hit the hills in Laos and head north until we were just southwest of Hanoi. Then our dash from the hills over the delta to the target would only take about six minutes. The A-4s were "slick" without bomb racks and could keep up with the F-8s. Our route out would be every man for himself, striving to reach the water as quickly as possible where airborne tankers waited.

Two diversionary strikes from other carriers were scheduled for the Hanoi area with target times just prior to ours. We had planned it so the enemy gunners would presumably expend most of their ammunition and loaded missiles prior to our arrival. The primary goal was to knock out the capital city's electricity, and we would do almost anything to optimize our chances.

After the details were worked out, the air wing "heavies" assigned the mission pilots. Of course, Homer Smith would lead his two Skyhawks. He picked as his wingman one of the steadiest young pilots in the squadron, Mike Cater. Bobbie Lee, my roommate, and yours truly were chosen to fly the mission. Bobbie would lead a flak suppressor section and I took the escort section with Lieutenant (JG) Bill Metzger as my wingman.

After the planning was done, B.C. and I went to bed the night of 18 May full of anticipation. We had been in our sacks over an hour, each not knowing that the other was still awake. Not a word was spoken until I muttered, "Bobbie, you awake?"

"Yeah, just thinking." We were both married with two children. "Bobbie, you scared?"

"Nah, I just hope I don't screw up my switchology over the target." There was a saying – better to die than look bad.

We lay there and reviewed together our switch positions for the different weapons we would carry. Little did we know that next day such details would make the difference between getting that coveted MiG kill and being an also-ran.

The launch, rendezvous and refueling overhead the ship were uneventful. The weather enroute was beautiful – a requirement for the TV-guided Walleye. As the strike group coasted in south of Than Hoa we could see 57 and 85mm flak puffs bursting around us. As we entered Laotian airspace we started hearing the "Big Eye" radar calls, informing us that MiGs were scrambling from their bases around Hanoi. I thought to myself, "I hope those *Kitty Hawk* and *Enterprise* fighters don't shoot them all down – leave some for us!"

Twenty minutes from target we heard the activity around the two diversionary strikes just ahead of us. Someone had really pissed off the enemy gunners. We later learned that the North Vietnamese had determined that our flight was Air Force F-105s out of Thailand. It was logical. Based on radar information, we were coming at them from the southwest along typical USAF routes. This misinformation would prove costly for the North Vietnam Air Force. They directed their MiGs to engage our F-8s – a very different breed of cat from the "Thud" – to say nothing of the dozen very aggressive fighter pilots flying those Crusaders, each one starving for a MiG kill.

Flying PAGE BOY 405, I had the lead of the middle section of fighters, trailing the A-4s about a half mile. The six suppressor F-8Cs were out front. Just as we entered the Hanoi plains area, Bobbie called over the tactical frequency, "Hey roomie, we got a single MiG at our ten o'clock."

I looked left and saw a Grumman Intruder from *Kitty Hawk* heading south, exiting the target area. "Negative, that's an A-6," I replied.

B.C. wouldn't be denied. "I'm telling you PAGE BOYS there's a MiG at nine o'clock." He couldn't chase the MiG because his role was to lead his element into the target area and suppress the AAA. The other two escort sections obviously didn't see the Bandit either, as they maintained their position. I looked left again and spotted the MiG-17 closing rapidly on the A-6's tail. I called "Tally ho", and broke hard left.

The MiG and Intruder were opposite my heading, but when I completed my turn I was at the MiG's eight o'clock, outside gun range. But I was starting to get the Sidewinder missile's "growl", indicating that the seeker head also had a bead on the bandit's tailpipe.

In afterburner, I was closing rapidly. God, was I excited. I squeezed off a missile. Stupid. I had too many "Gs" on the aircraft, putting the 'winder outside the firing envelope. The missile tracked the MiG initially, but couldn't turn the corner. It went ballistic behind the target but definitely got the MiG driver's attention. Seeing me in pursuit, he broke hard right from the A-6 and dove for the deck while lighting his afterburner. I was now too close to shoot a missile even though I had a steady-state solution. He was heading down a valley, limiting his maneuvering room, and as he appeared in my gunsight I fired two long bursts from my cannons. I detected no hits even though I seemed too close to miss.

All this time I was being led off into the boonies by the lone MiG. My primary mission was to protect the bombers, and this guy no longer was a threat so I broke off the chase and headed downtown, now about five miles behind the strike group. My wingman, Bill Metzger, was nowhere in sight.

I proceeded in burner toward the target, trying to catch the strike group. As I arrived over the outskirts of Hanoi all hell had broken loose. The scene reminded me of a front-page cover of an air action comic book. In living color! The black and gray puffs of AAA explosions, air-to-air missile trails, the white streamers of SA-2 SAMs lifting off their pads and silver MiGs. The Skyhawks were rolling in on the power plant and I could see two MiGs falling in flames. Later I learned that my roommate had bagged one of them.

While looking for my wingman I observed and heard cannon shells whizzing by my canopy. Stick hard left, burner on, pull like hell. I could now see a MiG-17 1,000 feet behind me but I didn't feel his rounds hit the fuselage aft of my canopy. I kept pulling, "getting angles" on the guy. He could no longer shoot at me and I was gaining the advantage.

He must have been an inexperienced pilot. He bugged out by reversing course, which allowed me to park behind him at 2,000 feet. He was running for his life. I switched my stick's armament switch to "heat" and put him in my gunsight. The missile growl was loud. I pulled the trigger and the half-second before my second Sidewinder fired seemed like the proverbial eternity. As the 'winder left the rail it dropped and appeared to have no guidance on the target. But moments later it started a gentle climb toward the bandit's tailpipe. The missile impact and expanding-rod warhead cut his entire tail off. The MiG lazily pitched nose-over and decelerated rapidly.

As I approached from the rear the pilot ejected and his seat took him up and clear of the aircraft. I passed him so close I could tell the color of his skin and see the patches on his flight suit. The pilot got seat separation, but I observed his main chute was a "streamer" and, sadly, knew that he would fall to his death. The only consolation was that he could not be punished by his superiors for losing his aircraft.

By now all the strike aircraft were outbound toward the Gulf. My wingman, Bill Metzger, plus Kay Russell from VF-211, had been shot down. We had bagged four MiGs, but that's a terrible exchange rate and the A-4s had inflicted minimal damage on the power plant.

The fun and games were just starting. I still had eighty miles of indian territory between me and the Tonkin Gulf. The four shells that had entered the top of my aircraft had passed through my avionics package before carrying through the engine intake, fodding the engine.

I was now low on fuel and couldn't make it back to *Bonnie Dick* without a drink from the duty tanker. My radio was out and my engine was only developing ninety-four percent of full rated power. Use of my nonfunctioning afterburner was academic because I didn't have enough fuel for it anyway. Finally, the cockpit air conditioning had been ripped apart and I became extremely hot inside my capsule.

As I egressed toward the water I could see the aircraft ahead of me taking fire from the gunners around Nam Dinh. If the tankers weren't right off the coast there would be some "hurtin' puppies" nearing fuel exhaustion.

Another F-8 was nearby and I joined on his wing. It was Commander Paul Speer, the executive officer of VF-211. As I joined, I passed the visual signal that my radio was out and I was low on fuel. Thank God for small wonders. Paul got on the horn and had the A-3B tanker, driven by John Wunch of VAH-4, meet us as we went "feet wet" over the Gulf. Wunch was a great aviator, widely respected by air wing pilots. He put the Skywarrior right in front of me with the refueling probe already extended. Since there were so many low-state aircraft he could only give me 1,000 pounds of JP-5, only enough to get me out to sea and find any carrier recovering aircraft.

My navigation aids were gone – shot to pieces by the second MiG – and Paul had headed north with the tanker to take his own drink. I had to proceed east.

I was alone, flying into the Tonkin Gulf. I knew that besides *Bonnie Dick*, *Kitty Hawk* (CVA-63) and *Enterprise* (CVN-65) also would be recovering strike aircraft. At this point I didn't have the luxury of choosing which one I landed aboard. I was down to 400 pounds of fuel – not much even for a single-engine fighter.

I was starting to get a little smoke in the cockpit when I spotted a big deck below. Traditionally it is a sign of poor airmanship and headwork to land on a carrier other than your own. But in this case, with my battle damage and diminishing fuel, to hell with tradition. I didn't want to get wet!

My descent began from the starboard side of the carrier – the wrong side – but it would draw attention to my arrival. I had to land on the first pass or use my ejection seat. As I entered the groove in the landing pattern I saw A-4s ahead of me, but the interval was too short for both me and the nearest Skyhawk to land. I hoped the LSO would give me enough credit to avoid the wrong ship unless I had a genuine problem.

Apparently the LSO did just that. On short final, the A-4 was waved off and I saw a clear deck. God, it was big compared to *Bon Homme Richard*'s. I got the green "cut" lights on the lens to indicate the LSO had me and I was cleared to land. As I hit the deck and shoved the throttle to full power, the afterburner section exploded and my Crusader rolled to a stop. I was aboard but where?

I looked up at the huge white numbers on the island and saw 63. My first landing aboard *Kitty Hawk*! If someone had told this young lieutenant that I would be captain of that big, glamorous carrier seventeen years hence, I'd have concluded they were smoking pot.

My Crusader was a strike, damaged beyond economical repair. As a yellow tractor towed the F-8 clear of the landing area so other planes could come aboard, local Air Wing 11 maintenance crews began "zapping" PAGE BOY 405. It was another tradition; any aircraft that lands on the wrong carrier is bedecked with emblems of the resident squadrons before return to its own ship. By chance, the embarked admiral and visiting Governor Love of Colorado observed the ritual and the governor asked about it. He was perplexed that an aircraft that had just returned from Hanoi, shooting down an enemy fighter, would be so treated. My airplane was repainted even though it was headed for the "boneyard" at Litchfield Park, Arizona.

Later we learned that the Vietnamese had launched sixteen MiGs out of Phuc Yen and Kep airfields. Four never returned. But though there was elation in "Bonnie Dick", the strike had failed to accomplish its mission. Because of scattered clouds over the target, the Walleyes were released low at 3,000 feet in a twenty degree dive – outside optimum parameters. One bomb did minimal damage to the boiler house of the power plant and the other hit the administration building.

It got worse. Next day Commander Smith, skipper of VA-212, was shot down on a strike against the Bac Ghiang thermal power plant. He was captured but never appeared among the POWs released in 1973.

However, on 21 May two Walleyes inflicted major damage to the original target. The lights went out all around Hanoi. But the loss of two pilots on the first mission was a poor trade.

Years later our returning POWs told us that the 19 May mission had a residual effect. They related how excited they were; how their hopes were regenerated when they heard and even saw Navy aircraft overhead, with bombs falling on Hanoi. Unfortunately, such positive efforts were on-again-off-again thing under the Johnson-McNamara regime in Washington.

Author's Note: Phil Wood retired as a Navy Captain, works full time for the Northrop Corporation and resides in southern California. This account of his MiG kill was written by Captain Wood and is printed verbatim.

43

LIEUTENANT COMMANDER MARION "RED" H. ISAACKS VF-24

"Red" Isaacks is one of the most respected gentlemen in the history of carrier aviation. His account of the events of his successful MiG engagement on 21 July 1967 and of another one later on in the next deployment are surprisingly candid and, consequently, carry the powerful ring of credibility despite the passage of twenty-seven years.

Red had just joined the squadron (VF-24) on Yankee Station on U.S.S. *Bon Homme Richard*. Although, in a short time he had gotten his share of Barrier Combat Air Patrols (BARCAPS), he had only been over the beach over North Vietnam two or three times. He considered himself "fresh caught" and not heavy on experience over "Indian Country."

The target for that day was a petroleum storage facility at Ta Xa about 20 miles north of the harbor of Haiphong. Red was the division leader of four F-8Cs from VF-24 which were serving as MiGCAP. Since there hadn't been any MiG activity in over a month there was a general sense in the strike group that this mission was probably going to be a milk run. Red's division of four F-8s was assigned the strike call sign of Pageboy.

Everything had gone as planned and it was fairly quiet until the strike group reached the target area and roll-in point. Suddenly, Red spotted three or four MiG-17s at his 10:30 position and a little below them, at perhaps 8 or 9 thousand feet. It took just a minor turn to the left to put himself in a firing position on one of them. Red got a good "growl" with his Sidewinder and fired it. It didn't guide so Red tried to fire the second missile but it didn't leave the rail.

In retrospect Red assumed that he had not waited enough time for the armament selector to step over to the number three station. On his second attempt to fire the second Sidewinder, the missile came off the rail, seemed to be guiding, then flew right up the tail pipe of the MiG. There was a tremendous ball of fire and the MiG simply disappeared. Red felt fairly certain that the pilot could never have survived the explosion and was genuinely surprised to learn later that the pilot had, in fact, ejected and that a parachute had been seen on the ground.

This was when Red admits to violating the principal commandment of every fighter pilot. It had been a relatively easy kill and Red, lulled somewhat by his success, was still watching the fireball on the ground where the MiG had crashed when he noticed a stream of tracers whip by his airplane and seemed to disappear into the space at one o'clock. Glancing quickly over his right shoulder as he simultaneously heard the sound of cannon fire, Red was horrified to find himself looking into the intake of a silver MiG-17 with a red painted nose and its cannons flashing menacingly. What shocked him most was how close the MiG was. Red said he could probably have "counted turbine blades" in the MiG's intake duct.

As an instinctive reaction Red "reefed it into him" and saw the MiG roll inverted just as it passed under his Crusader. It was so close that Red felt the "WHUMP" as the MiG passed. He estimates it was a "matter of a few feet, or maybe even inches." Commander Isaacks never saw that MiG again. But, he immediately took stock of his situation and saw that the MiG had indeed hit him in the starboard wing. (Afterwards the maintenance people found four holes from cannon fire in the starboard wing fold area and around the trailing edge of the starboard flaps).

Cdr. Red Isaacks, C.O. VF-24, after his MiG kill mission.

CHAPTER 43: LIEUTENANT COMMANDER "RED" ISAACKS

Four VF-24 F-8s preparing for launch in the Tonkin Gulf, 1967 – U.S.S. Bon Homme Richard (CVA-31). (U.S. Navy photo)

Red was so sure that the damage was serious that he headed immediately for the beach. He made the necessary MAYDAY calls and alerted everyone that he had been hit. He was fairly certain that he was about to join the ranks of his shipmates in the Hanoi Hilton. During all of this he saw fire breaking out in the wing fold area and was expecting to be confronted abruptly with a requirement to eject. Other downed Crusader pilots had experienced battle damage and had the airplane suddenly go out of control so violently that a safe ejection was put at hazard. Red said he spent the next forty-five minutes with his right hand on the control stick and his left grasping the alternate ejection seat handle.

About the time he reached the beach, Lieutenant Commander Bob Kirkwood joined up on him to look him over and watch the progress of the wing fire. To his astonishment, Red noticed that Bob's entire starboard UHT (unit hydraulic tail) was gone right up to where it joined the fuselage. Bob was unaware of this and never noticed any difference in the flying qualities of his Crusader with one half of his horizontal tail plane missing. In the post flight assessment it was undetermined whether the lost UHT was the result of battle damage or the result of an over stressing of the airplane.

(Bob Kirkwood got a MiG that day too. In fact, he had already fired his guns at the MiG which Red finally downed).

When Commander Isaacks finally crossed the beach, he had the distinct sensation that he had a damaged but flyable airplane. Meanwhile, he noted, the strike group remained totally unflustered by the MiG calls, MAYDAY calls, MiG activity and general activity in the air directly over head of them. They rolled in, dropped their bomb loads and egressed the target area "without ever batting an eye." Red, a relatively new guy in the air wing was deeply impressed by the general air of cool professionalism shown by his shipmates in the strike group.

Red makes the point that he experienced a tremendous sense of relief when he crossed the beach and went "feet wet." Now, at least, if anything went wrong he wouldn't end up in a prisoner of war status. About halfway back to the ship the fire went out and with it went the terrible apprehension which fire generates in any pilot. The only two things remaining to worry about were whether the wing would fall off; and whether he would be able to raise the wing, lower the wheels and "dirty up" to the landing configuration without utility hydraulic pressure.

Of course, there were emergency air bottles for doing all of those things . . . and they all worked. The recovery aboard the "Bonnie Dick" was relatively uneventful. There is an unwritten code among landing signal officers (LSOs) that when a battle damaged airplane returns to the ship and makes a successful carrier arrested landing, the pilot gets an automatic "OK,#3" landing grade. Commander Red Isaacks got an "OK, #3" that day.

Subsequent assessment of the aerial engagement by the various people who were listening in and watching on radar established that there were at least 8 MiGs airborne that day. They had probably originated from Phuc Yen, a major tactical jet base located a few miles north northeast of Hanoi, and staged out of Kep, an airfield a few miles farther north.

Because of the dearth of MiG activity in recent weeks, the four MiG killers of that day, Lieutenant Phil Dempewolf (probable), Lieutenant Commander Bob Kirkwood, and Lieutenant Commander Tim Hubbard and Red were all flown to Saigon for the usual press conference. A big thing was made by the press of the events of the day.

In the subsequent cruise on board U.S.S. *Hancock* on 17 September 1968 Red and his wingman, who shall remain unnamed, got into an extended engagement with two MiG-21s. This time the event took place farther south, just a few miles north of the demilitarized zone (DMZ); and Red and his wingman were flying newer F-8Hs.

They were vectored in on two "blue bandits" who were operating at relatively low altitude. The event, probably one of the longest single engagements of the war, lasted for eight minutes until it was finally broken off by both sides for lack of fuel. Neither Red nor his wingman got off a single shot . . . nor, does he believe, did the enemy airplanes. The engagement began at low altitude and consisted off a series of close in attacks, counter breaks, unloads and re-attacks which slowly bled off energy until both MiG-21 and F-8H were operating in a dogfight regime suited to neither.

In Commander Isaacks' opinion, it was a much more significant aerial event in that the level of performance of the MiG pilots was substantial. Red's wingman, described by him as "rather headstrong", remained in after-burner almost the entire event. And, as a

Bob Kirkwood, Phil Dempewolf, Tim Hubbard, and Red Issacks – "Four MiG Killers." (Photo courtesy of M. Isaacks)

consequence, he ran out of gas before he could refuel; despite the fact that there were tankers waiting off-shore to help and Red was frequently calling him on the radio to come out of burner. The pilot was recovered, but the loss of the airplane put quite a damper on the events of the day. I am sure the North Vietnamese calculated that aerial engagement as a win for their side.

Red remembers that the cockpit was "hot as hell" and they were pulling so many "gs" for so long that his oxygen mask worked its way down off his mouth and over his chin. In order to call his wingman about using burner he literally put the stick between his knees long enough to use his right hand to hold the mask over his mouth while the left pushed the microphone button on the throttle.

Commander Red Isaacks' story includes a relatively easy and quick MiG-17 engagement where a minor case of first engagement "buck fever" might have played a part, but didn't effect the outcome. The second story serves to highlight the broad spectrum of the aerial engagement environment into which the antagonists fought toe-to-toe. It was certainly an environment for which neither the F-8 and the MiG-21 were designed.

Author's note: Red retired as a Navy Captain and lives with his family in Corpus Christi, Texas.

44

LIEUTENANT COMMANDER ROBERT L. KIRKWOOD
VF-24

Lieutenant Commander Bob Kirkwood's MiG-17 was downed with twenty-millimeter guns. He was Red Isaacks' section leader. Flying as the leader of the second section of two F-8Cs employed in the target combat air patrol mission on the strike, against the petroleum storage facility, Bob was a part of the melee which ensued when the MiG-17s arrived on the scene. When the division turned to engage the MiGs, Bob took a Sidewinder shot at one MiG then became engaged with same MiG that Isaacks was fighting. Bob fired a second Sidewinder at that MiG-17 only to observe Isaacks' Sidewinder arrive a few seconds earlier and kill the airplane.

A third MiG appeared and Kirkwood fired his last Sidewinder only to observe it track and detonate but apparently do no serious damage to the airplane. Now armed with only his guns, Lieutenant Commander Kirkwood maneuvered for a kill with those lethal weapons.

Just after Bob's missile detonated the MiG pilot, apparently realizing his extremis circumstances, turned hard right and into Kirkwood. With what he thought was a good tracking solution, Kirkwood fired at an estimated range of 600 feet . . . perfect for a lethal gun kill. Kirkwood described seeing the sparkling effect of his high explosive incendiary rounds as they detonated on impact with the MiG's right wing and after fuselage. Flame sprang out of the after fuselage area and the MiG pitched up, at which point the canopy came off and the ejection seat fired. Kirkwood got a good look at the MiG pilot in the ejection seat as he passed close aboard. Then it was all over.

When Kirkwood joined up on Isaacks' airplane they assessed one another's battle damage. Both airplanes had suffered battle damage, both from enemy aircraft gunfire. But, Kirkwood's airplane, Isaacks noted, was missing its entire starboard unit hydraulic tail.

The two airplanes returned to *Bon Homme Richard* and recovered aboard safely. Just about every pilot in both fighter squadrons went up to look over the two airplanes. Mo Wright, VF-211's maintenance officer described the damaged UHT as having been cut off as with a cutting torch. VF-211's Commanding Officer, Paul Speer also looked it over and came to a similar conclusion.

The engineers who designed the Crusader never would have guessed that one could land the Crusader safely aboard a carrier with one UHT missing . . . but Bob Kirkwood did it.

Lieutenant Commander Kirkwood's MiG kill was one of only a few "gun kills" of the aerial war over Vietnam. Although guns were employed numerous times by F-105s and F-8s against MiGs they were rarely listed as the killing weapon of the engagement. This is often a misunderstood statistic of the war. Guns were, on many occasions, fired to intimidate the target into turning away or breaking off an attack. The net effect of the gunfire was often to bring the target into the lethal envelope of another weapon . . . a missile. Gun effectiveness needs to be evaluated as an important, contributing part of a fighter plane's quiver of weaponry. For his part in this major aerial engagement, Lieutenant Commander Bob Kirkwood was awarded the Silver Star medal.

Author's Note: I was unable to locate Lieutenant Commander Kirkwood despite extensive efforts among former associates and even advertising in several trade media periodicals. The above account represents the distillation of accounts of others who were in the same engagement and some unclassified and declassified operational reports.

45

LIEUTENANT COMMANDER RAY G. "TIM" HUBBARD
VF-211

"Tim" Hubbard's entry into the annals of MiG killers occurred on 21 July 1967. As a member of Fighter Squadron Two Hundred Eleven, in Air Wing Twenty-One on the U.S.S. *Bon Homme Richard*, he was assigned as "Iron Hand" escort for a strike group of eight A-4s on the Mei Xa petroleum storage facility north of Hanoi on the main highway (Route 15) to China about halfway between Hanoi and the Chinese border. There were two pouncer (Iron Hand) sections assigned to this particular mission because of the high surface-to-air missile (SAM) threat in the area. The lead Iron Hand pilot was Lieutenant Commander T.R. Schwartz, a former F-8 pilot turned light attack pilot. Tim Hubbard was T.R.'s escort.

Each Iron Hand section was made up of an A-4 equipped with anti-radiation missiles (ARM) designed to go against the fire control radars which comprised the extensive air defense network ringing the Hanoi/Haiphong area. Guarding the A-4 from enemy fighter interference, and providing a second set of eyes against all threats, was the fighter escort's function.

Also in the strike group were four F-8Cs from VF-24 which comprised the MiG combat air patrol (MiGCAP). The strike group

Hanoi Headaches – Left to right: Cdr. Speer, LCdr. Wright, Lt(JG) Shea, LCdr. Hubbard, and Cdr. Issacks. (Photo courtesy of P. Speer)

CHAPTER 45: LIEUTENANT COMMANDER "TIM" HUBBARD

Left to right: Major J.A. Hargrave; LCdr. T.R. Schwartz, VA-76; Cdr. Paul Speer, VF-211; "Mo" Wright, VF-211; Lt(JG) Joe Shea, VF-211; LCdr. Bobby Lee, VF-24; Lt. Phil Wood, VF-24, on board U.S.S. Bon Homme Richard, July, 1967. (Photo courtesy of P. Speer)

had launched, rendezvoused, tanked and proceeded up the Tonkin Gulf to a point north of the port of Haiphong and gone west via the "back door" enroute to their target.

It was the mission of the pouncer, in Tim's words, "to troll for SAMs." The Iron Hand's weapon, in those days was the Shrike anti-radiation missile. If, as was often the case in 1967, the ship's magazines ran low on Shrike missiles, they would simulate Shrikes by firing Zuni or 2.75" rockets from the A-4 or his escort. SAM radar operators, seeing the rockets fire, would often assume they were Shrikes and shut down their radars.

As soon as the pouncer airplane detected the enemy radar looking at him he went into an attack mode and, pointing toward the radar site, dove against it, launching his missile when he thought he was in the Shrike's envelope. If the enemy radar stayed on, the Shrike theoretically would home on it and damage it when it flew within lethal radius of the radar antenna. Most fighter pilots didn't enjoy Iron Hand escort missions. The reasons are several: the A-4s flew much too slow for the fighter to comfortably maneuver; there were rarely opportunities to kill other MiGs on these missions; and the Iron Hand shooter and his escort were usually the target of the SAM which often was launched in the Shrike shooting sequence. It was not a sexy mission. Most fighter pilots took their share of the Iron Hand missions as a necessary evil associated with combat operations. No one said that they had to enjoy it.

As the strike group ingressed into the target area at a relatively low altitude, T.R. Schwartz and Tim Hubbard wove back and forth about 2 or 3 thousand feet higher, in full view of the SAM acquisition radars, and several miles in advance of the strike group. As the strike group approached the turn-in point, eight MiGs appeared out of the north and flew right into the strike group. They appeared out of the north and came in between the Iron Hand sections and the strike group, pickling off their external fuel tanks as they dove against the strike group.

When T.R. and Tim briefed their own mission, they agreed that if jumped by MiGs the fighter escort would take the lead of the mission. This was standard procedure. As soon as the strike group tally ho'd the MiGs, Tim Hubbard took the lead, lit his afterburner and headed toward the closest MiG section. In so doing, he crossed over the top of T.R.'s airplane from right to left. T.R., as briefed fell into a supporting wingman position.

At this point, it is worth noting that Hubbard's armament consisted of one AIM-9D Sidewinder on the port fuselage missile station, four 5 inch Zuni rockets on the starboard fuselage stations and a full load of twenty millimeter ammunition. He had run into MiGs on one previous mission as an Iron Hand escort and learned that the air-to-air lead solution for Zunis (an air-to-ground weapon) was to hold the pipper "somewhere around the rudder pedal adjust lever." This, of course, was a fighter pilot's droll way of saying that the kinematic lead angle for firing a Zuni rocket against another airplane was so enormous as to make any meaningful aiming impossible. Having pickled off several Zunis (one of which actually detonated when the influence fuse sensed the proximity of the MiG) he finally managed to get a few bursts of twenty millimeter gunfire off, inflicting enough damage that the MiG departed the scene trailing smoke.

This, having happened on a previous Iron Hand mission, Tim Hubbard was at least experienced in the problems associated with air-to-air employment of Zuni rockets. He watched the first section of MiG-17s come through the formation from the north and turn back in a descending left turn toward the bomber group. Tim slid in behind the MiG wingman, much too close for a Sidewinder shot, and closed for a gun kill. The MiGs saw Tim at this point and tightened their left turn defensively into him. He increased his lead angle and began firing his twenty millimeter cannons seeing his tracers and high explosive incendiary rounds striking the wings of the MiG when he observed a few green balls going by him laterally. Since the MiG-17's huge thirty-seven millimeter cannon was known to carry tracer ammunition which left a green trail, this told him that another MiG was firing at him from the left rear quarter with a fairly large lead angle. Tim executed an extremely hard climbing left turn into the attackers causing a section of MiG-17s to over-

Above and below: VF-211 F-8Es with six MiG decals on ventrils and rudder. Above NAS Miramar, 1973, below NAS Miramar, 1971.

CHAPTER 45: LIEUTENANT COMMANDER "TIM" HUBBARD

shoot their turn and slide by him. Tim completed what was essentially a high "g" barrel roll defensive turn into the MiGs and ended up sliding into a gun attack against the wingman of this second section of MiG-17s.

Again, Tim saw his rounds begin sparkling all over the MiG and he observed, in his taped interview, that he had a "real good gun solution." This meant that this tracking run on the MiG was smooth and he felt assured that he was inflicting major damage. Then, to his dismay, he ran out of ammunition. The realization came over him that he had nothing left but the Sidewinder and four Zuni rockets. The Sidewinder had not seemed operable in his post-launch system check. Remembering his previous experience with the rockets, he decided to try this particular form of unconventional aerial warfare again. He pulled the enormous lead angle which he assumed was roughly right and pickled off two Zuni rockets. They undershot the target so he pulled even more lead angle and pickled off his two remaining Zunis, noting with pleasure that one of them detonated close aboard the MiG.

The MiG rolled out of his left turn toward the right, reversed the turn twice more and appeared to be smoking. Then the MiG, now turning left again, began to decelerate causing a large closure rate. Tim, realizing he was about to overshoot and slide in front of the MiG, executed what he called the "world's tightest high 'g' barrel roll" to gain nose to tail separation. He ended up above and behind the MiG, still in a left turn, watching as the fuselage seemed to grow several times in diameter and, as he watched, it simply disintegrated before his eyes. Just as this happened the pilot ejected and he watched, fascinated, as the parachute deployed and the pilot appeared to be safely clear of danger.

About this time Tim Hubbard looked around and saw that the A-4 bomber group was continuing on toward the target and there were no other airplanes in sight. He called his wingman, who was also no where to be seen and told him he was "AMMO ZERO" (out of weapons) and was departing the area to return to the ship. Later in the debrief he learned that his wingman, T.R. Schwartz, the Iron Hand A-4, had gone to the aid of Red Isaacks who was being badly hammered by a MiG-17 he couldn't shake. T.R. slid in behind the harassing MiG-17 and fired a few Zuni rockets at him causing him to desist in his attack on Isaacks and depart the scene.

It was during this same strike that three other MiGs were killed; one by Bob Kirkwood, (VF-24) with guns, one by Red Isaacks, (VF-24) with a Sidewinder and Phil Dempewolf, (VF-24) got a probable kill with a Sidewinder. For this mission, Tim Hubbard was awarded one MiG-17 kill with Zunis and guns and a probable MiG-17 kill with the same weapons as well as a probable MiG-17 kill from the other event also with the same weapons. He was awarded the Silver Star medal for this action.

Author's Note: Tim Hubbard is retired and resides with his family in Corpus Christi, Texas.

46

LIEUTENANT COMMANDER R.W. SCHAFFERT VF-111

As histories like this are assembled there is a tendency to focus perhaps too much on the discrete subject and miss the rich pattern of peripherals with which the Southeast Asian air war abounds.

One such peripheral is the story of Dick Schaffert's ten minute air battle with four MiG-17s on 14 December 1967. Although he downed none of the MiG's, Dick Schaffert displayed more skill, guts and airmanship in that long ten minutes than most airmen display in an entire career. Also, what came out of that air battle are some of the most unusual aerial combat photographs of the entire air war.

On that day Dick Schaffert had the unenviable mission of Iron Hand escort to perform. The Iron Hand mission was an anti-SAM mission in which the attacking airplane (an A-4 in this case) carried anti-radiation missiles (Shrike, in this case) and fired them at SAM sites to suppress their fire against strike groups. I always felt that we were trolling for SAMs when flying Iron Hand escort missions. That is really what it was. The launching airplane had to fly into the SAM envelope, get the SAM site to come up on the air and perhaps even fire a SAM. At that point, the launch airplane would launch its anti-radiation missile at the SAM site and hope to put it out of operations long enough for the strike group to hit its assigned target.

To help the Iron Hand pilot do his job, a cockpit display system was devised to help him determine when he was within lethal firing range of the SAM site. The pilot would usually drop his nose to pick up airspeed, then pull up, center the cockpit display and, when in envelope, fire the missile. If a SAM had been fired, it was usually at the Iron hand airplanes. This meant that the launching pilot had to have nerves of steel. While the SAM was accelerating to four times the speed of sound directly at him, the pilot had to look into the cockpit to center the display then fire. Only then could he look out. With the SAM coming at him at Mach 4.0 there was often only tenths of seconds to go into a SAM evasion maneuver. That was the reason for the escort.

The escort was supposed to be the Iron Hand pilot's eyes. He had to keep looking out for the SAMs and other dangers, and warn

Dick Schaffert, VF-111, after his MiG engagement. (Photo courtesy of R. Schaffert)

CHAPTER 46: LIEUTENANT COMMANDER R.W. SCHAFFERT

the pilot when to maneuver. The Iron Hand pilot and his escort were engaged in a dicey line of work. This is what Dick Schaffert and Chuck Nelson were doing on 14 December 1967. The following are his own words, unabridged.

"This is a verbal description of a mission that took place over North Vietnam on 14 December 1967 late in the afternoon off of the U.S.S. *Oriskany*. It involved aircraft from Air Wing Sixteen with A-4E attack squadrons VA-163 and VA-164 and F-8E fighter squadron VF-162 and F-8C fighter squadron VF-111. The target area was midway between Thai Binh and Hanoi in the Red River delta. The specific target was the Canal De Ambouze. The purpose of the mission was to deposit Mark 82 Snakeeye bombs with delayed action, metal-detector fuzes. These were to be used against barge traffic that might go through the canal during the bombing halt that was anticipated to come in the near future.

The attack force was composed of six A-4E bombers to be escorted by two F-8E MiGCAPs from VF-162. Pilots of the two MiGCAP airplanes were Commander Cal Swanson and Lieutenant Dick Wyman. The two Iron Hand teams for surface-to-air missile suppression were launched. They were composed of A-4E Shrike-equipped aircraft accompanied by F-8C fighters. Their call signs were Pouncer. The Pouncer 1 and 2 section was led by Commander Bob Rasmussen in an F-8C from VF-111 and he escorted an A-4E whose pilot's name, I'm sorry, I forget. They were to cover north of the strike group against surface-to-air missile sites in that area. Pouncer 3 and 4 consisted of an A-4E flown by Lieutenant Chuck Nelson and an F-8C flown by (me) Lieutenant Dick Schaffert. Our mission was to suppress surface-to-air missile sites in the Nam Dinh area which was generally to the south of the strike group. We were, of course, supported by the usual array of aircraft and ships. RED CROWN was on station and lent radar support to the mission.

One of the features that made this strike different from previous ones was that the fighters were carrying the new AIM-D Sidewinder, the newest, super-cooled Sidewinder air-to-air missile. Then, to confuse the North Vietnamese air defense efforts, it was decided that the Pouncer airplanes would use the same identification friend or foe (IFF) that had been designated for the main attack force. There was some theory at that time that perhaps the North Vietnamese were able read the IFF codes and knew where the attack force was, so they could deploy their missiles against them. The two Pouncer section would actually be at least 15 or 20 miles away from the attack force.

The 1600 launch went as scheduled and the strike group headed for the mouth of the Red River which was the coast-in point for the mission. As the strike group went "feet dry", the Iron Hand sections split away and the MiGCAP continued inbound with the strike group. Shortly after we went "feet dry", RED CROWN notified everyone that bandits were airborne at Bullseye. This meant that MiG-17s were taking off from Phuc Yen (a VPAF airfield north of Hanoi). A little while thereafter RED CROWN gave the strike leader a call telling him that the bandits were 25 miles northwest heading 150. A little later he gave another call to the strike leader telling him that the MiGs were on his nose at 15 miles.

As the strike leader was just about to start rolling in on the target he got another call from RED CROWN that the MiGs were now just west of him at three miles. As this was happening Bob Rasmussen was working over some missile sites at Hung Yen. In our section Chuck Nelson was working on some missile sites in the Nam Dinh area. The radar was attempting to lock on us according to our APR-27s (missile launch warning receivers). When RED CROWN issued its last warning Chuck Nelson was in the midst of what appeared to me to be a Shrike missile launch maneuver. He had begun a pitch-up to a firing position. This maneuver slowed him down. I was flying in a loose deuce formation so I performed a barrel roll maneuver to displace myself away so as not to over run him. I was in an inverted attitude and was looking back and down to the south to see Chuck. At the time my airplane was at about 15,000 feet. We were headed just about directly into the sun. It was then that I noticed two MiG-17 heading directly at Chuck Nelson. I recall realizing that RED CROWN must have been calling the MiGs relative to our Pouncer section and not the strike group. I don't know if our squawking strike group IFF codes messed up the North Vietnamese air defense radar network that day but it sure messed up RED CROWN.

I tally-ho'd the MiGs and warned Chuck that they were going to pass by his right side. I also told him that I was at his three o'clock high, that I had the lead. At that point in time I was on my back and indicating about 330 knots so I pulled the nose down and commenced a descending left turn to come around onto the MiGs. I bottomed out of what was a low yo-yo which put the MiGs high and to the north of us starting a left turn back toward us. My speed was about 400 knots and I was in burner. I was looking up into a brilliant blue sky background and the number two MiG seemed to have been thrown a little bit outside his leader's turn. I had a good (Sidewinder) tone on him, was closing, and at a range of about 4,000 feet, coming up through an altitude of about 11,000 feet I called a "Fox One" and fired the first of three missiles. When the missile came off the rail there was about a thirty degree angle off the tailpipe of the number two MiG. The missile tracked normally, did a little zig-zagging and continued toward the MiG. There was about a sixty degree angle off the tailpipe when it got to the MiG which continued a hard downward turn into the missile. The missile went by at a distance of about half the width of his aircraft ... about 15 to 20 feet away from but it failed to explode. We know from the pictures that this is a problem that Bob Rasmussen had a little bit later in the fight. I continued into a nose high left turn as the MiGs continued their left hand turn below me. At this point in time Chuck Nelson called that he had lost sight of me. I was climbing through about 15,000 feet and looked back over my left shoulder trying to see him. It was at that moment that I saw another section of MiG-17s in a fighting wing (formation) about 3,000 feet behind me coming out of the west above the sun and rapidly approaching a firing position. I still had enough energy left in the aircraft to execute a hard break into them and they overshot.

I called Chuck that I didn't see him, that we now had four MiGs (in the vicinity) and that he should get out of the area. I also tried to contact the Pouncer 1 and 2 section to come down and give us a hand. But they were interested in a SAM site in the Haiphong area and were considerably north of us at the time.

I came back down in a low yo-yo toward the MiG section that I had already shot at once already. They were starting a nose up pull and were at a distance of about two miles from me to the east. The low yo-yo took me across the circle toward them as they continued turning rather level. I closed to a range of less than one mile and had climbed to about level with them at an altitude of about 13,000 feet. I had about a forty degree angle off and although I was not able to get my nose on the (section) leader, I was able to get it on the wingman. I unloaded my airplane down to about two "g"s and fired my second Sidewinder at him. As the missile came off the rail I wanted to watch the shot but had to immediately increase my turn to about four "g"s in a left hand turn because I saw tracers coming by over the top of the cockpit. Looking back over my left shoulder I saw the other MiG section. They were both firing at a range of about 2,500 feet. Since they were already over-leading me I didn't dare increase my turn and pull into their line of fire. So I pushed the stick forward, unloaded the aircraft, pushed bottom rudder and went to a vertical nose-down position. With an airspeed of about 350 knots I started my nose back up, still in afterburner and, passing a heading of about north, went past them as they continued their turn down toward me. They reversed their turn and I stayed in burner going over the top at about 20,000 feet and as I looked back down, saw that they had attempted to follow me up but were falling off about four thousand feet below me. They dropped their noses and continued around to the west in an attempt to regain some speed. Continuing over the top I again had them in my gunsight but the sun was now low in the west and I had some difficulty getting a tone with my Sidewinder. I was closing distance rapidly and, at a distance of about three thousand feet fired my third Sidewinder. As I fired my third, and last, Sidewinder in a nose-down attack to the west I noticed in the upper right hand side of my wind screen two MiGs coming toward me from a northwesterly direction. They were at a distance of about one mile. I couldn't watch the missile I had just fired because I had to pull my nose up toward the two MiGs that were coming in from the northwest. As I was pulling my nose up they were in a line abreast with about one thousand feet between them, they each launched two missiles that I assumed were Atolls. All four missiles continued in a straight line, were never a threat, and passed well astern of me about three thousand feet. I went into a vertical nose-up maneuver to get up out of the fight area to see what was going on while those two MiGs ran right on by and continued out of sight to the east. I turned my attention back to the two MiGs that I had fired on with the nose-down shot. They had bottomed out and were now climbing back toward me from the west. I did not see Chuck nor did I see the section of MiGs that I had fired my first and second Sidewinders at.

At that time I heard Chuck Nelson talking to the MiGCAP trying to vector them over to our position. But, of course, the MiGCAP's duty was to (protect) the strike group and there had been MiGs seen in the area so they announced their intention to stay with the strike group until they were "feet wet."

By this time Pouncer 1, Bob Rasmussen was headed down toward our position and Chuck Nelson had visual contact with the one MiG which eventually was shot down. Chuck attempted to bring Bob into that area. Chuck was using the numerous Sidewinder and Atoll smoke trails as reference points in trying to vector Rasmussen into the area.

As I went over the top at about 20 or 21,000 feet, I looked back down at the section of MiGs that I had just fired at from the west. They couldn't get up to me again and they went over the top at about 15,000 feet so I came down again behind them accelerating in burner. The MiG leader again executed a nose up climbing turn and I cut across the circle as they did that closing for a gun shot. Again the MiG wingman was in about a 1,500 foot trail so I concentrated on him. I was pulling about six "g"s when I came within range of him. I always kept my gunsight set on a fixed range of 1,500 feet because the radar in our F-8Cs was not all that reliable. So, I put the pipper in front of him and decided I would fire a stream of bullets and let the pipper drift back through the target. So, at a range of about 1,500 feet I commenced firing and, unfortunately, firing at a "g" loading of six, I only got one round (counted after the mission) out of each gun before the feed mechanisms jammed. Historically, F-8 guns didn't work very well under high "g" loads. Realizing that I was now out of missiles and guns I eased off on the turn, went outside the MiG's turn and went high.

The leader did a very well-executed reversal climbing turn back toward me and we ended up in a vertical scissors maneuver. For the rest of the fight I didn't see any other MiGs except for the section leader I was fighting. The wingman got thrown out of the fight very early.

We continued the vertical scissors going up and both ran out of airspeed at the same time so we continued the same sort of cockpit-to-cockpit fight as we were in a vertical position going down. This fellow was very good.

Later I was able to fight "Tooter" Teague who was flying a MiG-17 around Nellis. "Tooter" had a pretty good reputation as a fighter pilot, but this fellow was much better. We were closely engaged, never more than 1,500 to 2,000 separating our airplanes, for about four minute's time.

During this four minute period the MiGCAP was able to escort the strike group "feet wet" and turned back, in response to Chuck

CHAPTER 46: LIEUTENANT COMMANDER R.W. SCHAFFERT

Nelson's vectors to the position where he was trailing (and taking pictures of) another MiG-17.

I had the very uneasy feeling that this MiG pilot was beating me. He was slowly gaining the advantage. He was getting inside of me just a little more with each succeeding maneuver. Frequently I had to unload the aircraft and do a negative "g" roll underneath. After a time during which our altitude was gradually decreasing I executed one of those negative "g" roll under maneuvers and after pulling out level I was looking horizontally at a pagoda. We were literally at treetop level. I had pretty good speed coming out of that maneuver and went vertical again topping out at about 6,000 feet. By that time he was right back on me again so I had to "break" down and into him. As I broke into him we passed almost directly head-on. After he passed me I looked back over my shoulder I saw him level his wings and head back in the direction of Hanoi.

My own fuel gauge was a little under 1,500 pounds and I needed to head back toward the ship. I headed about due east and watched him heading west. When I finally lost sight of him I began a climb. The low fuel level warning light was on and I remember when I finally got my first TACAN lock-on of *Oriskany*, the indicator said 68 miles. Shortly thereafter I crossed the beach and got a steer from RED CROWN which corresponded with my TACAN distance. I then switched to *Oriskany*'s strike frequency and asked for a tanker. I was told that there were none available at the time. I continued at maximum range cruise toward *Oriskany*, learning along the way that they had already finished their recovery and were holding a ready deck for me. The ship was steaming into the wind in a southeasterly direction. I was able to enter the landing pattern in a modified entry at about the ninety degree position with an indicated 300 pounds on the fuel gauge.

During the debrief I made a strong pitch to the CAG Shepherd that we escort Iron Hand mission with a section rather than a single fighter. He responded that he had such serious aircraft problems (having lost so many airplanes recently) that my suggestion was a luxury we simply couldn't afford. Some of the early reports on this fight had some errors which were later corrected. One of the errors was a reported time of engagement of about six minutes. A check of RED CROWN's tape recording of the incident showed that from the time the engagement began until I asked them for a vector to home plate was exactly ten minutes and forty-five seconds. Added to that was the time that Lieutenant Wyman and Commander Rasmussen were engaged with the MiG-17 that Chuck Nelson directed them on to, another three minutes, which makes the MiG engagement, between thirteen and fourteen minutes, the longest of the Vietnam war. Also Zalin Grant quoted me in his book (Over the Beach) as having said that Cal Swanson and Dick Wyman, when they made their victory roll over the carrier, nearly had a mid-air collision. I never said that.

Enough can't be said about Chuck Nelson and his pictures. He certainly had ice water in his veins. As I recall he had an offer from *Life* magazine for those photos and he could have made a couple thousand on them; but he kept them in the family. They became an important part of the slide presentation that CAG Burt Shepherd took back to Washington and showed around town. I don't know if you are aware but when Burt was back in Washington he appeared on the Ed Sullivan show and was acknowledged as the most decorated Naval Aviator at the time.

An interesting note in closing. As we began working more and more with the MiG-17 the Navy Fighter Weapons School (TOPGUN) announced that the MiG-17 had a flight control problem that made it difficult to unload the airplane and get a good rate of roll, such as we were doing during our scissors maneuvers. It just so happens that, by sheer coincidence, the unloaded roll under maneuver that I was forced to execute several time turns out to be the perfect maneuver to capitalize on this particular weakness.

One of our conclusions during the debrief was that my observation of the MiG-17 gun muzzle blasts as being evenly spaced would indicate the use of 23 millimeter guns only. This was the case in the MiG-17 C version. Also the launching of Atoll missiles and the fact that the airplane had to abruptly depart the area would indicate the airplane probably had an afterburner. All indications were that I was engaged with a MiG-17C model."

Although he didn't shoot down a MiG, Dick Schaffert became something of a legend. He had participated in one of the most difficult aerial engagements of the entire war and had taken on incredible odds with a reckless abandon which makes me proud to know him. As one listens to his tape account his tone of voice and the words chosen make it sound like a Sunday picnic. I found my palms getting sweaty just listening to his story. I am absolutely certain that the four MiG-17 pilots sat around their version of the ready room table that evening and all agreed that they had just tangled with the world class wild man of naval aviation. Certainly there had to be more than a few comments about the fact that certain unnamed parts of his anatomy had to be made out of solid brass. Lieutenant Commander Dick Schaffert was awarded the Distinguished Flying Cross for this mission . . . a richly deserved recognition.

PART V: AERIAL COMBAT

The series of eight photographs which were taken by Chuck Nelson while he was maneuvering with a MiG-17 which was trying to shoot him down remains one of the most startling photographic records of aerial combat ever taken. It is worth the time to examine them with a detailed explanation.

Photograph #1 (above left) was taken about five minutes after the engagement started. The MiG is seen from above as it makes a hard turn trying to get into position to fire on Nelson. More Crusaders were on the way to the scene and Nelson was guiding them in. The Navy pilots were uncertain, at the time this photograph was taken, whether the other three or four MiGs had fled or were returning.

Photograph #2 (above right) was taken only seconds after the first photograph. The MiG is shown as it continues its hard turn in an attempt to get into position to fire on Nelson.

Photograph #3 (below left) was taken about a minute later. Both Nelson and the MiG are at lower altitudes. The MiG is still turning and climbing to get into a position of advantage on Nelson. As this photograph was taken, Nelson was turning to the left to cause the MiG to overshoot him.

Photograph #4 (below right) shows that Nelson has completed a diving left turn to improve his position relative to the MiG. As he begins to sweep down to position himself behind the MiG's tail, the MiG is put on the defensive. By staying behind the MiG, Nelson forced the enemy pilot to concentrate on shaking him off. This kept the MiG from escaping. Two or three minutes later the other Navy Crusaders arrived.

CHAPTER 46: LIEUTENANT COMMANDER R.W. SCHAFFERT

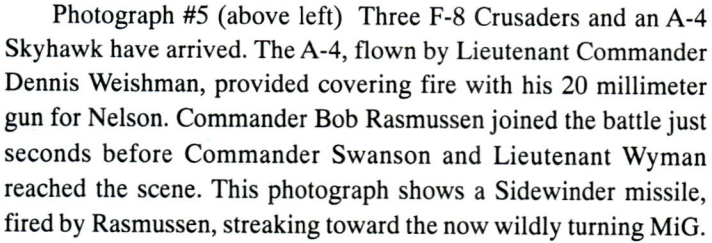

Photograph #5 (above left) Three F-8 Crusaders and an A-4 Skyhawk have arrived. The A-4, flown by Lieutenant Commander Dennis Weishman, provided covering fire with his 20 millimeter gun for Nelson. Commander Bob Rasmussen joined the battle just seconds before Commander Swanson and Lieutenant Wyman reached the scene. This photograph shows a Sidewinder missile, fired by Rasmussen, streaking toward the now wildly turning MiG.

Photograph #6 (above right) shows Rasmussen's Sidewinder missile passing near the evading MiG. The MiG evaded the missile by about 25 feet . . . the very edge of the missile's killing radius.

Photograph #7 (below left) shows that a second missile, fired by Dick Wyman finds its target. The residual trail of Rasmussen's missile is still seen across the photograph, cut by the diagonal trail of the lethal missile fired by Wyman from behind the MiG. Smoke from the detonation hangs in the air as the MiG plunges earthward in a fireball.

Photograph #8 (below right) shows the fireball turn into a second explosion as the MiG strikes the earth in a rice paddy 12 miles south of Hanoi.

Author's Note: Dick Schaffert retired as a Navy Captain, got a PhD in International Relations and now owns his own consulting company on international terrorism and international security. He has done well in this field and has published important articles in this critical field. He lives with his family near Whidbey Island.

The passage of the years have permitted Dick Wyman to do some retrospective thinking about the incident which is worthy of noting at this juncture in the narrative.

Lieutenant Commander Dick Schaffert, the Iron Hand escort, never got adequate credit for doing what Wyman could only describe as incredible flying! For what must have seemed like an eternity, until the cavalry arrived, he fought off four MiG-17s keeping both himself and the A-4 Iron Hand airplane alive by dexterous use of attack and counterattack tactics, and by making full use of the Crusader's advantages over the MiG-17. He certainly rewrote the book on Iron Hand escort tactics.

Lieutenant Chuck Nelson, the A-4 pilot, had "nerves of steel" while another part of his anatomy must have been made of solid brass. He stayed in the fight without any air-to-air weapons, gave short counts to F-8s enroute to the rescue and even had the chutzpah to take a dramatic series of photographs with a hand-held camera of MiG-17s even when they were attacking him. Of all of the Navy pilots in the engagement the only two who didn't lose sight of the MiG-17 were Wyman and Nelson. The green camouflage paint job on the MiG-17 made it exceedingly difficult to see; especially down low against the background of the rice paddies. Like his escort, Chuck Nelson never got adequate recognition for doing an incredible job! Some weeks later, *Oriskany* put in at Hong Kong for some well-earned rest and recreation over Christmas and New Year. Dick Wyman and Bob Rasmussen got together and compared notes on the aerial engagement. Both believed that the MiG-17 pilot was a hell of a tactician. Both were saddened that the man never survived the engagement.

They concluded, however, (as did the intelligence folks) that the North Vietnamese made very good use of their air assets. They used the MiG-21s as bait to lure escort fighters away from the strike group; then they brought in the MiG-17s from another direction (often by surprise using pop-up tactics) to attack the relatively defenseless strike groups.

Also while in Hong Kong, Dick Wyman met with the commanding officer of "Red Crown" (U.S.S. *Coontz*) who provided him with an audio tape recording of the engagement. The tape served to corroborate Dick Wyman's version of the engagement. Lieutenant Wyman was awarded the Silver Star for his performance on this mission.

Author's Note: He later resigned from the Navy and resides at this time with his family in Kittery Point, Maine working in the golf course maintenance business.

48

COMMANDER LOWELL "MOOSE" R. MYERS VF-51

The Story in his Own Words

"The morning of June 26, 1968 arrived like most days when you are deployed to the Tonkin Gulf aboard the USS *Bon Homme Richard*. The weather was beautiful, the gulf was peaceful, and I was enjoying the company of the most dedicated men, both young and old, in the United States.

The ship's crew and air wing of the "Bonnie Dick" were certainly not anything unusual from other Navy combatant ships with good leadership and excellent training. We were all products of background grounded in the belief that the United States was the greatest nation in the world and we were here in Vietnam to carry out the mission as dictated by the leaders of our country. That mission was to deter North Vietnam from invading and taking military and political control of South Vietnam. I understood this mission well as during an active 11 year career as a naval carrier aviator, I had developed a hobby of studying military and political history. The underlying purpose of my hobby was to continue to gain the strength of leadership experience that I knew would work when I was selected to be a leader in naval aviation.

The experience I am referring to is the study of the leaders that have counted in history and brought the rewards (good and bad) back to their country. I will not bore you with a long dissertation on how well schooled I am in political and military history. However, I want to point out the consistent connection between good leaders – success, and bad leaders-failure. Number one – a good leader never went anywhere without the backing of his countrymen. He certainly used devious means when he had to convince a reluctant nation to follow his lead, but he never disappeared over the horizon without a total commitment of the power of his people and resources. This approach sounds a little dictatorial and it is, because when you go to war, there is no alternative but to win and win convincingly.

This autocratic approach to managing a nation's wars is also good for the people doing the fighting. The military leaders chosen for their prowess in winning wars prosecute the war with only the political goal provided for direction. This leads to execution of strategy using maximum force to win while keeping the exposure of

"Moose" Myers describing his MiG kill mission. (Photo courtesy of L. Myers)

American fighting at the lowest level practical. This sure makes sense to me especially since I am a member of the professional military corps expected to lead men into battle.

What a terrible disappointment the Vietnam war became to me. Anyone reading this book must have some interest in military history, therefore, has an opinion of the nation's performance during the Vietnam conflict. Whatever that opinion is, I am here to tell you that the wartime conduct of our political leaders beginning with Kennedy and ending with Nixon was disgraceful and forever cast a pall on this country's ability to conduct foreign affairs the way our forefathers intended. The American dream has at its core the freedom and safety of our people to live their lives in a free and open society. The failure of political leadership to consider this their primary objective has brought us to the point where oil is more important than our own peoples lives, saving face is more important than using maximum force, political dominance at home is more important than good leadership abroad.

PART V: AERIAL COMBAT

"Moose" Myers after his MiG kill mission. (Photo courtesy of L. Myers)

"Moose" Myers, "There I was . . ." (Photo courtesy of L. Myers)

During my first cruise to Vietnam in 1966-67, the on-scene military leadership ran the show in the Tonkin Gulf. We were free to execute the military strategy reducing the capability of the North Vietnamese to prosecute the war against South Vietnam. We dropped every bridge in Indian Country (territory of North Vietnam north of the Red River), few, if any anti-aircraft guns or guided missiles were left in operating condition. We put out of commission every electrical power plant we knew existed, and generally had complete control of the skies over cities like Hanoi, Haiphong, Kep, and other important logistic centers. There wasn't much road or canal traffic moving South that we didn't monitor and destroy when necessary. Needless to say, we made life miserable for the logistic planners of Ho Chi Minh.

Things sure changed when I returned to the war in Vietnam in January 1968. President Johnson suddenly ordered us to stop bombing strategic targets in North Vietnam. We were not allowed to fly over important industrial areas like Haiphong or Hanoi for an extended period of time. The only flying we did was reconnaissance missions over these targets verifying that the bridges were being rebuilt, anti-aircraft guns were restored and an array of surface-to-air missiles was being deployed by Russia into North Vietnam at a level never seen in modern warfare.

You should have been along when we returned to bombing the North. President Johnson cleared us back into the North, however, he placed a no-fly zone around Hanoi, restricted us from bombing the harbor of Haiphong, and turned all mission planning over to the geniuses in the Pentagon. From then on things went from bad to worse. Our casualties went up, our effectiveness went down, and we were using the great strength of the aircraft carrier to blow up dirt bridges, and chase trucks and barges all over North Vietnam. It was a frustrating day when you would be proceeding toward some insignificant target and witness 6-10 surface-to-air missiles fired at the strike force from inside the no-fly zone and knew you were forbidden to counter-attack the missile sites themselves.

This sorry display of leadership from Washington forces the doubts to start creeping into the daily conversations in ready rooms of the tactical squadrons aboard the aircraft carriers. Subjects are discussed like "Why are we wasting our energies on such insignificant targets and not going for the jugular of North Vietnam?" Much of wartime activity and flying off aircraft carriers is a head game. It is really important to be committed to defined and attainable goals to be successful in this environment. Not to be committed body and soul to being the best there is under these conditions is often fatal. Therefore, when you start to question the motives of your political leaders, things can get out of hand real fast.

Every job on the carrier is important and critical to safe and effective operations. You are mixing steam power, gun powder (lots of it) airplanes filled with fuel, and PEOPLE. Every sailor from the

CHAPTER 48: COMMANDER "MOOSE" MYERS

18 year old son of a midwest farmer who fires the catapult when all the checks are complete, to the Commanding Officer of this powerful warship must do his job completely and unselfishly. Everyone is at risk and everyone depends on one another. You don't seek out support from others, it is always given fully without question or dependence. You think this is the mind game of the century! You can't play games aboard an aircraft carrier at sea. It is dangerously fast living and you can't go back and do it over.

This may explain why we navy men are so dedicated to sticking to our job. If we for one moment begin to think like politicians, we would all be dead. You must make a decision based upon the facts, plan a course of action and carry it out. Once the momentum is started, it must play itself out. Have you ever started a project and brought it to a halt because of some unplanned event? If you have, and have tried to re-start it, you will note that more bad things happen than good because of the doubts you have created in the minds of your men and the lack of commitment you have communicated to the people responsible for carrying out the action.

With this background, my division manned our airplanes in the early afternoon of June 26 for a barrier combat air patrol. I say "my division", because we flew together as a team at all times possible. Only sickness or other duties would shift the priority away from always flying with the same people. My wingmen knew my habits, they trusted me to lead them to and from the target, and had the confidence that as a team we had the best chance of successfully completing the mission and returning to the carrier. One of the four aircraft went down during engine start-up and launch, so the remaining three proceeded toward our assigned station under the communications and radar control of a picket line guided missile frigate.

We no sooner arrived on station than we were informed that there were four "bandits" heading south toward central North Vietnam. The four bandits were identified as two MiG-21's leading with 2 MiG-17's in a two mile trail. They were heading towards the area we were responsible for and if they continued, we would be vectored to challenge their approach to our assigned area. My 11 years of carrier aviation experience in fighter aircraft had prepared me for this kind of situation, and needless to say I was getting eager for the battle. Watching the strategy and tactics of the North Vietnamese for the previous two years of extended deployments, I had formulated a plan just for the situation that was fast developing. The North Vietnamese pilots like the Russian pilots were closely controlled by ground supervisors. They would not venture into unknown areas without good radar control and support from their ground control stations. If any situation would develop where they didn't have the clear tactical advantage, they would immediately turn back toward home base and discontinue the mission.

My plan was to get between them and their home base undetected by their controllers and force them to fight their way through me to get back to home base. With this in mind, the three of us were vectored toward the bogeys by our shipboard controllers as the MiGs continued to penetrate deeper into the southern area of North Vietnam near the town of Vinh. I took my division down to treetop level and increased speed up to 600 knots as we crossed the coastline heading straight toward the oncoming MiGs. I was heading for a position below and behind their left side to gain the tactical advantage before they detected our movements. We were risking heavy small calibre arms ground fire by enemy troops by flying at this low level. However, our high speed would make us difficult targets to shoot at as we were traveling nearly at the speed of sound.

As we approached the MiGs, my wingman on the right side called a "tallyho" on the bandits locating them just where we wanted them. He called the MiGs out at 10 o'clock high at 5 miles. This would place them 20 degrees left of the nose of my aircraft and well above the horizon. I quickly focused in on the direction, as by custom, if the leader did not make immediate acquisition, he must turn the lead over to the wingman with the visual lock-on. I acquired them immediately calling out my tallyho and identified two MiG-21s in a one mile trail. I informed our controller of the situation and asked the whereabouts of the other two MiG-17s. He observed on radar their turn away from the formation returning to home base at this same instant. I then took control of the situation and assigned the trail MiG to my section leader and pitched up and into the oncoming MiG.

The moment I changed direction in my Crusader, the MiG saw the flash of my wings in the sunlight and turned down and into me to counter my initial move toward him. The MiG-21 was an advanced fighter plane manufactured by Russia and noted for its high maneuverability and quick acceleration. The aircraft had two bad characteristics: rearward visibility was very poor and airspeed dissipated rapidly under high "G" conditions due to the high drag of the delta wing configuration. My strategy was to quickly take advantage of these two weaknesses. I must get close to him quickly and get into his rear quarter making it difficult for him to keep track of my flight path.

We passed nearly head-on into our first maneuver and began to turnback into each other. Here he made a near fatal error by turning back into me in a nose low maneuver. This increased the distance of his turn and allowed me to come back into him in a near vertical maneuver slowing my aircraft and slipping into his rear quarter. He saw the error of his ways and immediately pulled into an extremely nose high turn and got back just about all he had lost in the first turn. We traded reversals for several turns and I would fire my 20 mm cannons at him each time I would cross his tail at a distance of about 200-300 feet. Of course I was pulling 6-7 G's at the time and had little chance of hitting him with any of my armor piercing shells.

I was continually gaining the advantage over him by getting closer to his line of flight each time we would cross flight paths. However, I vas much too close to fire one of my 4 Sidewinders at

"Moose" Myers landing after his MiG kill mission. (Photo courtesy of L. Myers)

him as the distance was too close for the missile to leave the launcher, arm itself and track the target. After the fourth or fifth scissor maneuver, and seeing he was steadily losing the advantage, he attempted to detach from the dog fight by pulling into a very steep climbing maneuver and accelerate away from me. It was just what I was waiting for and I fell in behind him letting the range to his aircraft increase to a distance suited for my Sidewinder air-to-air missiles. The target presentation that I now had before me was a MiG-21 in full afterburner, in a 60-70 degree nose high climb with a clear blue sky background for reference and contrast.

I switched to missiles on the armament panel and shifted the select switch to my number two missile. During the early part of this flight, I had tested the condition of all four missiles and had determined that the number two missile was operating a degree better than the other three. I placed the tailpipe of the MiG in the center of my sight and listened for the tone indicating the missile saw the target and was seeking it with its infrared homing device. When the tracking sound diminished slightly, I knew that the target was in the null (or center) of the tracker and I pulled the trigger firing the Sidewinder at the MiG. As the manufacturer guarantees when you do everything right, the Sidewinder tracked the MiG perfectly, flew up the tailpipe and blew the back of the airplane away from the fuselage. The MiG pilot successfully ejected from his aircraft and floated down into the rice paddies and his friends watching from below.

I turned my attention to my section leader and the second MiG. Apparently, as soon as he realized that he was opposed by three Navy Crusaders, the second MiG immediately turned for home and abandoned his leader. Unlike the MiG pilot, my section leader observing him turn away from the fight, remained in a protective cover position to his leader and wingman as he was expected to do. We quickly regrouped and headed for the coastline and out of harms way from the many anti-aircraft guns positioned in and around Vinh.

Back at the ship, the Commanding Officer of the "Bonnie Dick", "Big Daddy" Dankworth gave me permission to make a victory fly by over the ship telling the crew that we had scored another victory for the good guys. Thus, the flight ended with a very challenging event of landing aboard a moving aircraft carrier. The trick is to follow the bouncing ball reflecting a light path through the sky indicating the right track to land in the middle of the landing area and catch one of the four wires that will snag the airplane and bring it to very quick stop. Having made hundreds of carrier landings, none of them the same, this particular landing was routine when viewed with the events of the previous hour.

The opportunity to engage an enemy aircraft and best him in an aerial duel is what helps you justify to yourself the position that our political leaders have put us in by prosecuting the Vietnam War in such a weak and indecisive way. This event boosts the morale of the shipboard crew and gives us all something to cheer about. The loss of one aircraft to North Vietnam will not influence the outcome of the war, and Russia will quickly replace the fallen aircraft anyway. But what it does for all of us is make it a little easier to live with the frustration of this war of attrition and the loss of our comrades to enemy guns and missiles. One small victory goes a long way in raising the spirits of the good guys and keeps us aggressive and at our best in continuing to carry the honor of the American fighting men regardless of the futility of the mission when you have to fight with one hand tied behind your back."

Author's Note: "Moose" Myers retired as a Rear Admiral and resides in Valley Center, California, working as a baby eucalyptus farmer. He was awarded the Silver Star medal for this action.

49

Lieutenant Commander John B. Nichols VF-191

Lieutenant Commander John "Pirate" B. Nichols' MiG kill day began as a relatively normal flying day in the Tonkin Gulf with a photo reconnaissance escort mission with Lieutenant Bill Kocar from the VFP-63 photo detachment on board U.S.S. *Ticonderoga* (CVA-14) in Fighter Squadron One Ninety One. "Pirate" manned up in Feedbag 181, an F-8E, almost a brand new airplane, having arrived in the squadron just before deployment. Bill Kocar was in Corktip 602, an RF-8G, photo reconnaissance version of the Crusader.

This was not a post-strike mission. Rather, it was a mission to reconnoiter half a dozen targets which had been identified as likely ones for the ongoing Rolling Thunder bombing campaign. John recalled that the weather was perfect with clear skies and almost unlimited visibility. He also noted that he carefully checked out all of his weapons system features after the running rendezvous with Kocar. To his satisfaction, all systems were working perfectly. Even in a brand new airplane, this was a pleasant surprise.

"Pirate" also commented on feeling lucky to be flying with probably the best photo reconnaissance pilot in the detachment. He had flown with Kocar before and had developed a great deal of respect for his all-around skill as a pilot.

Their flight was put under the control of the U.S.S. *Horne* (DLG-30) a guided missile destroyer. A Lieutenant (junior grade) Kozma was on duty in *Horne*'s combat information center that day and he was alert enough to turn on a tape recorder as the MiG engagement began to unfold and preserve most of the radio transmissions on tape.

"Pirate's" mission load-out was two AIM-9 Sidewinders (heat-seeking air-to-air missiles) and 600 rounds of twenty millimeter ammunition. The Crusader's four guns were reasonably reliable but they were known to jam every once in a while when heavy "g"s were being experienced at the time of a firing attempt. The more ammunition in the belts, under those circumstances, the greater the load on the gun feed mechanism. For strafing and photo escort missions, which were usually one "g" firing circumstances, a full load of 150 rounds per gun (total 600 rounds) was appropriate. If, however, there was a reasonably high likelihood of aerial combat the usual practice among fighter squadrons was to down load some of the ammunition. Fighter Squadron Fifty-Three, for example (the author's squadron) always down loaded to 60 rounds per gun under those circumstances.

The flight, under the lead of Lieutenant Kocar, coasted in at the conjunction of several small waterways about fifteen miles south of the port city of Haiphong. Their airspeed was 450 knots and altitude about 2,500 to 3,000 feet of altitude. John had his master armament switch set on, the left hand Sidewinder selected and the mode switch in GUNS!

Gunfire erupted at almost every target they went to that day. They were headed south toward the city of Vinh with John Nichols changing sides from time to time to keep himself generally looking at the photo plane with the threat sector beyond. He just happened to be on the right side and 1,000 to 1,500 feet higher that the photo plane as the approached their target about 15 to 20 miles inland. It was a clear day and he could clearly see the *Horne* (their radar control ship) just off the beach. "Pirate" suggests that the proximity of *Horne* gave him a false sense of security in that he felt there could be no enemy aircraft in his vicinity or the ship would have seen them with their radar.

Perhaps the photo plane pilot shared the sense of security because he wasn't doing any "jinking." This troubled John. But he rationalized it by assuming that Kocar was concentrating on the most important target at the moment.

"Pirate" was looking back over his left shoulder clearing the area behind the photo plane and spotted the MiG at two or three miles. He recalled being pleasantly surprised that he was able to recognize the airplane easily as a MiG-17 bearing a camouflage paint scheme. He also noted that the MiG had a fairly high closure rate and estimates that it must have been doing 600 knots.

He called out sharply, "Corktip, you've got a MiG on your ass. Break left." He was also pleasantly surprised that not a split second elapsed before Lieutenant Kocar had that photo plane in a hard, left nose-down break turn. What astonished him was that by the time

Cdr. John "Pirate" Nichols after his MiG kill mission. (Photo courtesy of J. Nichols)

the photo plane had turned 90 degrees and was essentially headed out to sea, the MiG had only accomplished about 30 degrees of turn and had slid way outside of Kocar's turn radius.

Simultaneously, John rammed the throttle to full military power (no after-burner) and began a high "g" left nose-down turn. At precisely that instant he noted tracers passing to the right and below him. The first reaction was to assume it was from ground fire. But, a second later, he decided that, from the flatness of the tracers' trajectory it must have come from an airplane somewhere behind him. But "Pirate" was in a turn which put him in a favorable position behind the first MiG. So, he "gave little thought" to the second airplane, "put it out of his mind", and immediately got a good Sidewinder tone on the first MiG. So, he pickled off a Sidewinder, saw it try to turn with the MiG and then appeared to explode close aboard but outside the MiG's turn.

The Sidewinder didn't appear to do any damage and the MiG reversed his turn and lit his after-burner. "Pirate" was now directly behind the MiG and was closing fast. Both airplanes were now in a hard right turn. He got a good growl on his second Sidewinder and pickled it off at what appeared to be close to minimum missile range. He had a good angle up, a good tone and practically no "g" on his airplane when he pickled off his second Sidewinder.

The Sidewinder came off the rails, appeared to be tracking the MiG and exploded in a very large shower of what appeared to be tinfoil all around him. He emerged from the cloud and saw that the MiG was still flying, but his after-burner was out and he seemed to be trailing a small cloud of white fuel vapor. "Pirate" believes that the Sidewinder explosion either blew out the after-burner or flamed out the engine. He had suddenly developed an enormous overtake rate and, without knowing exactly why, he chopped the throttle and popped the speed brakes. He barely had time to reach down and turn on the gun switch, noting the pleasant sound of the "thump" as the gun breaches closed. By this time the MiG was filling up his windscreen and was presenting a plan view to him with both airplanes in a hard right turn. John's gunsight pipper was ahead of the MiG but was drifting rapidly back through the MiG so he fired the guns.

"Pirate" immediately saw five or six large flashes on the upper part of the MiG then, "all of a sudden it began coming apart." He saw a wing come off then there was a large explosion of debris and it scared hell out of him for fear that his engine would ingest a lot of that stuff. In actual fact, John had relaxed on the stick and flew aft of the MiG and through the debris cloud. The MiG continued in a right, downward trajectory and impacted the ground.

Almost immediately, he remembered the other MiG and, becoming concerned about it, ran his throttle to full military. After the first ninety degree turn Kocar had reversed his turn and called out.

"Attaboy, Feedbag. You got him. Your six is clear", as the first missile exploded. The engagement looked like a 90-270 turn with the fight ending up about the time they were headed north again. John's neck was on a swivel as he searched for the second MiG. He was nowhere to be seen.

When the first MiG impacted the ground, "Pirate" transmitted, "I've got it!" Kocar rogered the transmission. Meanwhile, John's Air Wing Commander, Phil Craven was calling, asking for their position. John responded to Phil's first request with, "I can't talk now." When Craven called again, John told the photo pilot to "Tell him our TACAN position."

Craven told him to settle down and informed him that he was inbound with a flight of Crusaders. John continued north a little longer and realized he was out of missiles and had some unknown amount of gun ammunition left.

"Let's get out of here", he told the photo pilot and the two airplanes went "feet wet." Once clear of the coast and outside of SAM envelopes, they climbed to twenty thousand feet for the return flight to the ship. Of course, the ship were well aware of the aerial engagement and went out of their way to be accommodating. Would he like a tanker?

John had plenty of fuel and didn't need a tanker. However, years of ingrained training kicked in. Never turn down fuel was one of those unspoken carrier pilot maxims. So, the two airplanes joined up with a tanker enroute to the ship and took on a token amount of fuel. By the way, as an indicator of the adrenalin still floating around in his system, John admitted to having "a little trouble" with the tanker.

CHAPTER 49: LIEUTENANT COMMANDER JOHN NICHOLS

When he finally checked in with the ship fifty miles away, they announced they were ready to take him aboard. How accommodating! That had never happened before! Carriers always told you to wait. Did he have any special requests? This was almost too much! So, John asked if the ship would let him make a victory fly-by! Of course, be our guest, was the amazing response!

So, "Pirate" put Bill Kocar in a right parade echelon and descended to 200 feet, flying up the ship's wake at 650 knots.

"Two. You go right. I'll go left," he told his wingman sotto voce as they rocketed past the ship. Pulling up into the vertical in full after-burner John rolled left into a series of aileron rolls that took the pair all the way to forty thousand feet. At 200 knots John came out of after-burner and descended to a normal approach and uneventful carrier arrestment aboard "Tico." The date was 9 July 1968.

John was whisked off to Saigon shortly thereafter and appeared on what we called the "five o'clock follies", a press conference. Although he had been advised by an Army officer on the Joint Staff at Saigon not to discuss his unit, the kind of mission they were on or the type airplane he flew, John realized that was silly advice. He responded candidly to reasonable questions and the whole thing went fairly well.

It was on a later deployment that John and three other pilots (1 from VF-24 and 2 from VF-211) were sent to the Thailand Air Base at Udorn at the request of the U.S. Air Force. With 300 combat missions, over 3,000 hours in the Crusader and a MiG-17 under his belt, "Pirate" was a good choice of Navy fighter pilot to go to Udorn to examine their modus operandi and perhaps find an answer to why Air Force exchange ratios against MiGs were essentially one to one. While at Udorn, John gave some lectures on Navy fighter tactics, flew with the Air Force on a few combat missions and shared his frank views with them on fighter tactics. Ultimately, the bureaucracy got to him and he returned, troubled and dismayed to the ship. The straws that broke the camel's back were the issue of a camouflage paint job and the Navy IFF.

The local Air Force folks insisted that they be allowed to put a camouflage paint job on John's two F-8s before they could continue going on combat missions. John asked what that would add to the airplane's weight. When he was told that the paint scheme would weigh several hundred pounds to the airplane's weight he properly refused. The F-8 had already gained so much weight with combat modifications (RHAWS gear, ECM boxes, chaff dispensers and smoke abatement fuel additives) that its maximum arrestment weight with bring back weapons became a serious problem. The net effect of the above-mentioned items meant that the F-8 could arrive at the ramp with only enough fuel aboard for three or four landing attempts.

The Air Force were unclear about their objection to the Navy IFF (identification friend or foe) except to insist that it jeopardized their air crew's survivability. In their daily forays to Route Package 6 the Udorn aircrews were being regularly jumped by North Vietnamese MiGs. Their losses to MiGs continued, in "Pirate's" opinion (and the author's too) to be unacceptably high.

"Pirate" summarized the "lessons learned" from his aerial engagement as follows:

- "Loose Deuce" tactics work!
- He could, and should, have done better than he did (maybe have gotten the other MiG).
- He must have just begun his turn when the second MiG opened fire. (it probably saved his life).
- The proximity of the *Horne* lulled them both into a false sense of security.
- The MiG pilot did not take full advantage of his airplane's advantages.
- U.S. Navy tactical training is probably a decisive factor in its success against North Vietnamese MiGs.
- U.S aircrews generally have better air-to-air weapons.
- Overall, the total effect of sound tactics, good airplanes, good weapons and a superior training philosophy will always make a winning combination.

Author's note: John Nichols retired after having completed a tour of duty in command of a fleet fighter squadron. His activities since have included co-authoring a book about Navy combat operations in Southeast Asia, titled *On Yankee Station*, a novel about aerial combat in the middle east titled, *The Warriors* and there are more ongoing writing projects. He resides with his family in West Palm Beach, Florida.

50

COMMANDER GUY CANE VF-53

When solicited for an account of his MiG kill, Guy Cane submitted the following operational report.

"At 11:32 AM local time on 29 July 1968 four F-8E fighters from Fighter Squadron Fifty-Three were vectored over the beach of North Vietnam. The controlling radar ship was the U.S.S. *Long Beach* and the flight leader was Commander Guy Cane.

The radar ship had detected a flight of MiG fighters proceeding south past 19 degrees north latitude in North Vietnam. Commander Cane, anticipating a possible engagement, had shifted his flight to the new secure voice radio frequency to discuss tactics with the ship. The secure voice system had just recently been installed in the squadron's F-8Es and had yet to be proven in aerial combat. Commander Cane asked the ship to position them a little closer to the North Vietnamese coastline than they were normally stationed.

Upon receiving the initial vector, Commander Cane immediately dove his division to low altitude to confuse enemy radar and take advantage of surprise. In spite of the heavy concentration of anti-aircraft artillery and automatic weapons in the area, the division of F-8s crossed the beach at 18 degrees 56 minutes north latitude and 105 degrees 38 minutes east longitude at extremely low altitude and 630 knots. Commander Cane positioned his second section of F-8s about 2-3 miles in trail of the first. The flight had penetrated about 7 miles inland when two MiG-17s were sighted at 10 o'clock high at 2-3 miles heading about 130 degrees magnetic at about 400 knots. The two opposing sections of airplanes met almost head-on and turned into one another.

At this time the second section of F-8s sighted two more MiG-17s descending from a cloud to the west and turned to engage them. Thereafter the entire fight took place in the same 2 to 3 mile diameter airspace between the surface and an altitude of about 6,000 feet.

After his initial turn, Commander Cane executed a maximum performance port reversal into the bandits which placed him directly aft of the MiG leader after only one turn. An AIM-9D air-to-air missile was fired in envelope on a good tone from the MiG's

Cdr. Guy Cane, VF-53, after his MiG kill mission. (Photo courtesy of G. Cane)

tailpipe and it appeared to guide initially. However, the missile missed as the MiG leader increased his rate of turn. Executing another hard port reversal, Commander Cane maneuvered to a position behind the lead MiG's wingman and fired his second missile which guided well and detonated immediately aft of the MiG tail section.

A portion of the starboard wing was seen to disintegrate and the MiG rolled off into a steep, nose-down left spiral. After a quick check of their six o'clock position, Commander Cane and his wingman observed a huge fireball on the ground directly below them where the MiG impacted. Reversing again into the fight a guns attack was then pressed on a third MiG to a range of 200 feet but no hits were observed.

CHAPTER 50: COMMANDER GUY CANE

The aerial engagement continued for a total of five minutes with intense close in fighting. Commander Cane's division, demonstrating superb airmanship, and providing magnificent mutual support, achieved the tactical advantage initially and never relinquished it throughout the fight. Despite the fact that almost all ordnance had been expended and the flight was low on fuel, Commander Cane continued the engagement until the enemy fled north.

Disengaging, Commander Cane detached his two wingmen as a section and then joined with his second section leader to sweep back through the area in order to cover their departure. Completing the sweep, Commander Cane headed feet wet and regrouped his division as they crossed the beach. Following expeditious in-flight refueling all aircraft returned aboard ship undamaged."

Author's Note: Captain Guy Cane, USN (retired), lives in Annapolis, Maryland where he resides with his family. He is working in the general aircraft brokerage business.

51

Lieutenant Norman K. McCoy
VF-51

Norm McCoy's victory over a MiG-21 on 1 August 1967 was the third MiG kill for the Air Wing Five fighters in a matter of five weeks... but it was the wing's first MiG-21 kill.

There were some distinct differences between the F-8s flown by the two fighter squadrons in Air Wing Five. Therefore there were some differences between the weapons load-outs for the two airplanes. There is no doubt that the sequence of events in the 1 August engagement was heavily influenced by these differences. It is worth a few lines to explain them.

There is an unwritten law about the acquisition and ownership of tactical aircraft which, for lack of a better name, I call "Gillcrist's Law." I tested it out before articulating it when I was Director, Aviation Plans and Programs (OP-50) in 1982 and it is absolutely unassailable! Simply stated, it says that during the lifetime of a tactical (from the first assembly line roll-out to the last retirement from the fleet) it gains weight at the rate of roughly one pound per day! My test was simple. I took ten tactical airplanes in the U.S. Navy inventory in 1982. Taking the weight of the most recent model of each airplane, I subtracted from it the weight of the airplane when the first model rolled off the assembly line new. I took the time between those two events (expressed in days) and divided that number into the difference in weights. The answer for each of the ten airplanes came out uncannily the same (within a few percentage points).

In 1967, when VF-51 and VF-53 were flying F-8s in combat operations in Southeast Asia, Crusaders had already been in the fleet for eleven years. VF-53 was flying F-8Es which represented the fourth growth version of the original F-8A (F-8U-1). Each growth version saw the addition of important improvements. In accordance with existing accounting rules for maintaining the proper weight and balance records, each change was recorded in terms of the added weight and the distance of that weight from the airplane's center of gravity. However, some of the alterations occurred as a result of what was called "quick reaction" requirements imposed by the needs of our operating squadrons in combat.

As a consequence of their urgent need, the scrupulous process of the weight and balance accounting system was by-passed temporarily. In some of these cases it took years for the process to catch up. Some of the changes, like the addition of a fuel additive to prevent black exhaust gas smoke (a visual signature enhancement), or chaff dispensers (to foil enemy radars) or the "Alaska Patch" to beef up the fuselage or a few minor improvements to the AIM-9D Sidewinder missile were all changes which added weight.

The author, as Commanding officer of VF-53 during this deployment, took the extraordinary step of weighing each of the squadron airplanes on the scales at Naval Air Station, China Lake, California. I knew it was a step beyond my authority, but I also knew that the lives of my pilots might very well depend upon knowing what the real weight of the F-8E was. It was not the least surprising, then, that the average weight of all twelve squadron airplanes came out about six hundred fifty pounds greater than what the aircraft logbooks said they weighed.

This was serious! It meant that if one of the squadron airplanes were loaded with a full inventory of ammunition (600 rounds), two dual missile pylons, all of the electronics countermeasure equipment, a full load of chaff and four Sidewinders, its empty weight would be so great that it could land aboard the Bon Homme Richard with only 900 pounds of fuel. This was several hundred pounds below that at which the low fuel warning light would illuminate and allow only one or two carrier passes before a decision would have to be made to execute a barricade arrestment. It certainly did not allow our pilots any operational flexibility or safety. Any more fuel than that would exceed the maximum carrier arrestment weight specified in the handbook.

After much serious discussion, it was decided that the combat load-out for VF-53 airplanes would be only 60 rounds per gun (240 rounds versus 600) and two single missile pylons versus two duals which meant only two missiles instead of four. It was a bitter pill to swallow but far better and safer than returning to the carrier with only 900 pounds of fuel for the first landing attempt. That would be cutting it too close. The reduced weapons configuration permitted

CHAPTER 51: LIEUTENANT NORMAN McCOY

Aboard U.S.S. Bon Homme Richard, intelligence officer Cdr. D.D. Ritchey, debriefs Lts. George Hise (VF-53) (center), and Lt. Norm McCoy (VF-51) (right) after their MiG engagements. Cdr. Harry Blake, C.O. VF-53 (left) and Cdr. Guy Cane, X.O. VF-53, stand in the background. (Photo courtesy of N. McCoy)

Lt. George Hise, and Lt. Norm McCoy discuss their MiG engagement. (Photo courtesy of N. McCoy)

a pilot to bring back his weapons and still be able to make his landing approach with 1,500 pounds of fuel.

The other squadron, VF-51, was flying F-8Hs which weighed considerably less than the F-8E. The F-8Hs were essentially overhauled F-8Ds with a few minor modifications like: an avionics upgrade and a new radar antenna with a smaller nose cone. The 800 pound difference between the two models of the F-8 permitted the F-8H to carry a full load of ammunition and four missiles. This represented a significant difference in the war fighting capabilities of the two airplanes.

In the spring of 1968 the *Bon Homme Richard* and the other carriers on Yankee Station were operating under the heavy constraints of the bombing halt. This meant that there could be no strike missions conducted north of the 20th parallel. In order to ensure that there would be no inadvertent violations of that restriction, the U.S. Navy imposed a further buffer zone of another degree of latitude. No strikes were to be conducted by Navy airplanes north of the 19th parallel.

Principally as a result of the 19th and 20th parallel restrictions, "Bonnie Dick's" combat operations settled into a rather routine cyclic operational schedule involving eight launches per day of one and one half hour duration (ninety minutes between successive carrier launches). It went on day after day and the routine, if not boring by any means, became predictably stultifying. Missions such as coastal reconnaissance, interdiction of north-south lines of communication, targeting of railroad yards, bridges, highways, motorized convoys of war materiel, power plants and so forth.

Meanwhile, the North Vietnamese, quick to take advantage of such a gratuitous respite, began upgrading the small airfield at Bai Thuong, just north of the 20th parallel. When the construction was finished, the small airfield became a staging area for MiGs from the major airfields at Kep and Phuc Yen. Able now to launch from Bai Thuong with impunity, the MiGs were employed to harass the U.S. Navy's interdiction program in North Vietnam south of the 20th parallel.

On occasion the Tonkin Gulf radar control ship RED CROWN and the Southern SAR destroyer *Alcoa*, were able to observe these operations through radar and radio communications intelligence. Thus it was that "Moose" Myers, on 26 June and Guy Cane on 27 July were able to take advantage of the situation to cut off the harassing MiGs, getting between them and their staging area at Bai Thuong. This all served to set the stage for Norm McCoy's aerial engagement.

On 1 August Lieutenant Norm McCoy, VF-51 launched as a photo escort with a photo version of the F-8 to get photo reconnaissance coverage of the several targets around the city of Vinh. Later, in the same launch Lieutenant George Hise, VF-53, got airborne as leader of a flight of two F-8s to accompany a mini-Alfa strike against targets at Vinh Son. Unfortunately, George's wingman and the spare airplane both went down, after engine start-up, for mechanical problems and George got airborne alone. Because the Air Wing Commander, Bruce Miller had mandated that no A-4s go over the beach in North Vietnam without fighter escort (this because of the Bai Thuong development) the mission was put on temporary airborne hold while George tried to get assistance from Norm.

Lieutenant McCoy and his photo plane had been first to launch, had streaked into Vinh, taken their photos and were actually on their way back to the ship when George called Norm and asked whether he had enough fuel to go with him as his fighter escort wingman. Norm responded that he could do so. They effected a running rendezvous and the strike group was off for Vinh Son.

As the strike group approached the target Bandit calls were received from RED CROWN that told of MiGs headed south and closing fast. The strike leader decided to complete their work quickly and get out. As soon as the strike aircraft were feet wet and headed for the carrier, George shifted the section of F-8s to the fighter tactical UHF radio frequency under *Alcoa* control and identified themselves by Norm's BATTER UP and George's FIREFIGHTER call signs, as now, VIKING flight. (VIKING was George Hise's personal call sign).

Without any discussion the two airplanes went back over the beach, this time at an altitude of 9,000 feet and a speed of 450 knots, headed 270 degrees and in a combat spread hunting for MiGs. The combat spread put them abreast and about 1 to 1 1/2 miles apart . . . far enough for each to keep the other's six o'clock position clear from an ATOLL attack. At this time the two fighter pilots were more or less on their own because *Alcoa* couldn't see them on his radar. They gave *Alcoa* their positions periodically using bearing and distance cuts from *Alcoa*'s own TACAN station. This was done in hopes that the ship could eventually find them on its radar and provide even a modicum of control against the estimated six MiGs that were now clearly in their immediate vicinity. From the MiG cuts *Alcoa* was providing, Norm and George concluded they were on a fairly good intercept vector. The seconds ticked by tensely as both pilots kept their heads on a swivel . . . but no joy!

Finally, as the two began crossing over the foothills leading to the Laotian border, they concluded that they had passed the MiGs by somehow. They executed an in-place turn (a tactical maneuver intended to reverse course quickly but with continual mutual coverage). It took about fifteen seconds and our intrepid duo were now headed due east toward the North Vietnamese coast, still at about 9,000 feet, and 450 knots with military power and in a combat spread, this time with Norm now on the north flank of the formation and George to the south.

It was about this time that Lieutenant Tony Nargi checked in with *Alcoa*. They were from another squadron, VF-111 and another ship, but had heard the MiG activity and wanted in on the action. *Alcoa* was at first reluctant to vector them in over the beach . . . probably because their radar coverage was so bad they still couldn't see VIKING flight.

However, Tony Nargi, later to become a MiG killer in his own right, was very aggressive and very persuasive and finally conned *Alcoa* into vectoring them in over the beach in the same general vicinity. Now, we had two sections of very eager F-8s headed toward each other with a closure rate of almost 1,000 knots; and somewhere in between them were an estimated six MiGs! The radar control ability of *Alcoa*, at this point, was virtually useless.

Enter the MiGs! Norm saw it first. It was a small, silver MiG-21 at about 4:30 to 5:00 high, behind George at a distance of about 2 1/2 to 3 miles . . . and he appeared to be rolling in on an attack on George with a good closure rate.

"Tally ho. MiG-21. Break port. Break port, VIKING", was all Norm think of shouting over the radio. George's reaction was almost instantaneous, as he broke his airplane hard into Norm's position. At the same time Norm saw the MiG fire an ATOLL missile at George. George's uncanny reactions caused the missile to "go stupid" and fly wide outside of his tight turn. In order to keep George's six clear and stay out of his way Norm pulled up hard and rolled toward George keeping the MiG in sight.

At this point, Norm described the MiG as "doing an unbelievable thing." It broke hard right (away from George) and appeared to be headed home. Although it would have been in keeping with the "slash and run" tactics of the North Vietnamese Air Force, it was an incredibly stupid tactical maneuver . . . and spelled the beginning of the MiG pilot's eventual death.

At precisely this moment, both George and Norm spotted another MiG-21, presumably the wingman. But, this MiG appeared to be less than eager to either join the battle or to lend mutual support to his section leader. He passed directly overhead the three airplanes, at high speed, in wings-level flight, crossing from right to left and sped off out of sight never to be seen again. He left his comrade in harm's way.

Norm was the first to detect the MiG's fatal error and called George, "Come on starboard. Come on starboard, VIKING."

George reversed his turn high as Norm crossed his six o'clock high falling in above and behind the descending MiG. When George came out of his turn he tally hoed the MiG now headed west at high speed at 12:00 o'clock.

At this time Norm got a Sidewinder tone on the MiG's tailpipe and fired a Sidewinder. The MiG was now turning hard right and the Sidewinder never appeared to guide. On reflection, Norm opined that he may not have kept the gunsight pipper on the MiG during the 0.75 seconds it takes for the missile to leave the rails. In other words, Norm may have allowed the Sidewinder to loose track during the launch sequence.

George saw this and announced, "I think I can get a tone." He then fired a Sidewinder. Norm watching, thought it was a certain kill because it seemed to be guiding. But, unfortunately, it went right by the MiG and never detonated. It possibly never fuzed. At this time the MiG headed directly toward a small, fluffy cumulus cloud just as George fired his second (and last) missile. The MiG disappeared into the cloud and the missile, following it, apparently lost heat source/track because of the cloud. It never detonated either.

Almost immediately the two F-8s penetrated the same cloud momentarily losing sight of each other and the MiG. They popped out the other side in time to spot the fleeing MiG low and to the left but clearly distinguishable against the dark green jungle background. George was now out of missiles.

Enter Tony Nargi and wingman! Tony arrived overhead with his section of F-8s and called a tally ho on the MiG. The plot was

CHAPTER 51: LIEUTENANT NORMAN McCOY

thickening by the second. We now had four very talented and eager F-8 pilots all coveting a piece of the poor MiG pilot. He never had a chance! Momentarily, while Norm was tracking for another missile shot, Tony Nargi actually flew in between Norm and the MiG. Apparently, Tony's track crossing angle on the MiG was too great because he drifted past to the left and out of Norm's missile field of view.

The MiG was turning hard to the right and Norm was turning with him and got a good tone when he fired his next Sidewinder. In the initial part of its flight before it armed, Norm opined that the missile flew out of its own envelope and thereafter never seemed to guide on the MiG. At this time Norm was closing fast on the MiG and actually popped his speed brakes, lagging a little outside the MiG's turn to open the range a little.

It was at this point, Norm recalled, that he remembered his Skipper, Bill Parish, admonishing all of his pilots at a recent briefing, to continue to track the target smoothly throughout the launch sequence. He tried very hard to do so as he fired his third missile. Somehow, he had the sense that this might be his last firing opportunity. (Considering Tony's intrusive presence, he was probably right!).

Norm's missile was fired inside one mile, with a track crossing angle of about 20 degrees and not so high a "g" load as before. The missile tracked smoothly and impacted the MiG's fuselage just aft of the starboard wing root. There was a huge fireball and the MiG flew out of it apparently unscathed. A fraction of a second later, however, the airplane pitched up violently to the left. Then the nose fell through and the airplane flew almost vertically into the jungle. Norm popped up high and saw the impact followed by another huge fireball.

It was at this point that Norm did a peculiar thing. He had been carrying a small eight millimeter movie camera taped onto the top of his radar scope pointed directly forward through the windscreen. Earlier, when the engagement had begun, he recalled thinking about turning the camera on. However, so worried was he that the momentary glance into the cockpit to find and hit the on switch, might cause him to lose sight of the MiG that he decided against it. Now, however, he turned on the camera and made one pass over the crash site getting a low quality film sequence of burning debris which, in the final analysis, had no intelligence value whatsoever. Norm McCoy was a MiG killer!

Lt. Norm McCoy recovers aboard U.S.S. Bon Homme Richard after his MiG-21 kill mission on 1 August 1968. (Photo courtesy of N. McCoy)

In retrospect, Lieutenant McCoy had some very interesting observations which he obviously felt compelled to convey. In all of my interviews, I found myself asking the question (always of myself), are these afterthoughts developed in the hindsight of the intervening twenty-seven years? Or, are they the real thoughts of the pilot immediately after the fact? In most cases it has been impossible to tell.

Norm McCoy, however, made special note of several cogent points. One of them was the ease of integration of two pilots from two different squadrons who, not having briefed together, embarked on a combat mission over North Vietnam, engaged and shot down a MiG-21. They completed this highly complex task with hardly a hiccup. They understood each other's tactics and mutual responsibilities and placed total reliance upon one another with complete confidence in each other's response to demand. The fact served, in Norm's view, to emphasize the complete integration of the air wing as a highly skilled and well-coordinated team.

The second point was that, after the shootdown, the two pilots found themselves very low on fuel. In accordance with air wing policy the duty tanker pilot, flying an A-4 with a buddy store, headed for the coast of North Vietnam to meet the two pilots as close to the beach as possible. In fact, as he approached the beach, he was listening to the F-8s on their tactical frequency and surmised that not only didn't they have enough to get back to the ship; they barely had enough to get to the tanker. So, Lieutenant J. Douglas Meador ventured much too close to that hostile shoreline than dictated by squadron SOP (standard operating procedures).

Meanwhile, Tony Nargi recognized their fuel state condition and offered to cover the departure of the two fighters with his own two so they could egress North Vietnam at a more economical altitude and airspeed. This could mean the difference of several hundred pounds of precious fuel. They accepted and immediately climbed to 20,000 feet, slowing to 360 knots for best cruise to the coast. Nargi and his wingman took up an overhead weaving position such that any MiGs attempting to get to Viking flight would have to go through Tony Nargi first.

The tanker pilot executed the standard air wing tanking rendezvous. The three airplanes were approaching one another with a closure rate of over seven hundred knots. At exactly eight miles apart he began a standard rate turn to a course headed back to the ship. Halfway through that turn he chopped throttle slowing to tanker drogue deployment speed. At 275 knots he deployed the tanker basket and, with thirty degrees to go in his turn he spotted the two Crusaders.

"Vikings. This is Canyon Passage tanker. Your 11:30 four miles. Do you have me? Over."

"Tally-ho, Tanker. This is VIKING ONE joining." George Hise was the first to the tanker and even though he had more fuel than Norm, he immediately engaged the basket rather than waste precious minutes. When Norm reached the tanker, George backed out and let his wingman have a shot at the basket. This was no time for a missed attempt at the basket. It was country hardball and everyone knew it. When Norm saw the green light on the tanker store illuminate (indicating the start of fuel transfer) his fuel gauge read 400 pounds . . . enough for a few more minutes. That was cutting it pretty close.

Unfortunately, the tanker pilot had already given away some of his fuel so there wasn't enough left in his tanker store to get both Crusaders back to the ship. The ship knew this and had already scrambled their duty deck launched standby tanker right in the middle of cyclic operations. Lieutenant Meador made a decision without a second thought. He transferred a substantial portion of his own airplane's fuel supply into the tanker store and gave that to Norm McCoy also. Later on, after tanking from the second tanker, Norm and George executed an uneventful recovery aboard the "Bonnie Dick." Norm learned later that Meador had recovered aboard ship with a fuel state that was almost critical.

Routine sacrifices such as this were made all the time within the air wing team with never a second thought. It is what makes a fighting force greater than the sum of its parts!

Author's Note: Norm McCoy works for an outfit located north of San Diego. He enjoys his work in the civilian work force but recognizes that the loyalties and camaraderie of his shipmates are characteristics he will never find in the private sector but which he will never forget. McCoy was awarded the Silver Star medal for this action.

52

LIEUTENANT ANTHONY "TONY" J. NARGI VF-111

Each time I begin to describe a particular Crusader pilot as "colorful" I have to restrain myself. They are all, by definition, colorful! It's like using an adjective like spectacular to describe a nuclear explosion! It is pure inadequacy.

Tony Nargi was, nevertheless, a colorful pilot. Somehow, it seems fitting that he was the last Crusader pilot to down a MiG with ordnance. (I use that qualifier because Jerry Tucker downed his MiG with sheer intimidation . . . and, as a consequence, was not rewarded for doing so).

Lieutenant Nargi was a part of that peculiar experiment called "*Intrepid!*" Someone on COMNAVAIRPAC Staff thought up the idea of populating a carrier with nothing but A-4s (obviously an A-4 pilot). It was in essence what came later to be called a "low side" carrier (on the low side of the carrier high/low mix). At first, the Intrepid experiment was to be all A-4s; acting as both fighters and attack airplanes. Then a small detachment of F-8Cs, called Fighter Squadron One Hundred Eleven, Detachment Eleven (named for the U.S.S. *Intrepid*'s hull number), was assigned to provide a more credible fighter capability.

The fighter detachment, numbering at first only three airplanes, obviously couldn't perform all the functions of a full blown fighter squadron when *Intrepid* went into full-scale operations in North Vietnam in 1968. Later with only four airplanes, all the detachment could do was provide fighter escort missions for F-8 photo reconnaissance flights and target combat air patrol (TARCAP) missions in support of Alpha strikes. The step from A-4 fighters to an F-8 fighter detachment was followed shortly by a full ten plane fighter squadron and followed shortly after that by cancellation of the whole silly idea. *Intrepid* had been given a quick and dirty modification to change her from a CVS (anti-submarine warfare carrier) to a CVA (attack carrier), as a stop-gap to allow the fulfillment of an increasing carrier commitment to the Tonkin Gulf.

On 19 December 1968 Lieutenants Tony Nargi and Alex Rucker launched in a pair of F-8Cs from *Intrepid* in support of one of their Alpha strikes. RED CROWN, the U.S. Navy radar ship in the Gulf detected some unidentified bogies approaching the strike group and vectored the two Crusader to intercept them. Nargi, the flight leader, spotted one of the MiG-21s at "two o'clock high" and turned to attack. The MiG pilot apparently made visual contact with the F-8s at about the same instant and went into a vertical maneuver to counter Nargi's climbing turn into him. The higher thrust-to-weight ratio MiG outclimbed Nargi even though his F-8 was in full afterburner, forcing the F-8 to fall off from the overhead maneuver. However, the MiG pilot completed a conventional looping maneuver and probably lost sight of Nargi's airplane in the process. Nargi had a near optimum Sidewinder shot as the MiG came out of the loop.

The Sidewinder tracked and detonated near or inside the MiG's tailpipe, blowing the empennage completely off the airplane. The MiG pilot ejected safely and the total for the F-8 community was 18 MiG kills for three losses to MiGs. Both Nargi and his wingman Rucker fired missiles at another MiG-21. Although the missiles seemed to track and came close enough to detonate, they didn't seem to do any damage, and the MiG escaped. The resultant six to one exchange ratio was the highest for any airplane during the Southeast Asian air war. Lieutenant Nargi was awarded the Silver Star for his achievement.

Author's Note: Tony Nargi left the service to work for Grumman for several years and later for Beechcraft. Now, he owns his own consulting company and lives in Italy.

53

Lieutenant (JG) Phil Dempewolf (Probable) VF-111

One of the tragedies of aerial combat is, and always has been, the touchy question of confirmation. One only has to remember the intense scene in the aviation cult movie, "The Blue Max", to remember George Peppard's reaction to the lack of confirmation of his first "kill."

The need to confirm kills is a necessary evil of modern warfare... especially aerial combat. If one were to take, on face value, the sincere claims of aviators in past wars, the statistics books would carry more aircraft downed than ever existed in the inventories of the world's air forces. No doubt, in the process, some downings were never confirmed.

Doubtless, there have also been confirmations of claims of aircraft downings which never occurred. A good example is the list of Vietnam People's Air Force's claims of U.S. airplanes shot down by their pilots in the skies over Hanoi. The VPAF claimed they had 15 aces in the Vietnam War. If one totals up all of the U.S. airplanes shot down by just those fifteen pilots (106 airplanes), the number exceeds the total lost from both VPAF and the People's Republic of China Air Force action (91).

So, when Lieutenant (junior grade) Phil Dempewolf launched from "Bonnie Dick" on 21 July 1967 as strike escort for the Air Wing Twenty-One strike against Ta Xa, a fuel depot northeast of Hanoi he was destined to enter the history books as one of those "probable" killers. He was in the middle of the fur ball involving Tim Hubbard, Red Isaacks and Bob Kirkwood when he took a Sidewinder shot at a MiG-17. The missile appeared to be tracking the MiG when Dempewolf last saw it. However, simultaneous with his missile shot, two more MiGs appeared in the furball. Their untimely arrival, as well as the immediate threat they posed, prevented anyone from following the flight of Phil's missile. No one saw the result. Therefore it entered the books as a "probable", but unconfirmed, kill. For his part in the action, Phil Dempewolf was awarded the Distinguished Flying Cross.

What follows is his account of the mission.

"I was flying on an Alpha strike as fighter escort and was Bob Kirkwood's wingman. The four plane division was led by the Skipper, "Red" Isaacks whose wingman was Don McKillip. We were escorting a strike group made up of about 12 A-4s. My weapons load consisted of full ammunition and four AIM-9D Sidewinders. Also airborne was an RF-8 photo plane and his escort, an F-8 from VF-211.

This was a repeat of what we had been doing every day for a month and we followed roughly the same ingress route to our target in the Haiphong/Hanoi complex. Our approach route to the target was north in the Tonkin Gulf until we could turn in and stay relatively hidden by the karst ridge line which ran east and west and lay to the north of Haiphong and Hanoi. When we coasted in my section leader, Bob Kirkwood and I were located on the left flank of the strike group.

Our altitude was about 12,000 feet. As on previous strikes, the northern SAR destroyer was broadcasting the detection of Bandits (MiGs) taking off from Bullseye (Hanoi area airfields) and circling about 40 miles northwest. This was a standard tactic we had observed in recent weeks. It seemed to be a way of getting airplanes out of the area of incoming and egressing strike groups rather than a serious effort to intercept them.

This time, however, the call was slightly different, and I noted it. The location of the Bandits was northeast rather than northwest. I attribute this anomaly to a mistake on the part of the MiG flight leader; because, all of a sudden, there was a flight of five MiG-17s directly in front of us.

It was not a formation in the true sense of the word because they seemed to be straggling along in a general column. At this point the strike group was at an altitude of 12,000 feet and the MiGs were crossing from left to right at an altitude several thousand feet higher.

Our F-8s were slow, about 280 kts as our section pulled our nose up to pursue the MiGs. I slipped into a "loose deuce" formation on Kirkwood's airplane and saw that two of the MiGs had separated from the original flight. From this point on I lost sight of Isaack's section of F-8s. It wasn't at all clear to me that the MiGs had seen us yet. They were turning to the left and went by me so

226

close I could see the pilots' helmets. As they went by the second MiG was in trail of the first.

I was assuming that my section leader would attack but he didn't turn. All this time the ultra high frequency (UHF) radio channel was so clobbered with people talking on it that there was no opportunity to call my section leader. The MiGs were going by, close aboard, and I was about to lose the opportunity of a lifetime to kill one or both of them unless I acted immediately.

So, I broke hard left in pursuit of the MiGs and that was the last I saw of my section leader for the remainder of the engagement. At this time, I estimate that about fifteen seconds had elapsed since the first sighting. I armed my missiles and did a low yo-yo to put myself into position for a missile shot at the MiGs. I ended up looking down on them. The wingman was in trail of his leader by about six or eight plane lengths at a range of about 2,000 feet.

As soon as I selected a missile station I got a good missile tone and fired the first of my Sidewinders. The missile went straight into the ground. I was real slow about this time, about 200 knots. Still, I didn't use afterburner. Some of our fighters had gotten burned recently by getting low on fuel and being unable to find a tanker. So, we were all a little careful about using afterburner. In fact, I never used afterburner during the entire engagement.

The MiGs, by this time had completed about a 270 degree left turn to roughly a heading of north, toward their home base. I got my speed up by lowering my nose a little and dropped in behind the MiGs, launching my second Sidewinder from a position directly astern. The missile seemed to be guiding perfectly and, to my astonishment, detonated directly in between the two MiGs. Maybe, in the final portion of its flight the seeker head locked on to both MiG tailpipes and went after the centroid of the two IR sources.

The explosion of the missile seemed to galvanize the MiGs into action because they broke hard in a right turn to a heading of roughly east. The MiGs were now descending as I was climbing toward them and again I got a good missile tone from a position astern of them. I fired my third Sidewinder missile and saw it go straight into the ground.

It is worth noting, at this point in the narrative, that during my post-take-off missile checks (I did it on Kirkwood's airplane) I had failed to get a missile tone on the number four missile and assumed it was no good. by now, of course, the missile station selector had stepped over to the fourth and last missile station automatically.

The MiGs were still in a right turn and had passed through a southerly to a westerly direction. I was under the assumption that I had no good missile and heard the other three F-8s in my division calling "feet wet." With only guns for weapons I decided to get out of there and turned south. I was a lone airplane over enemy territory and was jinking all the time and looking over my shoulder for trouble.

All of a sudden a MiG-17 appeared directly in front of me, also headed south. I added throttle to close to gun range for a kill when I heard a good missile tone on missile number four. I was dead astern, in the heart of the missile envelope (about 1/2 mile) when I fired my fourth Sidewinder missile. The missile appeared to be guiding perfectly and I was already counting my kill while the missile flight seemed to be interminably long.

Suddenly, out of the corner of my eye I saw two MiG-17s at two o'clock high. They were about 1,000 to 2,000 feet above me and the same distance ahead, crossing from right to left in a descending left turn. They were close enough for a guns kill but, as I broke hard right and into their turn, it was obvious that the angle off was too high for me to do any serious gun tracking. Nevertheless, I continued the hard right turn and began firing the guns. I saw the tracers crossing well behind them as my guns quit firing.

Then, I reversed my turn back in the direction of the MiG at which I had just fired a missile and there was nothing to be seen . . . no fireball, no sign of an explosion. Just as I was beginning to conclude that my missile had somehow gone astray, I caught sight of a man in a parachute at an altitude of 10,000 feet. As I whipped by him I noticed that his parachute was square in shape and white. By now I had lost sight of the two MiGs and began to circle the man in the parachute.

About this time the A-4s from the strike group began to drift by my position. One of them called out the man in the parachute and asked whether he should be gunned down. The A-4 flight leader nixed the idea and directed his group to continue to "feet wet." I followed suit and trailed the A-4s out to the coast. The return to the ship was uneventful. Back aboard I learned that both Isaacks and Kirkwood had bagged a MiG."

Author's Note: Phil Dempewolf now works for USAir as a pilot and resides with his family in the San Diego area.

PART VI
THE SOUTHEAST ASIA AIR WAR: A PERSPECTIVE

"War is an ugly thing, but not the ugliest of things. The decayed and degraded state of moral and patriotic feeling which thinks that nothing is worth war, is much worse. A man who has nothing for which he is willing to fight, nothing he cares about more than his own personal safety, is a miserable creature who has no chance of being free, unless made and kept so by the exertions of better men than himself."

"Diamond of Diamonds." Sixteen F-8Es from VF-13 and VF-62, U.S.S. Shangri La (CVA-38), over the Aegean Sea, 1965.

This inscription, typed on a piece of eight-by-ten inch plain bond paper, was mounted on the steel bulkhead over my bunk in the U.S.S. *Hancock* and held there by a neat frame of one inch masking tape. I never found out who the author was but I liked it. It was a particularly appropriate statement for all of the poor devils who were "laying it on the line" day after day in a god-forsaken corner of the world for something they thought was right; at a time when "miserable creatures" at home were demonstrating and burning their draft cards in opposition.

During the early summer of 1967 the heightened war effort resulted in target taskings in Indian Country on a daily basis. Aviators going off on those strikes day after day were pumping themselves up by various means. I kept a reel permanently fixed on my tape deck, which I particularly liked. It was an excerpt from "Victory at Sea," titled "Under the Red Sea Sun." I would put on my headphones and listen to it late into the night. It was wishful thinking I suppose, but it would be a great deal easier for these young aviators to pump themselves up if the home folks treated them as patriots rather than bums.

My oldest son, Jim, a bright-eyed and handsome young ten-year-old, was bothered most by all of this protest. He even heard in grammar school things which worried him. Somehow, I got wind of his unease. I sat down a few weeks after *Hancock* had departed Alameda for the western Pacific and struggled with pen in hand for the right words. "What do I say to a ten-year-old whose father is headed for the Tonkin Gulf," I wondered, "to make him feel comforted?" The right words to make him feel that the deprivation, the absence of his Dad was somehow acceptable and bearable flowed like glue. It was a struggle, but finally I scribbled the following words to my first born:

1-25-67 *Hancock*

Dear Jim, It occurs to me that I should write you a short note and give you a "rationale" (look that up – it means "a way of thinking") to explain why I'm not home to throw the football with you in the backyard and go khayak surfing with you at the beach. Here it is . . . in a nutshell. You get up each morning from a warm bed, in a snug home, eat a good breakfast and go to the school that your Mother and I selected right? You serve Mass in a church (which your Mom and I picked out) – to a God in whom we all believe. Right? Now just think a minute about that!!! The bed is warm because there's gas in the furnace. The house is snug because it is a fairly expensive one and the breakfast is good because the icebox and cupboard are nice and full. The school is a good Catholic school because we all choose to believe that there really is a God!

There are millions of people whose beds are cold (no gas), whose homes are drafty (poor house) and whose stomachs are empty (no food). They go to an institution (not a school) run by the government which tells them what to think, what to read, what to believe and, worst of all, that God doesn't exist!

Now, the reason you have all these good things which many poor millions don't, is because about 190 years ago a handful of farmers said "goodbye" to their sons and daughters, took up their muskets and powder horns and walked across the corn and tobacco fields early in the morning to a bridge near a town called Concord. The reason for going to that particular bridge on that particular morning was to protect not only their own families, but also the right of you, Jim Gillcrist, to be a Catholic, and my right to earn my own living as I want in order to keep the house snug, your bed warm and the icebox full. Those farmers at Concord stood at the bridge to stop a company of British soldiers (come hell or high water) who were trying to tell them how to live.

It may seem funny, Jim, but when I left La Jolla to come out here I was just like one of those old farmers. The red-coats were coming again. But this time it isn't a village at Concord, it is a spot in the Gulf of Tonkin 9,000 miles away. But it's just as important. The road, this time is not the Boston Post road – it's the Ho Chi Minh trail. The geography is different but the principle is the same. Some 12 to 14 million people in South Vietnam want to live their own lives in their own way . . . but once again . . ." the redcoats are coming." Now it's Yankee Station instead of Concord but it's the same old story.

Don't brag about your Dad. I wouldn't want that! But never be ashamed or sorry that I left you to come out here. One good turn deserves another. Some calloused old farmer did it for me at Concord. Now I'm returning the favor for some little boy your age in a little hamlet in South Vietnam who doesn't have all the nice things you have.

Does this make any sense to you? I'd prefer to stand guard on Yankee Station rather than close to home. We don't want the redcoats any nearer do we?

Love, Dad

I knew this letter to my son was a little "hokey." But I also knew that buried beneath all the obfuscation, chicanery and political posturing there really was the basic principle of the right of a large number of South Vietnamese to determine their own future. I thought that the principle of gradually "upping" the ante to the point at which the North Vietnamese would find adventurism below the DMZ too costly, was just plain stupid. The fellow who dreamed up that theory and sold it to our policy makers had two strikes against him. First, he had never been in a war or been shot at. He couldn't have, or he would know how idiotic it is to gradually increase pressure and risk; giving the enemy a chance to forecast the next move and dig himself in. Secondly, the concept's author had totally miscalculated the North Vietnamese mindset, misread the history of the French

PART VI: THE SOUTHEAST ASIA AIR WAR: A PERSPECTIVE

involvement in Indo-China and misjudged the capability, mental resilience and the terrible resolve which we nourished in the North Vietnamese with each increase in pressure. I knew the U.S. war strategy in 1967 was doomed to failure.

In fact, I had written a Master's thesis at Naval Post Graduate school just two years earlier entitled: "Limited War and American Foreign Policy." One of the conclusions in my thesis was just that ". . . limiting a conflict by placing restrictions on one of the combatants ensures that he will never achieve his objectives." As early as 1967 I could see coming to fruition some of the dire forecasts contained in my Master's thesis . . . and that fact was cause for little comfort.

However, I was able to set politics aside, in my own mind, and focus on the mechanics of fighting "the only war we had." The sheer exhilaration of combat operations overcame all of the other minor irritations. I had been training all of my adult life to kill MiGs. Now, here I was on an aircraft carrier steaming off the coast of a country loaded with MiGs. Not only that; we had a war going on and I was going to be sent over the beach day after day with live Sidewinder air-to-air missiles and loaded guns. If I should be so lucky as to find MiGs, I would do the best I could to shoot down as many as I could. What more could a fighter pilot ask? To quote the Baron von Richthofen "all else was rubbish."

Certainly there was the frustration felt by everyone onboard the deployed carriers over not being allowed to win the war. U.S. battle forces possessed the tools and the talent to do just that if they could be turned loose to fight the way they'd been trained to fight and to use the weapons which they had developed against the targets for which they had been designed.

On the fighter side of the coin; the best, most cost-effective way to achieve and maintain air superiority over enemy territory is to destroy the enemy's anti-air defenses on the ground. Unfortunately, surface-to-air missile sites were not permitted as targets until it was too late, and the entire network of SAM sites too formidable. The best way to destroy MiGs is on the ground. But it was not until October of 1967 that air strikes against North Vietnamese airfields were permitted. Hot pursuit of North Vietnamese fighters across the Chinese border was never permitted. The North Vietnamese Air Force had a perfect haven just a few minutes flying time from their major bases anytime they chose to take advantage of that escape route.

Center stage for the aerial drama which unfolded in the skies over North Vietnam between 1964 and 1972 was taken up by three airplanes, the F-4 Phantom, the MiG-17 Fresco and the MiG-21 Fishbed. This is not to say that fighter planes like the Air Force F-105 Thunderchief and the Navy F-8 Crusaders did not play an important role. Far from it! The performance of the pilots who flew those airplanes should be appropriately recognized by anyone who tries to chronicle those years. For example, the F-105 was a tremendous strike airplane but its contribution, as a fighter, to achieving and maintaining aerial supremacy over Route Package Six was minimal. The F-8, on the other hand, had a superb record. Unfortunately, there were only ten fleet squadrons flying the fighter version of the Crusader and although it achieved the highest exchange ratio (how many we killed divided by how many we lost) of six to one in Southeast Asia, it only accounted for eighteen MiGs shot down.

What makes a good fighter aircraft? One could get different answers from different fighter pilots about the relative priority of the basic ingredients; but, the basic ingredients should all be there. The following are my priorities and ingredients:

First and foremost, a fighter plane should be agile; that is almost by definition. Those things that give it agility are low wing-loading and high thrust-to-weight ratio. Wing loading is the weight of the airplane in pounds divided by its wing area in square feet. The wing of an airplane with low wing loading can "lift" it better in turning and maneuvering flight. Thrust-to-weight ratio is the thrust put out by the engine at maximum power (expressed in pounds) divided by the weight of the airplane in pounds. By generally accepted convention the conditions determining that weight include two-thirds internal fuel and a full load of air-to-air weapons. An airplane with a high thrust-to-weight ratio can sustain its speed better in maneuvering flight and thereby sustain its agility. It also enhances its ability to accelerate. Agility is measured by turning rate, turning radius and by the ability to accelerate.

The second most important feature in a fighter plane is its weapons systems. Otherwise, why ever put it in the air if it can't kill other airplanes well? Weapons systems include not just the weapons, but also the sensors, displays and controls. Those systems must not only be lethal and reliable, they must also span the greatest possible spectrum of the air-to-air arena. An example of a bad weapons suite is the F-4 Phantom. Robin Olds, the wing commander at Ubon, a principal U.S. air base in Thailand, cursed bitterly a number of times when he found himself behind a MiG with no gun, inside minimum missile range. When I visited him in the summer of 1967, he was very outspoken on that subject. Back in 1956, as an aerial gunnery instructor in the Fleet Air Gunnery Unit, I had spoken out vigorously against the decision to omit a gun for the Navy's new F-4H-1 Phantom. Unfortunately for F-4 pilots the world over, the wisdom of engineers, analysts and acquisition specialists won out over the pleas of operators. I will remember to my dying day the patronizing tone of the brilliant young engineer who explained (to the dumb pilot) that at the speeds and altitudes that the Phantom would be flown a gun would be useless and therefore an unnecessary weight penalty. I left that meeting muttering to myself, "Jesus, if we let pimply-assed, twenty-eight year old PhD's in three-piece suits decide on our weapons systems, we deserve to lose the next war!"

An example of a good weapons suite is the present Navy fighter, the F-14A Tomcat. It can kill a target a hundred miles away with an

AIM-54 Phoenix air-to-air radar missile. Or, it can close to twenty miles and kill it with an AIM-7 Sparrow radar missile; or, it can close further to five miles and kill with an AIM-9 heat-seeking missile; or, finally, it can close to ten feet and kill it with a twenty-millimeter M-61 gatling gun. Now, that is a versatile weapons suite!

The fight to put a gun in the F-14A was waged bitterly while I was in the Pentagon after my last tour of duty in the Tonkin Gulf. The same old arguments were put forth by the same misguided kinds of acquisition experts, analysts and engineers. I had been brought to the Pentagon specifically to bring operational flavor to an office whose function was to analyze the lessons of Southeast Asian combat operations and ensure that they figure in decisions being made regarding future weapons systems. Fortunately, in the F-14s case, the operators view won over the engineers. Sadly enough, as plans were being finalized for the only major upgrade of the F-14A in fourteen years, the F-14D, the head of aviation acquisition questioned the need to upgrade the gun system, saying that it wasn't really a fighter. Unbelievable!

The third most important feature of a fighter is its durability in combat environment. Durability includes many things. Durability in the heavier, less agile F6F Hellcat was a big determinant in its success over the lighter, more agile Japanese Zero in the Pacific theater during World War II. A fighter plane, to endure in a combat environment has to have enough fuel to fight long enough to kill the opponent. Durability includes being able to absorb damage and continue to kill the enemy. This means a fighter must have high survivability and low vulnerability built into it from concept, and kept in all the way to production. No matter how agile a fighter may be, in a multi-plane engagement, if it stays in it long enough, it will sustain damage. We forgot all about survivability and vulnerability when we designed the fighters which engaged the MiGs over North Vietnam. A good example is the F-8. When it lost hydraulic power to the flight control system by catastrophic battle damage, Navy pilots discovered that the airplane pitched violently nose down. A similar problem with the F-105 caused the Air Force to put in a "Rube Goldberg" fix which did nothing more than enable the pilot to steer it roughly to some place where ejection would result in rescue rather than capture. Beautiful!

The fourth most important feature in a fighter is "flyability." The airplane must be easy to learn to fly so that high proficiency levels can be achieved and maintained with the least amount of training. It is generally appreciated that, other things being roughly equal, the best trained air force always wins. The Israeli Air Force has proven that several times in recent years. If a high level of proficiency takes a long time to achieve, even modest attrition rates will dramatically reduce the effectiveness of an air force.

The fifth most important feature in a good fighter is visibility. A pilot must be able to see out of his airplane easily. He should not have to look through four inches of bullet-proof glass to see ahead of him. His field of view around him, especially to the rear should be as good as technology can achieve. The old adage "trouble always comes from six o'clock" is still true. The visibility aft is so bad in the Phantom that the Israelis, copying from the Russian MiG-17, put a rear-facing periscope on top of the canopy. Visual enhancing devices like the television cockpit system in the F-14 are part of the visibility equation as well. U.S. Air Force aircrews found that jury-rigged Weaver rifle scopes mounted in the front cockpit above the instrument panel increased their combat effectiveness in an important weapons systems evaluation conducted in the mid-1970s. Night vision enhancements will be important parts of the visibility equation in future fighter systems.

The sixth most important feature in a fighter plane is its observeability. Not only should the physical dimension of a good fighter plane be as small as technology can achieve, but the vulnerable size is of equal importance. Survivability of a fighter plane against radar directed weapons is directly proportional to its vulnerable size. Visual acquisition has always been the key to victory in clear air mass air engagements. The one who sees his opponent first nearly always wins. The radar cross-section and the infra-red signature of a fighter must also be small as technologically possible. Camouflage and other means of reducing the "visual signature" are all important. Since, in high technology, the cost of the airplane is directly related to size and weight, this feature is vitally important to the taxpayers as well.

The list could go on, but the seventh most important feature is a good fighter is its reliability and maintainability. Although this feature may be more important to the force commander than to the pilot, the reliability of weapons systems means that more of them will be available at any given time to fight. High maintainability of a fighter means that it can be returned to the battle arena as quickly as possible if one if its systems fails. What is equally important to the fighter pilot and the commander as well is that subset of reliability known as graceful degradation. When a system fails, whether battle induced or otherwise, is the pilot out of business or is it just a little harder for the pilot to kill his opponent? This all important feature can only be designed into a fighter from first concept.

It is interesting, now that the features of a good fighter have been established, to see how many of them were embodied in the cast of actors who performed in the drama which unfolded in the skies over the Red River Valley in the mid-1960s and early-1970s.

MiG-17 FRESCO

The MiG-17 Fresco had been in operational Soviet fighter squadrons for eleven years when U.S. Navy airmen first engaged it successfully on 17 June, 1965. Two F-4B Phantoms from U.S.S. *Midway* (CVA-41) downed two MiG-17s by Sparrow radar-guided missiles in that engagement. A small airplane, with a length of forty feet and a wing span of thirty-six feet and a combat weight of 13,200

PART VI: THE SOUTHEAST ASIA AIR WAR: A PERSPECTIVE

MiG-17s with North Vietnamese camouflage, 1979. (Courtesy of Dave Bourland)

pounds. The mid-mounted wing was swept back, and cranked at 40 degrees and had a wing area of 282 square feet. This produced a very low wing loading of fifty five pounds per square feet. The airplane was powered by a Klimov VK-1 turbojet engine featuring an afterburner which was rated at seven thousand five hundred pounds of thrust. With a combat weight of 13,2000 pounds, the Fresco had a thrust-to-weight ratio of 0.57. MiG-17s carried 4,500 pounds of internal fuel which gave them an estimated combat radius of only one hundred fifty miles on a point defense mission. In full afterburner it had a sea level rate of climb of 36 thousand feet per minute and a service ceiling of fifty-seven thousand feet. It's one "g" acceleration from two hundred knots out to five hundred fifty knots was estimated to take 27 seconds. It was a subsonic airplane with a reasonable controllability limit of five hundred and fifty knots.

Rating the Fresco C against my seven criteria, it received a very high score in the agility criteria of turning rate, turning radius and acceleration from two hundred out to five hundred knots. Although the basic airplane was built to withstand eleven "g's." The foreign pilots whom I interviewed were unable to achieve more than eight "g's" even with both hands pulling the control stick.

The weapons systems of the Fresco gets a low score. But, what weapons it did carry were extremely lethal at short ranges. Early

MiG-17s with North Vietnamese camouflage and markings at Champlin Museum. (Photo by the author)

models carried no air-to-air missile but had three cannons. The big caliber weapon was an incredible thirty-seven millimeter cannon with a small eighty round capacity and estimated low rate of fire of 80 rounds per minute. The MiG-17 also carried two twenty-three millimeter guns with a total capacity of 250 rounds and a combined rate of fire of 1500 rounds per minute. The fire control system which aided the pilot in aiming those three guns was primitive by U.S. standards. However, a MiG-17 pilot could fire all three guns at once and if he ever hit another airplane it would be fatal. I always felt that a well-aimed MiG-17 strafing attack could have sunk the northern SAR (Search and Rescue) destroyer in one pass.

The Fresco's durability was low simply because of its small internal fuel supply. As for vulnerability, not many MiG-17s flew away from the battle arena after having been damaged. My review of all the engagements described in open literature substantiates the fact that the Soviets, in their striving for performance in the late 1950s, didn't place any higher priority on survivability and vulnerability than did U.S. aircraft designers.

The flyability of MiG-17 was superior. Those foreign airmen whom I interviewed gave rave notices about how quickly a neophyte airman could master the aircraft and the simple weapons systems. For this reason, it has always been a puzzle to me and many other tactical aviators as to why North Vietnamese MiG-17 pilots were not more aggressive and more effective.

Visibility from the cockpit of the MiG-17 was terrible in virtually all quadrants. The ability to see forward through a combination of a gunsight combining lens and a bullet-proof windscreen was very poor. The sideward visibility was limited by the high canopy sills. Rearward visibility was restricted by cockpit structure blocking but a rearward facing periscope provided excellent rear hemisphere visibility within the restricted field of view of the periscope. The optical clarity of the periscope was exceptional.

The MiG-17s small size and camouflage paint scheme made it extremely difficult to acquire visually. Up until the Vietnam conflict, Navy and Air Force tactical aircrews did their air combat maneuvering training against their own airplanes all of which had paint schemes which made them stand out rather than blend in with the background. It was possible to look over at one's wingman to visually check his rear sector and completely fail to see a MiG-17 simply because one's eyes were accustomed to looking for an airplane twice its size and decorated with contrasting markings and even squadron insignia. The MiG-17 Fresco got high scores in this category.

Probably one of the most surprising features about the MiG-17 was the ease with which relatively unskilled maintenance personnel in the third world countries were able to maintain it, refuel and rearm it and send it back to the combat arena in a remarkably short time with relatively austere facilities. The MiG-17 received very high scores in reliability and maintainability.

MiG-21 FISHBED

The MiG-21 Fishbed was produced, beginning in about 1956 by the Soviet Union as a follow-on to its predecessor the MiG-17 with the mission of point defense fighter. It is probably one of the most successful export fighters ever built. About 5,000 thousand were produced in the Soviet Union, and many more have been co-produced in other countries. The People's Republic of China has even produced its own version and is producing them for export. I often described the MiG-21 as the smallest possible airframe wrapped around the biggest single engine the Soviet designers had in that class. United States airmen first engaged it over Cuba in the early-1960s. In Vietnam the first lethal engagement with a MiG-21 occurred in April, 1966 was a kill by an F-4 from the 35th TFW.

The MiG-21 is a small, single-engined airplane with a length of fifty five feet, a wing span of twenty-five feet and a combat weight of seventeen thousand pounds. The mid-mounted wing was of pure delta geometry swept back at an angle of sixty degrees and had a wing area of 310 square feet. This resulted in a very light wing loading of 53 pounds per square foot. The airplane was powered by a single MK R37F axial flow turbojet engine featuring an afterburner which was rated at twelve thousand five hundred pounds of thrust. The Fishbed enjoyed a thrust-to-weight ratio of about 0.74. The airplane carried about four thousand five hundred pounds of internal fuel which gave it a combat radius of less than one hundred miles in a point defense role. In full afterburner the Fishbed had a sea level rate of climb of thirty thousand feet per minute with a service ceiling of 58,000 feet. It's one "g" acceleration from mach .87 to mach 1.2 is estimated to take only 32 seconds at thirty thousand feet. It's maximum mach number was in excess of mach 2.0.

Rating the Fishbed against my seven criteria gave it a very high score in agility. Its turning rates, turning radius and acceleration from one hundred twenty knots all the way out to mach 2.0 were comparable to, or better than any U.S. fighter in the combat zone.

The MiG-21 gets a good score in the weapons department because it possessed two thirty millimeter cannon and carried two

Above and below: MiG-21s with North Vietnamese camouflage and markings. (Courtesy of Chuck Parnell, EAA 1994)

PART VI: THE SOUTHEAST ASIA AIR WAR: A PERSPECTIVE

Atoll heat-seeking air-to-air missiles. Ammunition capacity is estimated to be 380 rounds with a combined rate of fire of 1200 rounds per minute. The Atoll missile possessed only a rear hemisphere capability against U.S. jet powered aircraft and therefore could not score with missiles in a head-on encounter.

Because of its small internal fuel capacity the Fishbed received a low score in my third criterion, durability. An assessment of its ability to absorb damage is difficult to make but a review of all the MiG-21 engagements with U.S. aircraft in Southeast Asia doesn't reveal many instances in which a damaged Fishbed departed the battle arena.

As far as flyability is concerned, the general consensus of those foreign pilots whom I interviewed, who flew the airplane, was that one could become fully proficient for basic air work in one or two flights. Proficiency in the finer aspects of air combat maneuvering could be acquired in as few as fifteen flights. The fire control system was rudimentary by U.S. standards and easy to master. A high level of combat readiness could be maintained by flying half-a-dozen times per month. The Fishbed would get high marks in this category.

The pilot's visibility from inside the cockpit of a MiG-21 is absolutely terrible. The ejection seat and cockpit aft bulkhead structure severely limited six o'clock coverage. High canopy side rails limited side downward visibility and a full four inches of bullet proof windscreen and gunsight combining glass restricted the forward visibility. It was worse than the MiG-17 visibility and furthermore had no rear view periscope.

The MiG-21s small size made it extremely hard to visually acquire beyond three or four miles, depending on background. The head-on profile was very small. Pilots in combat were having difficulty seeing it beyond one and one half miles in some background conditions. The clean head-on profile presented a very small radar cross-section.

The reliability and maintainability of the MiG-21 was, by all reports, even better than the MiG-17. More design effort had apparently been put into maintainability. Third world countries, even with lesser skilled personnel, were able to generate remarkable sortie rates with this airplane. Turn around times were normally determined by the time required to re-arm and refuel.

This simple, small, uncomplicated mach 2.0 MiG remains, even today, one of the most successful fighter planes in the world. It represents a signal achievement of the Soviet aerospace industry.

F-4 PHANTOM

The biggest player, on the U.S. side of the air war was the McDonnell Douglas, F-4 Phantom. Biggest in total numbers, it was also the largest fighter in the aerial arena. The airplane's wing was swept at an angle of forty-five degrees with twelve degrees of dihedral in-

U.S. Navy F-4J. (U.S. Navy photo)

troduced at about half the span. With a length of fifty-eight feet, a wing span of thirty-eight and an empty weight of thirty four thousand pounds, this twin engined, mach 2 plus fighter plane was almost twice the size of the MiG-21. The original design of the Phantom came from a strike aircraft concept optimized for the high-speed, low-altitude environment. As a consequence, the F-4's wing loading of eighty three pounds per square foot was much too high to achieve even a modest turning radius. The F-4 didn't turn worth a damn but it accelerated like a rocket. With two sixteen thousand five hundred pound thrust General Electric J-79 engines and a thrust-to-weight ratio of 0.76 the F-4 could out-accelerate all other aircraft in the Southeast Asian aerial arena. As a consequence, those aircrews who learned to operate the F-4 in the vertical plane were most successful against the MiGs. Although its ability to accelerate gave the Phantom many advantages in aerial combat, the high wing loading made it impossible to match the MiGs in a level turning engagement. Successful Phantom pilots soon learned to "blow through" engagements using their head-on Sparrow air-to-air missile capability. Once engaged, they soon learned how to successfully "bug out", disengage and then to pitch back into the flight presenting another opportunity to fire the head-on Sparrow missile.

The F-4 was equipped with a medium range, radar guided AIM-7 Sparrow missile and a shorter range AIM-9M Sidewinder heat

F-4Cs from the 8th Tactical Fighter Wing refuel from a KC-135 Stratotanker during the Vietnam war. (USAF photo)

An F-4C of the 433rd "Satan's Angels" Squadron of the 8th Tactical Fighter Wing is loaded with ordnance and ready to go. Ubon Royal Thai AFB, Thailand. (USAF photo)

seeking air-to-air missile. It was a tragic mistake that a gun was not built into the original airplane. During the fall of 1967, Air Force aircrews, fed up with the lost close-in MiG kill opportunities, finally began carrying into combat the gun pod. This bulky, two thousand five hundred pound store created drag problems but was a better alternative than venturing into the target-rich environment of Route Package Six without a short-range weapon. The Air Force finally corrected the Navy's mistake when it built the gun equipped F-4E!

The Phantom exhibited pretty good durability in its aerial engagements. Carrying eleven thousand pounds of internal fuel, it ran MiGs out of gas in a number of engagements. The airplane could take punishment and stay in the fight. Many a battle-damaged F-4 limped home to fight again another day; and sometimes on one engine.

The F-4s flyability score was not nearly as high as its opponents. There were a number of systems to master and then to manage. The training and procedures necessary to get the maximum combat effectiveness out of the F-4 were complicated and involved. The initial training took a long time and constant refresher training was necessary to maintain an adequate level of combat effectiveness.

The pilot's visibility is not very good from any of the U.S. tactical aircraft in the combat arena. But the visibility from the front seat of the Phantom was definitely better than that of its opponents.

In addition, the extra pair of eyes in the backseat made a big difference in the visual acquisition of enemy aircraft and SAMs; and just as important in maintaining sight of wingmen. Analysis of exercises and tactical evaluations shows a definite correlation of success rate in air combat maneuvering, to the extra set of eyes.

Not only was the Phantom the largest tactical airplane in the aerial arena, it also left a tell-tale trail of black smoke (except when the throttles were in afterburner) that could be seen for miles. The F-4 was easy to spot a long way off from almost any aspect and its distinct profile made it unmistakable.

The reliability and maintainability of the F-4 Phantom was not good by any standards. The failure rate of the avionics systems was high. The engine, flight control system and the hydraulics system were all complicated and their maintenance and inspections consumed a great deal of time and effort. The number of man hours required to get a F-4 turned around and back in the air, re-armed and refueled was easily three or four times that of its opponents. The North Vietnamese gunners didn't get to shoot at the same Phantom more than twice in one day.

F-105 THUNDERCHIEF

The Republic F-105 Thunderchief was not an agile airplane, but it could go like hell. It could ingress into a target area hauling a load

PART VI: THE SOUTHEAST ASIA AIR WAR: A PERSPECTIVE

Four photos this page: USAF F-105 Thunderchiefs from Royal Thai Air Base, Takhli, Thailand. (Courtesy of Jack Broughton)

of bombs at six hundred knots and egress at seven hundred. The Thud pilots used the airplane's speed to great advantage. With a wing span of thirty-three feet and a length of sixty-nine feet, it was physically longer than the Phantom but it had only seventy-two percent of the F-4s wing area. The Thud had a very high wing loading which made it ride well at high speeds and low altitude. But that very feature gave it poor turning rate and radius.

The F-105's weapons suite was optimized for the air-to-ground mission but its air-to-air arsenal was pretty impressive. I saw F-105s at Takhli Air Base in Thailand carrying AIM-4 Falcon air-to-air missiles, whose performance in combat was dismal. But when it carried the AIM-9 air-to-air missile, the combination of Sidewinder and the M-60 Gatling gun proved to be a lethal combination for some MiGs who were so foolish as to get in front of it. Unfortunately both the gun and the missile were rear quarter weapons and it was difficult for a Thud pilot to maneuver to the rear quarter of an alert MiG pilot whose turn radius was half of the F-105's.

The Thunderchief did not score well in durability. It carried plenty of internal fuel but its flight control system didn't degrade gracefully if battle damage caused a catastrophic failure in a hydraulic system. Because of its poor turning performance, the Thud's best tactic was to keep its speed up. Above five hundred knots a Thud pilot could completely negate the slower MiG-17 and if he stayed low and fast he made the MiG-21s job very difficult.

Like the Phantom, the F-105 took a lot of practice for its pilot to maintain a high level of combat readiness. The systems were complicated and the procedures involved. One didn't just "kick the tires and light the fire" to go on a strike mission over downtown Hanoi. Flying the Thud effectively in combat took a very well-trained pilot.

The pilot's visibility from the cockpit was worse than the Phantom's but not nearly as bad as either of the two MiG types in the theater. The same problems that bother the Phantom pilot applied equally or worse in the Thud except for the fact that he didn't have that extra pair of eyes to help keep him out of trouble.

The Thud was a monster! Sixty-nine feet is longer than a school bus! But the airplane was deceptive. It was a very clean design without bombs, and in the head-on aspect there really was not much to look at. Rick Harris and I found that to be a fact that afternoon when we passed through a "gaggle" of them over Cam Pha.

PART VI: THE SOUTHEAST ASIA AIR WAR: A PERSPECTIVE

The reliability and maintainability of Thunderchiefs were not quite as bad as the F-4s but were generally in the same ball park. Air Force ground crews became very proficient at rearming the Thuds and getting them back in the air. The fire control system in the F-105 was not quite as complicated as the Phantom's but it still took a great number of maintenance man hours for each hour the Thunderchief flew.

F-8 CRUSADER

The Vought-built F-8 Crusader had been around a long time when it first began to take on MiGs in the skies over the Red River Valley. It was an old airplane but it chalked up the highest exchange ratio of any tactical airplane on either side. F-8s shot down only eighteen MiGs in the whole war (1965-1973) but they only lost three to enemy aircraft. The "Gator" was a fairly agile airplane. With a length of fifty-four feet, a wing span of thirty-five feet and a combat weight of twenty-five thousand pounds the F-8 had a wing loading of sixty six pounds per square foot and a thrust-to-weight ratio of 0.64. Although the F-8 didn't have as high a thrust-to-weight ratio as the Phantom, it had a substantially lower wing loading and therefore its turning rate and radius were better. The F-8, a much smaller airplane powered by a single Pratt & Whitney J57-P-12, came closer to the agility of the MiG-21 than any other U.S. tactical fighter that entered the skies over Indian Country.

The F-8s weapons suite included four AIM-9 Sidewinder heat-seeking missiles and four twenty millimeter guns with a maximum load of six hundred rounds of ammunition. A maximum rate of fire of about three thousand five hundred rounds per minute was achieved when firing all four guns. The missile was a rear-quarter weapon so the F-8 did not have a real head-on launch capability. The combination of a short range missile and a close-in gun turned out to be effective in combat.

The Gator scored high in my durability criterion. With all of its eight thousand two hundred pound fuel carried internally it had excellent range and staying power. The F-8 could take battle damage and continue to kill airplanes. The engine was rugged and wouldn't fly to pieces if it ingested foreign objects. There were several engagements in which the F-8 held the MiG at bay until it ran low on fuel. When the MiG was forced to leave the fight the F-8 ran him down and killed him.

The F-8 did not score high on flyability. It took a good deal of practice to become a proficient air combat maneuvering pilot. It also took constant practice to retain that proficiency. The post-stall gyration and spin characteristics were absolutely awful. Spin recovery procedures differed for different kinds of spins.

The pilot's visibility was not good. Visibility to the rear was terrible by today's standards but over all it was far better than the MiGs. A movie was made by the Grumman Aircraft Corporation entitled "No Points For Second Place." It was thirty minutes long and was essentially a plug for the F-14A Tomcat. It contained some interesting color footage of an F-8 and an F-4 in a dog fight. The footage taken by a camera in the F-8 cockpit showed the gyrations which I had often unknowingly gone through to look over my shoulder to keep track of the opponent across the circle while under six and a half "g's." The pilot in trying to lever his upper body around to the right would take his left hand off the throttle and, reaching up, grab some part of the instrument panel in front of him and then push his torso around to the right using the left hand for leverage. Contrarily, in trying to look around to the left, the pilot, Tony Nargi, actually switched hands on the control stick and used his right hand to do the levering. Sometimes the blindly grasping hand came close to the canopy jettison handle.

VF-191 F-8E from U.S.S. Oriskany (CVA-34), 1971. (Photo by Redditt)

PART VI: THE SOUTHEAST ASIA AIR WAR: A PERSPECTIVE

VF-191 F-8Es from U.S.S. Oriskany (CVA-34), 1971. (Photos by Redditt)

I decided, after seeing that film, to design what the squadron pilots eventually called the "hassle handle." It had to be made of non-ferrous metal (aluminum) because of its proximity to the wet compass at the top of the canopy bow. Oddly enough, when I got into an F-14 eleven years after my last tour in the Tonkin Gulf I noted that the Tomcat had a hassle handle at the top of the canopy bow. The bottom line was, that visibility out of the cockpit was not one of the Gator's strong points.

The F-8 was a lot smaller than the F-4 but a good deal larger then the MiG. Its distinctive high mounted wing and long thin fuselage with the chin mounted engine intake duct made it easy to recognize. I never could understand how that Phantom pilot could have mistaken me for a MiG over the Do San Peninsula that day!

The reliability and maintainability of the F-8 was not good, but better than the Phantom. One reason was that its systems were a lot simpler and less sophisticated. Only half as many engines to maintain, a very rudimentary radar, only one missile system, all reduced the maintenance man hours required to turn the F-8 around on the flight deck and get it back into the air. In March of 1966 on Dixie Station, both VF-53 and the sister squadron VF-51 logged over one thousand hours, and that was only with ten airplanes. Each squadron had left two of its twelve Crusaders with the detachment at Naval Air Station Cubi Point in the Philippines for corrosion control work. That worked out to a sortie rate of 3.6. That meant that each airplane flew over three and a half times each of the intense flying days. Amazing!

If those basic ingredients of what constitutes a good fighter plane are roughly right, and, if the relative priorities are also about right, I thought it would be interesting to score the five fighter planes which engaged each other in Southeast Asia to see how they rated and if there was any correlation between their rating and the outcomes of the engagements. The results of this first cut, or "Poor Man's" analysis, done while sitting on an airliner enroute to Washington, D.C. were interesting. The raw scores and weight factors could doubtless be the subject for great and heated debate.

I took the above qualitative evaluations of the five airplanes and assigned scores based on a scale of one to ten. I then applied a weight factor to each of the fighter criteria based upon their relative priorities; also on a scale of one to ten. The resulting matrix showed that the F-8 Crusader should have scored the highest in aerial combat; the F-105 should have scored the lowest; and the F-4 Phantom should have fallen some where in the middle along with the two MiGs. The MiG-21 should have performed better than the MiG-17.

A review of the final box score in the early part of the war correlated very well with this prediction. Changing the scores or the weighting factors didn't alter the above "pecking order" as long as the priorities and the relationships of the criteria remained the same. Of course the unknown factor in all of this analysis was the level of training effort exerted by the competing air forces involved. Certainly, the training environment in the skies over North Vietnamese airfields was not conducive to training young eaglets.

As it turned out, the U.S. Navy and U.S. Air Force F-4 communities had not been training properly for the kind of aerial combat which they encountered over North Vietnam. The improvement in statistics, however, as the war went on, attested to the efficacy of the dramatic changes which the Phantom drivers made in their training programs. The small Crusader community needed to make only minimal changes to its training regimen . . . and they did so. Obviously, realism in training, and plenty of it, are the key ingredients to achieving and maintaining high levels of combat effectiveness in the air-to-air arena. This message was well understood as the air war ground to a halt in January 1973.

Analysis was being done on a daily basis both at U.S. Air Force and Navy headquarters as the results of the aerial engagements came in. By the fall of 1968 sufficient data had been analyzed to tell a grim story. The exchange ratio between F-4s and MiG-21s was about even. What a shocker that was to the decision makers in both services! "How," they asked, "could the biggest, most powerful Air Force in the world, flying the most sophisticated and most expensive fighter planes in the world, get held to a stand-off by a rag tag, third world fighter force flying export models of older Russian aircraft?" The answer was not a simple one. Experience in World War II and in Korea had programmed U.S. leaders to expect numbers like ten or fifteen to one. But one to one! That was hard to take!

There is another one of my laws of aerial warfare that says, "We build and buy airplanes for the war we think we're going to fight, not the one we really fight." The principle airplane, the F-4 Phantom and its primary weapon were designed in the mid-1950s to shoot down long-range Soviet bombers at high altitudes in down-the-throat, head-on encounters. Considering the requirement, the F-4 was really a very responsive design. Had the aircraft acquisition people known back in the mid 1950s that our Navy and Air Force tactical aircraft were going to be put into the constrained environment that existed in Route Package VI from 1965 through 1972, they may not have produced the F-4.

But, it was "the only war we had" and the task that U.S. Air Force and Navy leadership had ahead of them was two-fold, first they had to find out why they were only able to achieve a one to one exchange ratio. Then they had to fix it!

Along came U.S. Navy Captain Frank Ault. Frank was an intimidating person. He reminded me of a burlier version of Telly Savalas; a big, broad-shouldered, totally bald man with a gruff voice, piercing black eyes, Frank had a very pointed way of expressing his thoughts. He was sent from Naval Air Systems Command Headquarters in Washington, D.C. to find out the "why" to the above questions. Frank's investigation in the field was typically thorough. Simply summarized, it stated that U.S. tactical aircrews weren't training right and the maintenance and ordnance personnel were not testing and checking all the components of the fire-control system properly.

In all the engagements in which U.S. Navy aircrews fired air-to-air missiles at their opponents the success rate of achieving kills had been less than ten percent for the AIM-7 Sparrow radar-guided missile and less than forty percent for the AIM-9 Sidewinder heat-seeking missile. Probability of a kill (Pk) in other words, was less than 0.1 and 0.4 respectively. I thought both the one-to-one exchange ratio and the low Pks were a disgrace and needed to be fixed with the highest priority assigned to the program.

In the early years of the Korean War two schools were formed to teach U.S. tactical pilots the finer points of aerial gunnery. One was the fighter gunnery school at Nellis Air Force Base near Las Vegas, Nevada. The other school was the Fleet Air Gunnery Unit, Pacific at Naval Auxiliary Air Station, El Centro, California. The purpose of the schools was to train a cadre of talented young pilots the intricacies of weapons delivery so that they could return to their fleet squadrons and serve as Weapons Training Officers, passing on their newly acquired skills to other squadron pilots and also acting as the squadron weapons experts. In theory, the Navy school was supposed to keep one FAGU graduate in every fleet squadron.

In order to do this, the school was divided into groups of instructors, each equipped with six of every type of tactical airplane in the fleet. When I arrived at El Centro in 1955 to be a FAGU instructor I was understandably miffed at being dragged out of my first fleet squadron tour after only one carrier deployment to the western Pacific. The normal routine was for junior officers to do two WestPac deployments in their first fleet squadron. It didn't take very long, however, for the realization to sink in that selection to FAGU was a distinct honor. FAGU, as small as it was, provided an important central repository for expertise in aerial weapons delivery.

Unfortunately, in 1958 a decision was made to close down FAGU. The rationale given to justify the decision was bureaucratic and short-sighted. Simply stated, the two aviation type commanders for the fleets, Commander Naval Air Forces, U.S. Atlantic Fleet and his counterpart in the Pacific, were going to commission shore based squadrons which would take fresh-caught, newly "winged", young training command graduates and teach them how to fly fleet airplanes. The new training squadrons, called RAG's (Replacement Air Group) would bring each fledgling aviator to a state of one hundred percent combat readiness so that the recipient fleet squadron wouldn't have to train the new comer at all. For this reason, there would be no need for a FAGU mission. A stroke of the pen turned the Fleet Air Gunnery Unit, Pacific into history.

Ten years later Frank Ault's report recommended, among other things, that a unit be formed to train fleet fighter squadron aircrews on the finer points of air-to-air weapons delivery. It also recommended that the Navy complete the development of the air combat maneuvering range and procure some of them for the fleet. The Ault Report was probably one of the most important documents to come out of the early stages of the U.S. aerial combat experience in Southeast Asia. Published in January of 1969, it represented over two hundred man-years of effort and made two hundred and forty-two recommendations ranging all the way from the way weapons systems specifications are written, through their development, training, deployment and finally, how they are reworked and refurbished.

If acted upon, the recommendations would have cost the United States Navy over half a billion dollars. Most of them were acted upon and their effect manifested themselves in increased combat effectiveness throughout the remainder of the war. The U.S. Navy had come full circle. It had taken another war and another disastrous combat experience to relearn an old lesson. The Navy Fighter Weapons School was born, first as a detachment reporting to the F-8 RAG (now called Fleet Replacement Squadron) at Naval Air Station, Miramar, California, then at the F-4 RAG. An almost immediate change in the success rate of U.S. Navy F-4s against MiGs manifested itself as the new Navy Fighter Weapons School began pumping young, aggressive TOPGUN graduates into the fleet F-4 squadrons flying from the carriers in the Tonkin Gulf. The U.S. Air Force

PART VI: THE SOUTHEAST ASIA AIR WAR: A PERSPECTIVE

leadership, also keeping day by day updates of the aerial box score, were quick to note a change and, following suit, reinstated their own Fighter Weapons School at Nellis Air Force Base, the same place they had operated their old gunnery school.

Any analysis of the air war in Southeast Asia ultimately examines the performance of the various weapons systems and tries to explain why some did better than others. Now, the real message of my "Poor Man's" analysis is that no U.S. fighter which fought in the skies over Route Package VI was optimized to fight MiG-17s and 21s in that sterile environment. That may or may not be as it should be. The U.S. needs to design systems to counter the major threat. That hasn't changed; and MiG-17s and 21s did not pose a major threat to the United States in the mid-1960s. However, they did pose a serious threat to the aircrews who flew over the Red River Valley in those years and they will similarly threaten U.S. aircrews the world over whenever they attempt to seize and maintain air superiority over turf that Soviet built fighters think is theirs.

The F-8 Crusader did well against the MiGs by any standard. The reason is simple. Since the F-8 was not equipped with a forward-quarter weapon (like the head-on, radar guided AIM-7 missile of the F-4s) the Gator pilots spent all of their peacetime training maneuvering to place their airplane in the best position to employ their weapons while at the same time staying out of the MiGs similar weapons envelopes. That is a funny way of saying that Crusader pilots spent all their time "dog fighting." Naturally they did better at it. F-4s up until the late-1960s spent most of their training time practicing head-on tactics to make best use of their main weapon, the Sparrow missile.

A subset of that explanation is that the simpler, less complicated heat-seeking Sidewinder missile was also easier to use. The F-8 pilots spend all of their missile training time learning to recognize visually where the "heart" of the missile lethal firing envelope was. The F-4s spent very little time practicing with the Sidewinder for the very reason it was simpler, and therefor should not occupy a large percentage of their limited training time.

One final factor which caused the difference was the fact that the Sidewinder, being simpler and therefore more reliable, had inherently a higher probability of kill (Pk) than the larger and more complex Sparrow, regardless of how skillfully it had been maintained or how competently it had been employed.

In the final stages of the war the U.S. F-4 aircrews began to get the maximum combat effectiveness from their deadly machines. The engagements during the spring of 1972 were fought hard and by skillful aircrews on both sides. None were better typified by aggressiveness, skillful airmanship and sheer guts than the five MiG kills achieved by Lieutenant "Duke" Cunningham and Lieutenant (junior grade) "Willy" Driscoll, making them the Navy's only MiG aces!

These were sad lessons but fighter aviation learned them well, taking the necessary steps to fix them, and fix them well.

On the strike side of the coin, hard lessons were learned the hard way. Admonitions, like "Dutch" Netherland's "twenty-nine commandments" had been articulated in World War II, relearned in Korea and again in North Vietnam. The evolution of the Alpha Strike, as a way of focussing the most fire-power on a pin-point target in the shortest time with the least loss of aircraft proved a well known

Lt. Jack Ternune, VF-154, ejects from his F-8E over the Tonkin Gulf, 8 November 1963. He ran out of fuel twenty miles from the U.S.S. Coral Sea (CVA-43). (U.S. Navy photo)

fact. As anti-aircraft defenses increased, a greater percent of total air assets was devoted to support. The threat of enemy fighters in the early years didn't concern strike planners. By the summer of 1967 strike escort fighters for Alpha Strikes were a way of life. By the summer of 1968, the real benefit of fighter sweeps was being demonstrated almost daily. As the surface-to-air missile defense threat in the north increased, anti-radiation missile attacks, called "Iron Hand" became more frequent until one or more sections of "Iron Hand" shooters were a part of every Alpha Strike. In the late spring of 1968, the *Bon Homme Richard* ran out of the AGM-45 Shrike anti-radiation missiles during one heavy line period. The A-4 Shrike shooters were loaded with 2.75 inch rockets. When the shooter detected a SAM acquisition radar emitter locked on him he would simulate a Shrike attack maneuver, pickle-off a 2.75 inch rocket and simultaneously transmit on his radio "Shotgun, shotgun." In every case the enemy radar site shutdown. In my view, pound for pound, the 2.75 inch folding fin aircraft rocket was the most effective anti-radiation missile ever employed.

Airborne jammers, to electronically blind their acquisition and fire control radars of the SAMs and AAA sites became a regular part of the Alpha strikes. Grumman E-2 airplanes from the bigger carriers and their predecessors, the E-1Bs from the *Essex* class carriers were used for vectoring strike leaders to their target coast-in points, warning strike aircraft over the beach of impending attacks by enemy fighters, vectoring egressing airplanes to waiting aerial refuelers and, finally, keeping score for the strike leaders of individual strike elements reporting "feet wet." The investment in support aircraft increased as the anti-air threat increased. But, this rule applied principally to Alpha strikes which were limited to day and clear weather conditions. A large part of the year, during the northeast monsoon season (November to May), the weather was generally too bad for Alpha strikes.

In addition, the night attack capability of the strike aircraft flying over the beach, was limited to those missions in which visual acquisition of the target was possible. Unless the moonlight was particularly bright, flares were the only alternative and they could be used only in the relatively benign environment of Route Packages I through IV.

The first real all-weather, night, overland strike capability for the U.S. Navy came with the advent of the Grumman A-6 Intruder. I first flew this airplane on 3 November 1960 at the Grumman plant at Peconic on Long Island. At that time the airplane was designated the A-2F-1 and its engine tailpipes could be rotated about twenty-five degrees downward to increase take-off performance. What ensued between then and 1965 when the Sunday Punchers of VA-75 initiated the Intruder into combat operations in the Tonkin Gulf from the U.S.S. *Independence* is a genuine success story.

The idea of the night, low-level bombing attack was not a new one. The mission was very attractive to the light and medium attack communities in Naval Aviation for a number of reasons. First and foremost, it offered protection to the attackers. By staying low the attacker avoided radar detection until so late in the approach that effective defense was minimized. By avoiding detection the bomber was safe from enemy fighters, anti-aircraft artillery and surface-to-air missiles until it was too late to respond.

If, however, detection did occur, protection was still provided against enemy fighters because their air intercept radar couldn't see the bomber through the ground clutter on the interceptor's radar scope. Even if enemy fighters detected the bomber and fired an air-to-air radar guided missile at him, the missile's fuzing function couldn't work properly so close to the ground. Furthermore, when finally detected close in by the SAM acquisition radar, and a SAM were fired at the bomber, the missile would have the same fuzing problem that its air-to-air cousin had. So, the low-level, night attack had tremendous potential but it needed a platform having special capabilities and carrying special equipment.

The capabilities needed in the basic airframe were the ability to carry a big bomb load, long range and a respectable low altitude maximum speed. The equipment needed was a large dish, brute-force radar to enable the crew to locate radar significant geographical features for over-land navigation and to attack radar significant targets. The airplane needed, in addition, sensors and cockpit displays to enable the crew to maneuver safely as close to the ground as possible without flying into it.

Other strike aircraft possessed some of these capabilities. The A-7 Corsair II, in fact, had a very impressive night attack capability. But it took a two man crew in that technological timeframe to do this demanding mission. The A-6 BN (bombardier navigator) was needed, devoting all of his attention to operating the weapons system, to do the job right. Possessing all of these features in spades, the A-6 was truly a mean machine. Mean and ugly! So ugly, in fact, that the irreverent aviators in the other communities on the carrier referred to it as the BUFF; standing for "Big. Ugly, Fat F#*@ers."

The BUFFs in the late-1960s wrote a new chapter in the annals of combat aviation. Streaking in below the radar horizon at 500 knots, the A-6 would cross the beach 200 feet over the ground carrying twenty-two, five hundred pound Snake Eye retarded bombs, release them as it passed over the target and egress at low altitude dodging SAMs and AAA all the way to the "Feet Wet" call. Intruders had essentially taken away from the North Vietnamese the "haven" provided to them by nightfall and bad weather. Night after night the black skies over Hanoi would light up from the fireworks display of exploding AAA, roaring SAM booster motors, SAM explosions and finally the fireballs of exploding ammunition storage depots or power plants. At night or in bad weather, the A-6 Intruder had become a one-plane Alpha strike.

A final lesson, again learned the hard way, was the need for stand-off weapons and precision-guided munitions which, in a sense, are one and the same thing. Many of the targets assigned to Navy attack aircrews were indeed pinpoint targets. There were cave entrances, narrow bridges, gun emplacements and other small targets which needed essentially a direct hit. An eighty-five millimeter gun is a hard target (and it is also a pinpoint target) but since one direct hit with a five hundred pound bomb will knock it out, it is not so hard to destroy. One attack plane on a single pass can drop a number of those bombs and have a reasonable chance of a hit. But, hard targets which require a direct hit with a large bomb are another matter. The Than Hoa bridge was such a target and because of that became a monument in the middle of a graveyard filled with destroyed U.S. tactical airplanes. That monument is mute testimony to the need for "smart" weapons.

On April Fool's Day, 1965, the "Rolling Thunder" campaign approved strikes against the Than Hoa bridge. Like most bridges which were a part of a major supply artery, feeding the North Vietnamese war effort in the south, Than Hoa was heavily defended with anti-aircraft artillery guarding the approaches to the bridge. The bridge itself was on the main coastal route leading south from Hanoi and included a railroad and a highway crossing of the Song Ma River about seventy-five miles south of Hanoi. The French engineers who built it had done a good job. It was constructed of heavy steel girders on large concrete abutments. Since it was a critical link in the flow of war material to the south, the strike planners eyed it covetously when it appeared on the target list in April 1965. About every two weeks thereafter it was targeted . . . and hit. However, the bombs chipped concrete and dented girders but never cut the span. What was needed was a direct hit by a large bomb. Between April 1965 and the bombing halt three years later, the bridge was targeted over six hundred times.

In late May 1966, I flew fighter escort on a mission against the Than Hoa bridge. But this mission was different from others which preceded it because there were two targets. The first, of course, was the bridge itself, a span fifty-four feet long and eighteen feet wide. The secondary target was a U.S. Air Force F-4 Phantom II. The F-4 had been hit during an attack a few days earlier. The aircrew had ejected and were captured. The unmanned airplane had circled for a short time then landed in a very flat, wings-level attitude in the Song Ma River. After it struck the surface of the water, it was observed to skip once and come to a stop on a sand bar in the middle of the river. Photo surveillance photographs indicated it was in a salvageable condition.

Strike planners, concerned that the North Vietnamese, with some Soviet technical help, might be able to use it to our disadvantage somehow, targeted the F-4. I had to pinch myself as we headed in toward the beach. My wingman and I were flying fighter escort with four "Tinkertoys." Two of them were to drop six Mark 83 1,000# bombs on the bridge. The second two A-4s were to drop twelve Mark 82 500# bombs on the F-4 hoping to damage it beyond repair or, if lucky, destroy it completely. "Here we are", I said to myself, "Heading for the Than Hoa again with a token effort directed toward the bridge, and really going to risk six more airplanes to bomb one of our own airplanes which failed to knock down the bridge in the first place. What a crazy war!"

When the strike group arrived at the push over point about thirty miles from the beach it was obvious that the weather was not going to be good enough. The strike leader aborted the mission and directed the group to its assigned alternate target, a small bridge south of Vinh. The F-4 was damaged by a later strike.

As a result of the six hundred plus strikes against the Than Hoa bridge the U.S. lost over thirty tactical airplanes over forty airmen lost their lives or were captured. In 1972 a U.S. Air Force strike group equipped with a new laser-guided precision weapon finally cut the bridge. The total U.S. investment in knocking out that little bridge in North Vietnam would have paid for the building of the Verrazano Narrows bridge at the entrance of New York harbor several times over. One chronicler of the U.S. Navy's air war against North Vietnam titled his foreword, "All For Naught." I don't like characterizing the magnificent effort, over nearly a nine year period, by so many talented aviators in such a negative way. But, in a sense, it was true.

In the air-to-air arena a bad beginning was turned around to a successful finish with a much-needed overhaul of tactics, training, and weapons logistics. The reestablishment of a Fighter Weapons school which would later report directly to the Deputy Chief of Naval Operations (Air) was essential to ensuring the perpetual health of that community.

Unfortunately, it took another twelve years and a disastrously botched strike over Lebanon on 4 December 1983, to get the Naval Strike Warfare Center (an attack counterpart of Topgun) established.

I never forgot a football game in which the U.S. Naval Academy team was beaten by a powerful Big Ten team. The score wasn't even close, and Navy's star quarterback was intercepted half a dozen times. I was playing second string defensive left tackle at the time and got to play most of the game because the first stringer injured his knee. The Navy's defensive squad played a long, hard, ass-kicking, pad-popping game. Playing perhaps a little over my head, I had the best football day of my life. I racked more ball carriers than I had ever hit before, and man-handled more big, fast, mean offensive linemen than ever before. As our team was walking dejectedly off the field after the game I felt a powerful hand on my shoulderpad. It was our defensive line coach, Ray Schwartz, who looked me in the eye and said, "Don't feel bad, Son. You've got nothing to be ashamed of!"

When I left the combat zone for the last time I felt the same way. When the U.S. forces pulled out of Vietnam in January 1973 I felt that I could say the same thing to the five thousand or so tactical aviators who for nine years carried the air war into the skies over the Red River Valley with as much courage, competence and elan as any aviators in any war. What I wish our own country had said to them would be simply, "Don't feel bad, fellows. You've got nothing to be ashamed of!"

PART VII
THE RESERVES

*"Now, here's to the Regular Air Force;
They have such a wonderful plan.
Just call out the god damn reservists,
Whenever the crap hits the fan."*

(Air Force ditty from the Korean War)

As Sir Winston Churchill might have said of the U.S. Naval Air Reserves during the Southeast Asian conflict, "It was not their finest hour!"

When President Lyndon Johnson decided to call up selected portions of the U.S. Naval Reserves in 1968; he probably had an inkling that he might be opening up the proverbial "can of worms." The seizure of the U.S.S. *Pueblo* by the North Koreans, while the announced "cause celebre" for the call up, was really only one of the reasons. The increasing demands of the Southeast Asian conflict were having their effect.

Commander Paul Speer (see Chapter 36) and the author were but two of a great number of senior carrier aviators who had been designated "combat limited" but whose names were thrown back into the hopper because of the lack of assets. Speer ended up as a wing commander in the Tonkin Gulf. The author ended up as a wing commander on U.S.S. *Saratoga* which, on the deployment subsequent to mine, was sent around the world to get into the Southeast Asian conflict. Both the Navy and the Air Force were running out of non-limited combat aviators and were forced to bring aviators from other communities and to break their own rules.

But, organizations were also being broken up in the frenzy to meet ever-increasing deployment rates. For example, Air Wing Five (of which VF-53 and 51 were a part) was deployed on three different attack aircraft carriers, on three successive deployments, each time with a slightly different air wing composition.

So, the arrival of the two reserve fighter squadrons from Dallas at Miramar was viewed by many of us at the latter base as another stop-gap solution to increasing demands on limited assets. Lyndon Johnson's own motivations for such a step can only be surmised. But, he obviously intended some political mileage from taking the extraordinary step of "mobilizing the reserves."

One can only assume that politics must have governed the selection of squadrons because common sense couldn't have played a factor. The Crusader, with its accident rate several times higher than any other community in the Navy, should have been the very last choice of communities to test the reserve concept in actual combat operations. In retrospect, the horrible selection had some long-term benefits. Shortly after the two squadrons were ignominiously returned, not combat tested, to Dallas, the whole system was reorganized. The general consensus among reserve experts is that it is now by far, a more workable system. However, whether it works or not remains to be seen. We will need another war to make that determination. The two key people within the system charged with the responsibility for getting these two squadrons ready for combat would probably be the COMNAVAIRPAC Training Officer at North Island and the commanding officer of the F-8 Fleet Replacement Squadron (VF-124) at Miramar. Both of these gentlemen, when interviewed agree that the logistic problems involved in getting the older reserve airplanes into a material condition for combat were not insurmountable.

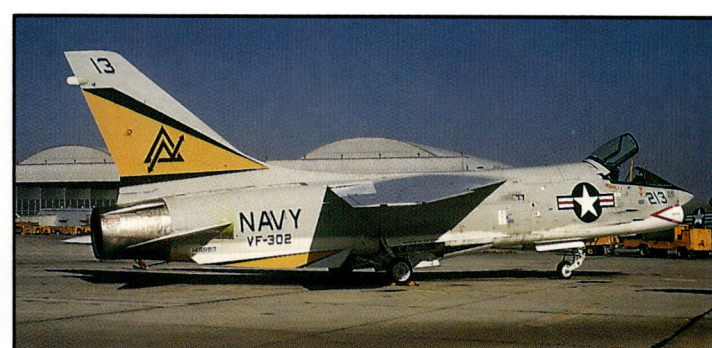

F-8Es from NAS Miramar's two reserve fighter squadrons (left VF-301, right VF-302). (Photos by Lewis)

PART VII: THE RESERVES

F-8Es form NAS Dallas' two reserve fighter squadrons (left VF-201, right VF-202). (Photos by Lewis)

The problem, in a word, was the pilots. They were airlines pilots, dentists, professional men; individuals whose own worlds had been turned upside down by Lyndon Johnson's decision. Businesses were left undermanned, medical practices abandoned and families disrupted and so on. The detractors were quick to point out that the pilots knew what they were getting into when they joined up. What then, was the big deal? If you enjoy the benefits of reserve duty in peacetime you ought to be prepared to pay the piper when the balloon goes up! Reserves had served honorably in other wars . . . what was so different about this one?

The list of Crusader pilots who had recently been shot down and had either been recovered or were prisoners of war or missing in action or killed was growing rapidly. Jim Stockdale, Howie Rutledge, Ron Dodge, Jack Snyder, "Tooter" Teague, Cole Black and a host of other Crusader pilots had been forced to abandon their airplanes because of enemy action. The first combat loss of a Crusader occurred in June 1964. From that event until the call-up of the Reserves four years later there were more than seventy Crusaders shot down by enemy fire in Southeast Asia. At the time none of us, I don't believe, knew that startling statistic. But, all of us knew that the Crusader was taking its share of combat losses.

There was, in all of us, I believe, a much greater fear of becoming prisoners of war than of dying. I was always able to reconcile death as a logical eventuality. It meant an end . . . a termination of a process. In carrier aviation one learns to accept that fact. But, there was no end in being a prisoner of war in Hanoi. That was the scary part . . . not the starvation, torture, privation and loss of freedom.

But, having made all of the above excuses for reserve pilots not wanting to deploy to Southeast Asia in 1968, I remain convinced that there remained one, single, fundamental, over-arching reason for their reluctance. That was the thought of facing the back end of the 27C class attack carriers at night in a Crusader.

This fundamental respect for the night carrier landing performance of the airplane was universally shared by all Crusader pilots, active duty and reserve. It is the basic reason for several requested transfers from the fighter to the other communities that I am aware of. The big difference was that the active duty pilots in general loved the airplane's performance in other areas so much that they were willing to put up with it around the boat at night as an unpleasant, but necessary, adjunct of a career as a carrier fighter pilot.

After all, carrier aviation was the chosen occupation of the active duty Crusader pilot. For the reserve Crusader pilots, it was a usually pleasant diversion. A chosen occupation demands commitment, a sideline does not.

After nearly two years of "standing up" the reserve Crusader pilots arrived aboard U.S.S. *Oriskany* ready for the final exam; day/night carrier qualifications. All of them were completed with their required day qualification arrested landings and the ship was steaming downwind to gain sea room for the first night launch in the operating area off San Diego. Much like the last minute telephone call from the Governor to San Quentin's death row, the message arrived from the fleet commander. It stated that a decision had been made by higher authority not to deploy the two reserve fighter squadrons. Instead they were to be returned to their home base at Dallas as soon as the carrier could arrange for their flight to the beach.

We will probably never know the motivation of the drafter of the message for the caveat that followed . . . but it must have been born of brilliance. The message contained an innocuous paragraph which acknowledged the tremendous effort that had already had been expended to get the reserve squadrons this far. Perhaps not to complete the effort, the message observed, would be a great disappointment to the pilots involved. Therefore, the message continued, all pilots who wanted to complete the night portion of their carrier qualification requirements (12 day/8 night arrestments) were free to do so.

To a man, the reserve pilots declined to continue. The only pilot who elected to continue was a TAR named Commander Abercrombie. That was the bottom line!

In a meeting of landing signal officers after the carrier qualification deployment it was generally agreed that the performance of the reserve Crusader pilots was "pretty average" when compared to other CQ deployments. There was no reason to believe that they would have had any more difficulty completing their qualifications than any other recent group of regular Navy Crusader pilots. That was the second bottom line!

PART VIII
NASA

NASA's digital fly-by-wire F-8 and Supercritical Wing F-8 in formation over the Mojave Desert. (All photos in this section are from NASA archives)

NASA's Supercritical Wing F-8 on dry lake bed at Edwards Air Force Base.

speeds for carrier landings, the single position supercritical wing caused excessively long landing roll-outs. This mandated use of the lake bed for recoveries.

First flight occurred on 9 March 1971 with McMurtry at the controls. The flight envelope was gradually expanded until, on 26 May McMurtry reached a mach number of 1.1 at 35,000 feet. But, the test program was really a supercritical wing proof-of-concept flight test program. Preliminary results from wind tunnel tests indicated that the promising characteristics of the design could also be achieved at higher Reynolds numbers in flight.

Specifically the tests were designed to determine gross indicators of design achievement such as transonic drag-rise Mach number, buffet and stability and control characteristics.

The basic idea of the supercritical wing is to delay what is known as drag-rise. All airfoils can be described with a drag coefficient curve. Conventional airfoils all possessed a drag coefficient curve which has a slope indicating increasing drag as a function of increasing indicated Mach number. As the Mach number approaches 1.0 the slope of the curve increases sharply to a point at, or close to, Mach 1.0 where the curve peaks and after which the slope decreases dramatically. The hump in the curve is the region known as the drag rise.

Most conventional fixed wing airplanes cruise at their most efficient point along their lift or drag-rise curves which is somewhere short of the drag-rise. Commercial airliners cruise at somewhere around Mach 0.8 to 0.9. Tactical military aircraft cruise gen-

NASA's Supercritical Wing F-8. Note empennage area rule fuselage blister.

CHAPTER 56: THE SUPERCRITICAL WING

NASA's Supercritical Wing F-8. Note area rule fuselage blister.

NASA's Supercritical Wing F-8. Note wing/fuselage blending.

erally at slightly higher Mach numbers 0.9 to 0.97. The penalties for operating in the high drag-rise region are fuel efficiency and the controllability problems associated with the region.

The supercritical wing concept attempted to delay the drag-rise, moving it to the right along the horizontal axis of the drag coefficient curve. The F-8 supercritical wing flight test program provided a design intended to permit economical cruise at a Mach number of 0.97. It was not until nearly after completion of the F-8 supercritical test program that the U.S. Air Force had developed enough confidence that the technology had no unpleasant surprises to embark on its military-oriented TACT program.

For his work on the supercritical wing program which was his brain child Langley's design engineer, Richard T. Whitcomb was awarded the 1974 Wright Brothers Memorial Trophy.

57
DIGITAL FLY-BY-WIRE CRUSADER

Another one of NASA's important technology programs which utilized an F-8 airframe was the digital fly-by-wire (DFBW) program. The original airplane also sits on display outside the visitor center building at the Hugh L. Dryden Flight Research Facility at Edwards Air Force Base, California. Again, the airplane lent itself to the test program because its design provided plenty of internal volume for instrumentation and because of its supersonic flight performance envelope.

In the initial stages of aviation, it was easy to connect a flight control actuator (stick or rudder pedals) directly to the control surfaces (ailerons, elevators and rudder) by way of a steel cable. That was known as direct control. As higher performance was achieved by airplanes, engineers tried to compensate for increased airloads on control surfaces by clever devices which provided mechanical advantage to the pilot in the form of mass balances and actuator lever arms.

With further increases in aircraft performance, flight control design engineers had to resort to hydraulically boosted controls. Back-up control for such systems was merely direct, brute force, no-boost control modes. Boosted systems managed to co-exist in modern tactical up to the MiG-17 which appeared in the late 1950s with direct control elevators and rudder and a boosted aileron control system driven by a 1500 psi hydraulic pump.

Finally, flight control system design engineers found the 1500 psi boost systems unable to cope and resorted to 3,000 psi hydraulic systems. Of course, the greater the hydraulic pressure, the greater the problem with seals and leaks. But, increased aircraft performance brought with it stability problems which challenged the pilot's ability to move the flight controls quickly enough to ensure satisfactory positive control.

This phenomenon ushered in flight control devices like yaw dampeners and finally stability augmentation systems about all three

NASA's Digital Fly-by-Wire F-8 over the Mojave Desert.

CHAPTER 57: DIGITAL FLY-BY-WIRE CRUSADER

NASA's Digital Fly-by-Wire F-8 landing at Edwards AFB.

control axes. Redundancy could no longer be provided by a degraded, brute force back-up system and triple and quadruple redundant stability augmentation systems began appearing in modern tactical aircraft powered by 6,000 psi hydraulic systems. The final step was, of course, the electronic, fly-by-wire system in which the pilot applied force to a control stick generating an electronic signal which went to a computer and thence to a control surface actuator. This also ushered in the ability of a flight control system which gave a pilot safe control over an airplane with negative natural aerodynamic stability.

As a part of the U.S. Naval Test Pilot School syllabus, each student was exposed to several flights in aircraft deliberately configured with negative longitudinal stability. In 1959 the author was assigned to a flight control project which featured a side-arm-control stick in a Lockheed T-33. The age of electronic fly-by-wire flight control systems as a normal feature of tactical airplanes and commercial airliners as well was rapidly approaching.

As might be expected, NASA led the way in pioneering digital, fly-by-wire flight control systems. There was a certain amount of support from NASA's Office of Advanced Research and Technology for space flight applications. A number of airplanes were examined for candidates as test bed airplanes for a digital fly-by-wire test program. The nod finally went to an F-8C acquired by the agency from the U.S. Navy.

On 25 May 1972 Dryden research pilot, Gary Krier made the first flight in what came to be known as the DFBW (digital, fly-by-wire) test bed. By using off-the-shelf items from space flight programs, like the digital flight computer and inertial sensing unit from the Apollo program, NASA was able to accelerate the test program by almost two years. The modifications to the F-8's flight control systems were substantial. Essentially, everything between the control column and rudders; and the control surfaces (unit hydraulic tail, ailerons, spoilers and the rudder) was removed. The systems were then connected electronically and a power pack fitted in at each of the control surface actuators to translate the electronic signals to the actuation devices.

Early test flights uncovered the need for software changes in the resident software. Accommodations were made for the operational problems associated with using the relatively coarse Apollo computer output signals to satisfy the minuscule demands of precise pilot stick inputs. Occasionally, the lateral-directional features of the system would "drop out", but never the longitudinal systems.

Some of the pilot induced oscillation (PIO) suppression filters developed during the DFBW program were employed in the space shuttle program for the landing phase of operations. The DFBW airplane flew a total of 42 test missions without incident and it was never necessary to resort to the system's emergency back-up flight control system.

It is worth noting that the NASA program totally eliminated a mechanical linkage and provided only system digital flight control system redundancy for back-up. This decision by NASA so impressed the General Dynamics flight control system designers that they decided to initiate their own F-16 program with a pure digital fly-by-wire system as well. In those days this decision was considered risky. It is a powerful vote of confidence in the new flight control system technology that General Dynamics would take on that level of program risk.

NASA's Digital Fly-by-Wire F-8 joins with VX-5 A-6 tanker for refueling tests with fixed IFR probe installed.

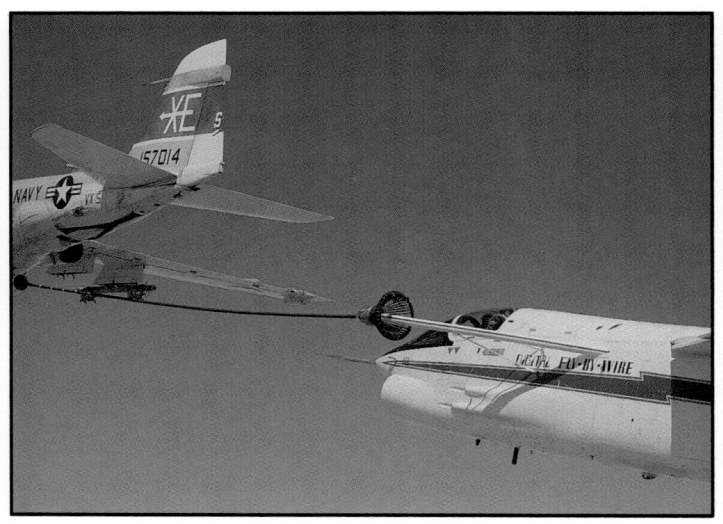

NASA's Digital Fly-by-Wire F-8 evaluates flight control system in refueling test.

But, DFBW continued to support the F-16 program. When company test pilot Bill Ostereicher took the F-16 on its now famous high speed taxi test, and inadvertently became airborne, it was consultation with NASA DFBW data that resolved the problem.

After the first phase of DFBW test proved that the special purpose flight control system hardware was reliable, and the system worked, system engineers began looking for a system with computer redundancy. As a result, a "triplex" DFBW system was installed using general purpose digital computers which was able to sustain a number of system failures and continue to operate. The program also introduced a viable fault tolerant system characterized by what is now called graceful system degradation. Not only was the F-8 supportive of the on-going lightweight fighter program, it was also used to test Space Shuttle computer software and to support the flight control system for the shuttle orbiter.

All in all, the F-8 DFBW program was an enormous success. When the author paid a visit to Dryden during the 1993 X-31 test program, there were two beautiful RF-8Gs being employed as chase airplanes. Those two airplanes came courtesy of the U.S. Navy's last operational squadron, VFP-206!

58

THE OBLIQUE WING

He had two strikes against him when he walked through the door of my office in the Pentagon. He was an acquaintance, not a friend, a retired, Crystal City Rear Admiral who admitted to be employed by Rockwell International and who was looking for a small amount of (10 million dollars) research and development money to develop the concept of the oblique wing. My office, Director, Aviation Plans and Programs (OP-50) had something to say about how and where Navy R&D dollars were spent. But, the artist's sketch he carried was the thing that did him in. And, the sad thing is that he didn't even know it!

Not surprisingly, the airframe of choice for the test bed for the oblique wing was the F-8. For the same reason as the supercritical wing flight test program (ease of removal and reinstallation of the wing) the F-8 was chosen. The concept of the oblique wing was not new then, in 1982, and it is, I believe, an off-shoot of the X wing concept. The wing pivots around the vertical axis, the tip of the left wing, as an example, would sweep aft while the tip of the right wing swept forward. It achieves the same effect as wing sweeping in the conventional sense with a greatly diminished engineering penalty in terms of weight, complexity and cost. Its only problems are that it looks like hell, it is asymmetric and it is terribly unconventional.

He must have caught me in a weak moment because I agreed to "donate" the dollars which I then had to steal from another funded program. My staff was furious at me . . . but then, that was not a new circumstance. The Navy money, along with a defense advanced research projects agency (DARPA) contribution permitted initial

Artist's conception of NASA's proposed oblique wing F-8 test aircraft.

PART IX
THE FRENCH CONNECTION

PART IX: THE FRENCH CONNECTION

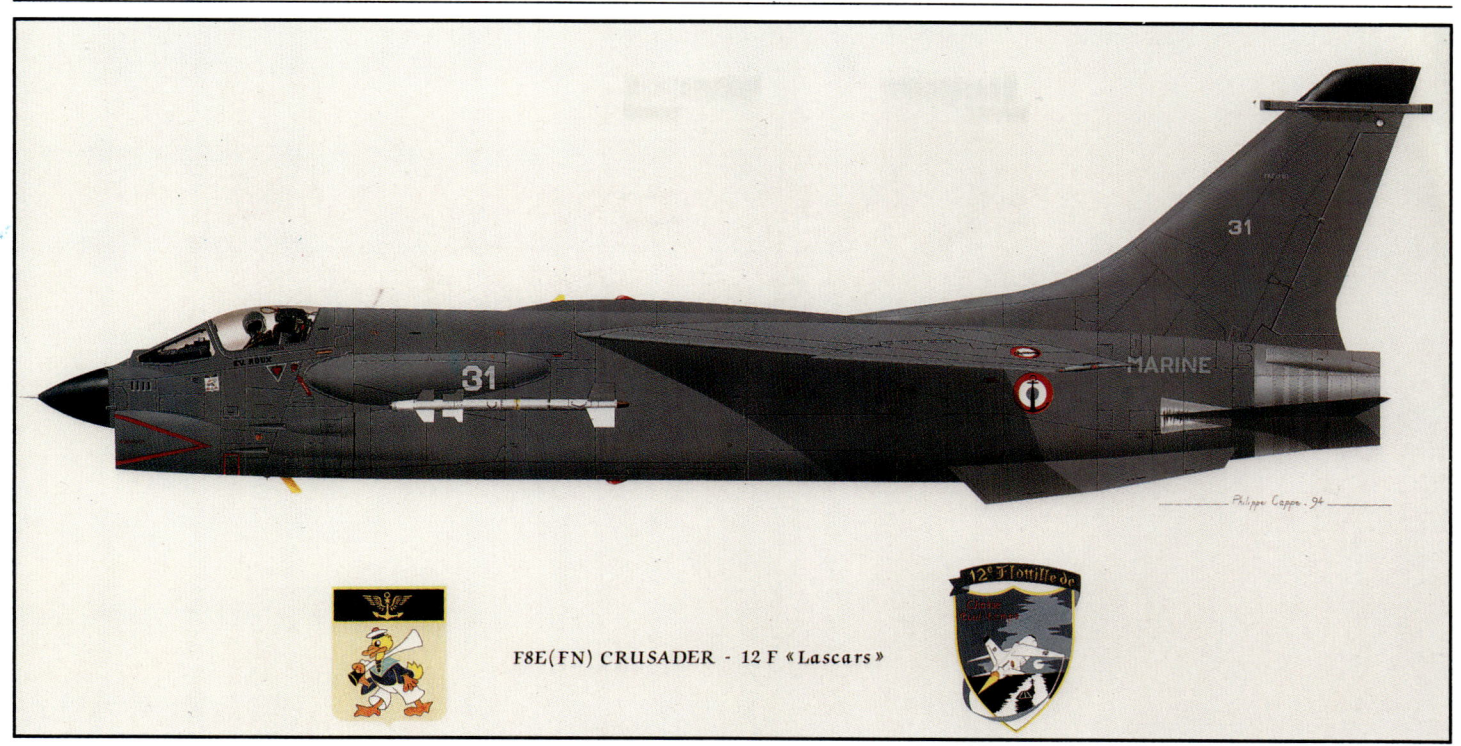

F-8E(FN) from Flotille 12F. (Official French Navy illustration)

The largest hurdle to overcome was not cost but politics. In addition to the balance of payments issue, there was also the issue of the Rafale program. This new fighter development program for the French Air Force would have been badly damaged by purchase of F/A-18s and the concomitant cancellation of the French Navy version Rafale M.

The idea of a French Hornet force officially died in 1989 when a decision was made to rejuvenate the Crusader one more time.

Although the Crusader has already gone through several substantial upgrades, it was now scheduled for one more . . . last Crusader improvement program. In 1969 the French Crusaders were all equipped with F-8J wings. Ten years later all French Crusader engines were equipped with new after-burners for their Pratt & Whitney J-57-PW-20A engines.

The new up-grade includes of course, a major overhaul; the replacement of the current escape system with a Martin-Baker ejection seat with a zero-zero capability, a complete re-wiring of the airplane; the replacement of almost the entire hydraulic system; selective replacement of many of the APQ-104 fire control system components with more reliable hardware; a new avionics suite to include the new global positioning system (GPS) and a Thomson-CSF TMV 011 Sherlock radar warning receiver.

The first two of the seventeen airplanes scheduled for this up-grading were accomplished at Dassault. The next fifteen are being accomplished at the French naval workshop at Cuers.

The first Crusader to return from this last, extensive up-grade program arrived onboard *Clemenceau* in early 1993. The second arrived in May. By the end of 1993, six Crusaders had been modified. By September of 1994, the number completed was planned at 12. The entire 17 airplanes will have been completed by the end of calendar 1995.

Unfortunately, the modernization plans of the French Navy hinge, to a large degree, on an aircraft development program (Rafale), which is a political minefield.

The planned acquisition of the 86 Rafale Ms should begin with the delivery of the first one to the R&D facility at Mont-de-Marsan in the summer of 1997. Nevertheless, if everything goes on schedule FLOTILLE 14F will be reactivated as the French Navy's first Rafale squadron in the fall of 1998 with an initial operating capability (IOC) of 1999.

The French Navy operated the "Crouze" in a totally different way from the way it operated in the U.S. Navy. It deployed only one squadron on a carrier; and the squadron deployed with only six to twelve airplanes . . . depending on the operation. Six was a standard number for most of the Crusader deployments. The remainder of the carrier's air group consisted of about twenty Etendarde strike and recce airplanes. These were ultimately replaced by the Super Etendarde. In addition there was also a detachment of Alize three seat anti-submarine warfare airplanes powered by turbo-prop engines. Finally, there were usually a two aircraft detachment of what we would call plane guard helicopters called Alouette III. These were later replaced by the newer Dauphin helicopters.

The first major deployment by the Crusaders was to the southern Pacific during a politically controversial series of French nuclear tests. There are photos showing French F-8s escorting U.S. Navy EP-3 and EC-135 intelligence gathering airplanes during these tests.

PART IX: THE FRENCH CONNECTION

The last Crusader squadron – Flotille 12F, French Navy, 1994. (Courtesy of Flotille 12F)

U.S. Navy aircrews saw the irony in these missions. In 1983 when the U.S. Marine Corps barracks was attacked in Lebanon, French forces in country were also subjected to attacks. Just as the U.S., the French retaliated with a strike on targets in the Bekaa Valley. Two French F-8s escorted eight Etendardes and were extremely successful in their strike tactics, with no losses.

In 1987 French naval forces were employed in escorting oil tankers through the Straits of Hormuz and across the Persian Gulf. For this operation *Clemenceau* was deployed for what was probably the longest carrier deployment F-8s were ever involved in. From 30 July 1987 to 16 September 1988 the carrier operated her complement of aircraft, including a detachment of 6 F-8E(FN)s in the Arabian Sea.

Presently Flottille 12F has 10 to 12 airplanes in an operational status with 14 pilots assigned. The squadron is planned to remain in its present draw-down status until decommissioning in 1998. The French F-8s still operate from the carrier at night but with a night team. Of the fourteen pilots, only about six are on the night team.

French F-8E(FN)s from Flotille 12F Lann Divisiau at sea. (Photos courtesy of J. Miottel)

PART IX: THE FRENCH CONNECTION

Two Philippine Air Force F-8Es. (Photos courtesy of M. Grove)

Typically, the French navy has maintained a pilot seat ratio of 1.5:1. What is fascinating is the level of F-8 expertise in Flottille 12F. Over half of the pilots (in 1994) had over 2,000 flight hours in F-8s.

When the Crusader finally leaves service in the French Navy, it will have done what we in the U.S. Navy call, "stud duty" for over 33 years. What is even more startling is the fact that it will have been over 43 years since it was introduced into service in the U.S. Navy!

It should be noted that the only other foreign acquisition of the Crusader was by the Philippine Air Force which took receipt of the first of thirty-five ex-USN F-8Hs in 1979. Their employment by the Philippine Air Force was extremely limited for a number of reasons. The airplanes were finally withdrawn from service after the disastrous damage inflicted upon them by the eruption of Mount Pinatubo at Clark Air Force Base.

In the late-1960s a patch appeared at odd and sundry places in the U.S. Navy's fighter community saying, "WHEN YOU'RE OUT OF F-8s, YOU'RE OUT OF FIGHTERS!" By 1999, the French Navy will be able to say the same thing!

PART X
THE RECORD

VFP-63 RF-8G at Fallon, Nevada. (Photo by J. Jacobs)

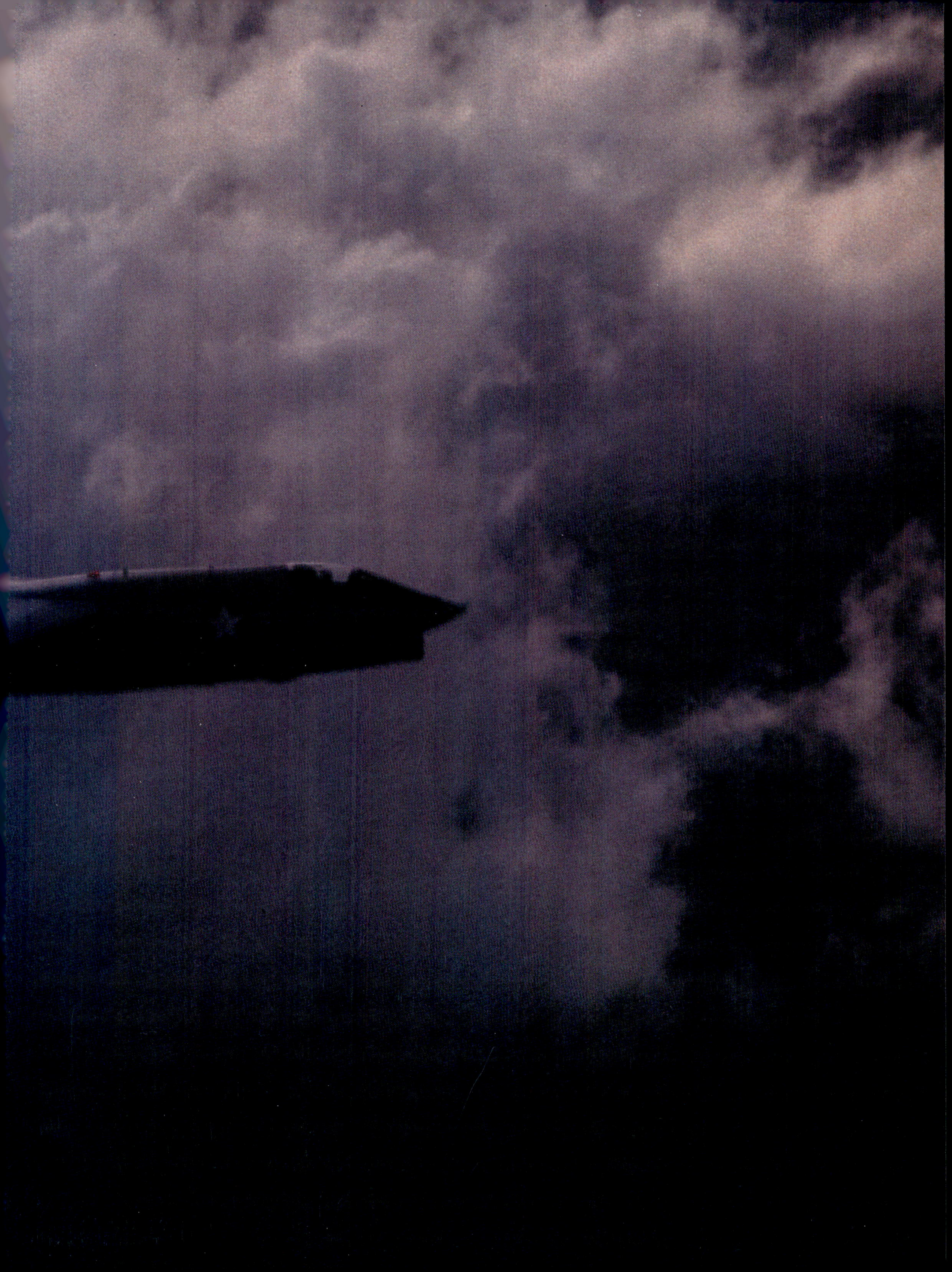

For as long as I can remember, the premier social event of the year for all of U.S. Naval Aviation, from a standpoint of sheer enjoyment, has been the annual Crusader Ball. For many years it was held exclusively at "Fightertown, U.S.A.", the Naval Air Station at Miramar, California. If you were a fighter pilot during the 1960s and 1970s and could attend only one professional event each year . . . it would have to be the Crusader Ball.

Finally, other sub-communities began to emulate the Crusader pilots. First among the emulators were those members of the fighter community who flew the newer F-4 Phantom II. They held their affair, called the Phantom Fling at the fighter base on the other coast, Naval Air Station, Oceana Virginia. Then, I believe, came the light attack community with their Corsair Ball (named after the A-7 Corsair II). Shortly thereafter the medium attack community came on line with their annual Intruder Ball (named, of course, after the A-6 Intruder). Then, in about 1972 the fighter community which flew the F-14 Tomcat decided that there must be an annual Tomcat Ball. This, of course, was also held at Miramar.

When the F-8 Crusader was finally phased out of the fleet, leaving only the photo reconnaissance version of the F-8 on active duty, the Crusader pilots still active, now in other communities decided to have what they called the Last Crusader Ball.

It was to be the last hurrah for this wonderful airplane, and Crusader pilots came from all over the world to tip their hats one last time to that wonderful airplane. If memory serves me correctly, Rear Admiral Paul Speer, Crusader pilot extraordinaire, and a Crusader MiG killer, himself gave the guest speaker performance of his life. I wish I had his words written down. It was a marvelous speech.

As a standard part of the ceremonies, the host always gave what I call the Crusader role call. He would stand up, drink in hand as though preparing to give a toast, and ask all present who had ever flown the Crusader to stand, and to remain standing. Of course, nearly every male in the room rose.

Then, after saying a few memorable things about them all he asked that only those standing, remain standing who had flown over 500 hours in the Crusader. This caused a significant number of the standees to take their seats. The next milestone was 100 carrier arrested landings, then two hundred arrestments. Then, after a few more interesting comments about the Crusader the host would ask only those standing to remain standing who had flown over 1,000 hours in the Crusader. The next milestone was usually one hundred night carrier landings. As might be expected, each successive sit down invitation was greeted with genuine awe by all present as they craned their necks to see who was still standing for the company, by this time, was getting fairly exclusive.

The 2,000 flight hour mark was then followed by the 500 carrier landings mark and usually there were but a hand full of gentlemen standing. About this time, the host asked all standees who had not shot down a MiG to please take their seats. I remember my first Crusader ball, I felt quite proud of my fifteen hundred hours and

Flight of VFP-206 RF-8Gs over the Grand Canyon. (Photo by D. Strong)

167 combat missions in the Crusader. I remember being quite in awe when the only man remaining on his feet was the only man present who had flown over 3,000 hours in the Crusader, had over three hundred combat missions and 1,000 carrier arrested landings in the Crusader and had shot down a MiG. The standing ovation he finally got went on for a long time. Other communities have tried to emulate this particular aspect of the evening's entertainment but never quite pulled it off. The man's name was John Nichols.

The last Crusader ball had been so well done and so well received that a poll was taken by mail as to whether people thought there could be one more Last Crusader Ball. The poll returns were overwhelmingly in favor of the idea. Of course, it would have to be labelled the Second Last Crusader Ball. Then there was the third, the fourth, fifth and on and on. The fourteenth Last Crusader Ball was held at Phoenix, Arizona in the summer of 1994.

The fourteenth Last Crusader Ball was hosted by Bill Hauprich, President of Thunderbird Aviation, the last operator of the Crusader in the United States. Bill announced, at the end of the evening's festivities, that the fifteenth last Crusader Ball will be hosted by the Naval Aviation Museum in Pensacola, Florida . . . who knows?

Naturally, the actuarial laws being what they are, the number of Crusader pilots is rapidly declining. It has been suggested that all living Crusader pilots be invited to throw a few bucks into a pool. When the number gets down to only two of them, the suggestion is that they pick up all the money accrued in the pool and take a cruise around the world. What a great idea! Each of us thinks he is bullet-proof so therefore, each of us thinks he will be one of the last two revelers. What a great fraternity to be a part of!

One of the most memorable of them all was the Seventh Last Crusader Ball, held at Miramar in 1988 in which one of the ball committee, one "Turtle" Redditt did two things for which he will always be remembered. First he read the following set of statistics which he had gleaned from the computer at the Naval Aviation Safety center in Norfolk, Virginia. "Turtle" had an uncanny sense of the comedian's timing and read the statistical report in such a way that nearly all present were weak with tears of laughter when he finished although not all of the data points he read to us were humorous. Here are the essentials of what I call the "record."

CRUSADER STATISTICS

F-8s flown by the U.S. Navy FY 1957 to CY 1985
Total F-8s manufactured for the USN - 1266
Total number of major F-8 accidents - 1106
Total hours - all years: 2,368,153
Most hours per year - 1964: 230,570
Most accidents - 1962: 106

EARLY ACCIDENTS

First loss of an F-8 (BuNo.140444): Contractor pilot named Brackett doing an air show at MCAS Mojave 1 Feb. 1957. Aileron hinge failure – fatal.

First military loss – a Marine named Felliton – Pax River. Tore the wings off in a pilot induced oscillation (PIO). 4 May 1956. First Navy accident was in Atlantic City, a runway overshoot. Landed in dirt short of runway. No injury. Pilot was Lt. Blake. On 21 December 1956, a pilot in VX-3 on his first (and last) flight raised the wing at 500 knots and lost the wing. First Navy fatality. First carrier accident was on the *Roosevelt* on 2 April 1957. Lcdr Miller from VX-3 sheared a landing gear on touchdown. On 5 April 1957 Cdr. Boyd from VX-3 lost power on take-off and executed a successful ditching but both he and the helicopter rescue crewman drowned during the rescue attempt.

OVERALL F-8 ACCIDENT RATE: 46.70 PER 100,000 FLIGHT HOURS

Maximum in 1957 at 243.9 per 100,000 hours (Before 1977 the term Major accident was used. From 77-85 the term Class A mishap-flight related describes what used to be the Major accident).

ACCIDENT RATE COMPARISON
Rates shown per 100,000 hours

Fiscal Year	Mishaps	Hours	Rate
1957	9	3,690	243.9
1958	59	33,569	175.76
1960	77	136,802	56.29
1961	80	171,042	46.77
1962	106	189.951	55.80
1963	97	199,558	48.61
1964	87	230,570	37.73
1965	99	218,865	45.23
1966	66	202,373	32.61
1967	81	173,158	46.78
1968	47	138,362	33.97
1969	71	112,516	63.10
1970	52	104,646	49.69
1971	34	84,393	40.29
1972	14	68,724	20.37
1973	18	60,382	29.81
1974	19	43,940	43.24
Jul-Dec 74	3	17,382	17.26
1975CY:	9	32,808	27.43
1976	6	9,704	61.83
1977	4	7,849	50.96
1978	0	8,316	0
1979	1	7,668	13.04

Year	Mishaps	Hours	Rate
1980	1	7,443	13.44
1981	2	6,502	30.76
1982	0	3,862	0
1983	0	2,506	0
1984	0	2,525	0
1985	1	1,453	8.826
TOTAL	1106	2,368,153	46.70

TOTAL FATALITIES: 186 MEN (not counting combat losses)

For other USN fleet aircraft

Aircraft	Mishaps	Hours	Rate
F-14	77	825,937	9.32
F-4	653	3,238,090	20.17
A-4 (less Ts)	1217	5,210,058	23.36
A-5	99	237,895	41.62
P-3	35	4,971,800	0.7
E-2	31	578.005	5.36

Earlier in this chapter I mentioned that "Turtle" Redditt did two important things at the 1988 ball. The second thing was to give "Bug" Roach's benediction. That, I am reserving for the epilogue to this book.

Author's Note: The French Navy provided me with the names of all of their pilots who have flown the F-8E-(FN) from 6 May 1963 until 24 May 1994 (the date of the survey) . . . a total of 236 intrepid fighter pilots. More will join their ranks before the airplane is phased out of the French Navy in 1998. As I ran through the roster I recognized several names like Lieutenant Goupil, now a retired Rear Admiral, who helped me with this book. But, by far the most startling figure in the list was the number who were killed in the service of their country. Thirteen pilots lost their lives flying the French Crusader due to operational losses – this is just shy of six percent of all of their Crusader pilots. These French statistics were received by the author at the eleventh hour and, therefore, precluded any serious effort to compile the total number of U.S. Navy Crusader drivers to derive a comparable statistic for the 186 lost in operational Crusader accidents.

PART XI
"It Ain't Over . . ."

There have been traditional ceremonies in the United States Navy when various kinds of weapons systems have been officially phased out. Whether ships or airplanes, the phase out ceremonies have always been appropriate to the occasion; and the amount of emotion expressed and tears shed have always been in direct proportion to the degree to which the fleet operators loved or hated the equipment.

In the case of the Crusader, it was the end of a sometimes tempestuous love affair! The ceremony was conducted at, of all places, the Naval Air Facility, located on Andrews Air Force Base, Camp Springs, Maryland. This was where the last Crusader organization was sited. It was a reserve light photographic squadron, flying the RF-8G, which had the honor of saying "Goodbye" to the Crusader . . . and the ending was carried off with panache!

VFP-206 RF-8G over U.S.S. Eisenhower (CVN-69) in the western Atlantic. (Photo by J. Cotton)

On 29 March 1987 Light Photographic Reconnaissance Squadron Two Hundred Six was officially decommissioned. It was, appropriately, a Friday afternoon in the Capital of our country. The evening before had seen a gala black tie ball and banquet, attended by a bevy of dignitaries. Legendary F-8 pilots like Charlie Tinker and "Snake" Morris were in attendance, almost as though it were a command performance.

The Smithsonian Institute was to be the recipient of one of the last airplanes in the squadron. To honor the occasion, an ancient F-8A which had been stored on display at the Paul Garber Aircraft Restoration Facility at Silver Hill, Maryland was towed (with full police escort) to Andrews Air Force Base. There it sat, parked on the ramp in front of Hangar #14 next to two of the squadron's RF-8Gs. After the decommissioning ceremony, one of the RF-8Gs would be flown to the new Smithsonian Facility at Dulles International airport at Chantilly. The other would join the rest of them at the desert storage facility at Tucson, Arizona.

The official Navy band from the Washington Navy Yard provided the musical tribute. The squadron (officers and enlisted personnel) were formed up in ranks and guests were seated in proper pecking order inside the hangar bay. On the speaker's platform were Rear Admiral Tommy Rinert, Chief of Naval Air Reserves, Vice Admiral Cecil Kemp, Director of Naval Reserves, Commander Dave Strong, Commanding Officer of VFP-206, the Reserve Air Wing Commander and, yes, even a Chaplain to give the invocation . . . but not the benediction!

No, the benediction was reserved for a wonderful woman, a pleasingly plump lady with a gifted singing voice. She had sung the national anthem at the opening of the ceremony and, of course, as these ceremonies go, everyone thought she would not reappear. Everyone except Dave Strong.

There was the reading of the decommissioning order followed by the reading of Commander Strong's orders and then there was a symbolic turning over of the "keys" to the airplane to a representative of the Smithsonian Institution by Cecil Kemp. The very last event was to be a traditional piping over the side of the airplane itself. The airplane had already been hooked up to a tow bar and a tug whose motor was idling. The audience rose as if on cue. Commander Strong called the squadron to attention. Two rows of side boys formed and everyone saluted. A bos'un piped the pipes shrilly and the airplane was slowly and sedated towed through the side boys and into the history books! It was a touching scene!

But, as Yogi Berra was famous for saying, "It ain't over . . .", Commander Strong held the salute, as though waiting for one more, final thing to happen . . . and happen it did! The rotund singer reappeared and strode purposefully to the microphone. She picked it up as the band struck up the opening bars of "God bless America." Then with a powerful voice, reminiscent of Kate Smith, she belted out the marvelous melody. As the last notes faded away, Commander Strong dropped his right hand to his side and quietly announced to the assembled crowd, "Now, it's over!"

EPILOGUE

There have been countless jokes over the years about fighter pilots... some of them (like Barry Goldwater's) have not been very complimentary. But, fighter pilots in general enjoy the characterization which most Americans apply, of a square jawed, steely-eyed killer with nerves of steel etc, etc. Given all of that, there is a distinct, and very unique subcategory of that genre called "Crusader pilot" which conjures up perhaps an extreme version of the basic category.

One only has to spend some time reading the statistics in Part X, The Record to realize that the principal attribute required of a successful Crusader pilot was a powerful sense of survival! As one who has flown every tactical airplane in the U.S. inventory over the last thirty years (except the F-111, B-1, B-2 and B-52), the author can state categorically that the Crusader is the most demanding of them all... and by a large margin!

For most pilots who flew the Crusader it was a love/hate relationship. The airplane had its wonderful features and there were some not-so-wonderful characteristics. Whenever a pilot exploited the good parts and achieved a modicum of satisfaction he loved that machine in a way that was almost juvenile. On the other hand, when one of the Crusader's bad features caused him embarrassment, he would curse it roundly. One of my best friends had a series of bad experiences with the Crusader at Patuxent River, and described the airplane to me as having the flying qualities of "... a barrel of turds"!

Once the wing was down and the speed was up the airplane was a fighter pilot's dream! However, when one raised the wing and tried to fly it in the landing configuration in a carrier environment it was horrible. Making good carrier landings consistently in the Crusader (especially at night) required a degree of personal discipline, concentration and flying skills that very few professional pilots ever achieve. There was only one exception in the landing configuration where the Crusader shone. It took an absolutely marvelous catapult shot. If one had the guts, a pilot could easily take a hands-off catapult shot. Nowadays with airplanes like the F/A-18 which "capture" the optimum angle of attack automatically on a catapult shot, hands off is no big deal. But, in the F-8's case the flight control design engineers achieved the same aerodynamic objective mechanically... something of an engineering miracle.

On the other side of the coin, the Crusader (with the wing down) had one very ugly characteristic... called the post-stall-gyration. Before settling down to a steady-state spin the Crusader went through post-stall excursions so violent (depending on entry conditions) that the airplane became legendary. So, the love affair among Crusader pilots with their airplane grew as it matured in the fleet from the late-1950s until its final departure from Miramar in 1975. It proceeded to set all sorts of records. Some were good... others not so good. Some records were intentional... others were not. For example, the Crusader enjoyed the highest exchange ratio for all U.S. tactical aircraft in the Southeast Asian conflict with a six to one ratio against Soviet-built enemy aircraft of the North Vietnamese Air Force. On the other hand, it established the worst safety statistics for any U.S. tactical aircraft... by a huge margin.

"Duke" Windsor, Bob Dose and John Glenn and Billy Lawrence all set speed records in the Crusader. These were the intentional marks. Other Naval Aviation luminaries (whose names will remain unmentioned) set all sorts of unintentional records, six of them with the airplane's wings folded... one for greatest distance flown, another for highest altitude achieved, a third for highest airspeed achieved and yet a fourth for the greatest payload lifted (in bombs). The Crusader pilots who flew combat missions over Vietnam and who operated off the smaller (more hazardous) carriers did so with a degree of panache that established them, forever, as a breed apart. So, if one of them occasionally set his hair on fire at the Cubi Point O'Club or performed some other comparable dido; it was often viewed with condescending nods of understanding because: after all, he was a Crusader pilot!

So, even today, Deer Valley Airport in Phoenix, Arizona (where the last Crusader is still flying), the peculiar sound of the J-57 engine buried halfway down that thirty foot duct causes people to stop and stare. Then, when the long, pencil thin fuselage of the airplane appears in the break it conjures up images of an airplane which North Vietnamese MiG pilots feared above all others. It is the last of the gunfighters.

But, besides fearless pilots, the Crusader also bred another individual, equally colorful and courageous . . . the F-8 landing signal officer (LSO). This small body of experts were truly a breed apart . . . from other LSOs. They were more familiar with stark terror, carnage and mayhem than their peers who "waved" conventional airplanes.

One of the most colorful of this colorful group was Commander John J. "Bug" Roach III, USN. His recent untimely death during a flight as a Navy adversary pilot made all Crusader pilots pause to reexamine their own mortality. For all of the "guidance" he gave to Crusader pilots over the years about their landing performance; and the carrier approaches he salvaged; and the lives he saved; and the careers he salvaged; and the egos he bruised then massaged; he may well be best remembered for something entirely out of character. At one of the Last Crusader Balls, "Bug" Roach rose and gave the following benediction, with his own unique signature. It represents a fitting benediction for the F-8, the "Last of the Gunfighters"!:

"This evening, rather than pray as individuals, I would like to offer this prayer as a people – this prayer of the nation we have created . . . in order that we here may better understand who and what we are. Let us pray!

Lord, we are the nation! We celebrate our on July 4th, 1776, with the Declaration of Independence as our birth certificate. The bloodlines of the world run in our veins because we offer freedom and liberty to all who are oppressed. We are many things and many people. We are the nation.

We sprawl from the Atlantic to the Pacific; to Alaska and Hawaii. Three million square miles throbbing with industry and with life. We are forest, field, mountain and desert. We are the wheat fields of Kansas, the granite hills of Vermont and the snow capped peaks of the Sierra Nevada. We are the Brooklyn Bridge, grain elevators in Nebraska and the Golden Gate. We are the nation.

We are 213 million living souls – and yet we are the ghost of millions who have lived and died for us. We are Nathan Hale and Paul Revere. We are Washington, Jefferson and Patrick Henry. We are Lee, Grant, Abe Lincoln and George Bush. We are Presidents. We are paupers. We are the nation.

We stood at Lexington and fired the shot heard 'round the world. We remember the Alamo, the Maine, Pearl Harbor and the Persian Gulf. When freedom calls we answer. We left our heroic dead at Belleau Wood, on the rock of Corregidor, on the bleak slopes of Korea, in the steaming jungles of Vietnam and under the rubble of Beirut. We are the nation.

We are schools and colleges, churches and synagogues. We are a ballot dropped in a box, the harmonious voice of a choir in a cathedral, the crack of a bat and the roar of a crowd in a stadium. We are craftsmen, teachers, businessmen and judges. We are laborers and nurses. We are parents and we are children. We are soldiers, sailors and airmen. We are peaceful villages, small towns and cities that never sleep. Yes, we are the nation and these are the things that we are.

We were conceived in freedom and dear God, if you are willing, in freedom we will spend the rest of our days. May we always be thankful for the blessings you have bestowed upon us. May we be humble to the less fortunate and assist those in need. May we never forget the continuing cost of freedom. May our wishbone never be found where our backbone should be. May we possess always the integrity, the courage and the strength to keep ourselves unshackled, to remain always a citadel of freedom and a beacon of hope to the world.

We are the nation. And this is our wish – this is our hope – this is our prayer. Amen."

Sources

The major primary source material in this book comes from interviews. Many were conducted personally. A few were conducted by telephone and supplemented by tapes sent by mail in cases where it was impossible for the author to travel to the source. The few cases in which documents were the primary source of information are identified, with particulars in the following pages.

PART I, Chapters 1 through 5 and PART II
The material in this part of the book comes from the personal experience of the author and his direct involvement in the formulation of fighter requirements and the operational employment of U.S. Navy tactical weapons systems over a thirty-three year career. For example, in Chapter 5 "FAGU and Topgun", the author was a weapons delivery instructor at the Fleet Air Gunnery Unit during the period under discussion; and, later, he exercised administrative command of Topgun during a portion of its applicability to the story.

Chapter 6
The material in this chapter was derived, in its entirety, from personal interviews with now, Senator John Glenn, Captain Robert Dose, USN (Retired) and Captain R.W. Windsor, USN (Retired), the principle characters.

Chapters 7, 9 and 10
The material in these chapters is derived from the personal records of the author who was a Navy test pilot at the Naval Air Test Center at Patuxent River during the period under discussion.

Chapter 8
The material in this chapter comes from numerous primary sources. First, there was Mr. George Spangenberg who was the chief design engineer for the Naval Air Systems Command during the time when the competition between the Phantom II and the Crusader III went on. Then my research included Vice Admiral Don Engen, USN (Retired) who was one of the principal members of the pilot evaluation team. Don flew both of the airplanes. Vice Admiral Billy Lawrence, USN (Retired) was the principal F-8U-3 pilot on the evaluation team. Captain Bob Elder, USN (Retired) headed up the evaluation team and flew both airplanes. Mr. Hal Andrews provided some fascinating primary source documents which date back to the great shootout and which he, as a NAVAIR engineer, collected and preserved.

Chapters 11 through 24
The material in this portion of the book was derived from the personal experience and records of the author who was the Operations Officer of Fighter Squadron of Fighter Squadron Sixty-Two, the fleet organization in which all of the events occurred.

Chapter 25
The material in this portion of the book comes entirely from interviews with two people: The first, John Miotell, is the first person to make an actual carrier barricade engagement in a Crusader. Strangely enough, Lieutenant Miotell also has the dubious distinction of being the only person to make two carrier barricade engagements in a Crusader. The second interviewee is noted aviation author/historian Peter Mersky, presently editor of *Approach* magazine and source of aviation accident statistics derived from the Naval Aviation Safety Center.

Chapter 26
The information in this chapter came from an interview with Ron Luther over thirty years after his accident, and from an account of the event written by him shortly afterwards. Because of the incredible nature of the story, tha author went to some pains to verify its authenticity. The records show that it is all true ... and just another incredible chapter in the history of the Crusader.

Chapter 27
The material in this chapter came from a series of personal interviews with Captain Jim Foster, U.S. Navy (Retired), the only person who ever ejected from an airplane underwater ... he did it in a Crusader.

Chapter 28
The material in this chapter came from several sources including its principal character. Several of his squadron mates recounted the story to me before I gave the principal character an opportunity to refute what I had written down. He is an old and dear friend, for whom I have the highest regard. One of the true legends of carrier aviation, (I will came him "Duke" to preserve his anonymity) "Duke" made a truly electrifying flight in the Mediterranean Sea as a young man. After a long telephone interview, he volunteered to provide a detailed account in a tape ... then changed his mind ... asking that his real name not be used. I have complied.

PART IV and chapters 29 through 31
The material in this part came from the personal records of the author who was, in chronological order, the maintenance officer, executive officer and, finally, commanding officer of the squadron in which almost all of the events occurred.

Chapter 32
The material covered in this chapter deserves special mention. To the best of the author's knowledge, after extensive research into available accident statistics, there have been ten recorded events of folded wings flight in the Navy and Marine Corps. Several of the individuals involved agreed to be interviewed only with the stipulation that their names not be mentioned. Such events, especially when involving senior officers have been extremely difficult to uncover. The author hopes that the reader will accept his assurance that his research efforts have been fairly successful, and therefore reasonably complete. But, who knows?

SOURCES

Chapter 33
The material contained in this chapter is the account of a single flight from U.S.S. *Bon Homme Richard* led by the author.

PART V
The information contained in this part was derived by the author during a tour of duty as an operations analyst in the Pentagon from 1968 to 1970 evaluating the results of southeast Asian aerial combat operations.

Chapters 34 through 54
The material contained in these chapters was derived from interviews with sixteen of the seventeen surviving confirmed Crusader MiG killers. In addition, interviews were conducted with other Crusader pilots who did not kill MiGs but who were in on the action in several significant aerial engagements. As explained in the preamble to this bibliography, these interviews were conducted personally (for the most part), or by telephone and backed up by written and taped accounts. The one Crusader MiG killer, Commander Dick Bellinger, who has since died, was the first pilot to down a MiG-21 in combat operations. His wingman during this engagement has also died. The description of Bellinger's MiG kill came from an interview with other pilots who were on the scene as well as personal discussions I had with Commander Bellinger before I decided to write this book.

Chapter 55
The material in this chapter came from an article written by Rene Francillon entitled; "MiGs' Kills" which appeared in the January-February Issue, Volume 1, #1 of the *Journal of Military Aviation*.

PART VI
The material contained in this part represents the distillation of the personal experiences of the author based on 167 combat missions flown in southeast Asia from March 1966 through April 1968. In addition, material gathered while the author was working in the Pentagon as a Southeast Asian air warfare analyst has been included in the chapter to fill in blanks.

PART VII
The material contained in this chapter came from a telephone interview with Rear Admiral Tom Hall, Chief of Naval Reserves, Captain Chuck Long, Chief of Staff, Commander Naval Reserve Forces and from material provided by him. Background was also provided by Vice Admiral Pete Easterling, USN (Retired), formerly the commanding officer of the fleet replacement squadron (VF-124) at the time and Rear Admiral Paul Peck, USN (Retired), formerly the COMNAVAIRPAC Training Officer during the same time period.

PART VIII and Chapters 56 and 57
The material contained in this portion of the book came from a series of interviews with Mr. Edward Schneider, NASA's Chief Test Pilot at the Dryden Flight Research Facility, Edwards Air Force Base, California and from Dryden flight test engineers Joe Wilson, Ken Szalai and Don Gatlin. In addition, Chapter 50 draws from material contained in SUPER-CRITICAL WING TECHNOLOGY, A PROGRESS REPORT ON FLIGHT EVALUATIONS, Scientific and Technical Information Office, a National Aeronautics and Space Administration, Washington, D.C. 1972. Chapter 51 draws from material contained in NASA TECHNICAL NOTE TN D-7843 DESCRIPTION AND FLIGHT TEST RESULTS OF THE NASA F-8 DIGITAL FLY-BY-WIRE CONTROL SYSTEM, February 1975.

Chapter 58
NASA engineer Ken Szalai's comments and his paper, NASA TECHNICAL MEMORANDUM 85913 ROLE OF RESEARCH AIRCRAFT IN TECHNOLOGY DEVELOPMENT, NASA November 1964 were the principal sources for the information contained in this chapter.

Chapter 59
The material contained in this chapter came from a series of interviews with Bill Hauprich, President and Chief Executive Officer and from Larry "Hoss" Pearson, Captain, U.S.N., (Retired), the former Chief Pilot and Director of Operations, Thunderbird Aviation, Phoenix, Arizona.

PART IX
The material contained in this portion came from official records from the French Navy through the French Naval Attache in Washington, D.C., a personal interview with Mr. Rene Francillon, noted author and historian of naval aviation matters and extensive correspondence with Admiral Guirec B. Doniol, French Navy (Retired) and Rear Admiral Yves Goupil, French Navy (Retired). Admiral Goupil was one of the first French Navy pilots to be trained as part of the initial cadre at Fighter Squadron One Hundred Seventy-Four. Admiral Doniol is presently Chairman of Association pour la Recherche de Documentation sur L'Histoire de L'Aeronautique Naval (ARDHAN) and graciously made records of the association available to me to complete the story on the French Crusader. The squadron operational history of Flottille 12F was provided by two squadron officers; the Skipper, Capitaine de Corvette Alain Levieux and the operations officer, Ensigne de Vassieau Phillipe Roux. A great deal of time was also generously given by Mrs. Evelyn De Bernardo, an employee at the French Embassy in Washington, D.C. in translating certain of the above documents from French to English.

PART IX
The material contained in this part came from records compiled by the U.S. Naval Aviation Safety Center, Norfolk, Virginia.

PART IX
The material contained in this part of the story came from Commander Dave Strong USN (Retired) the last commanding officer of VFP-206 stationed at the Naval Air Facility at Andrews Air Force Base, Camp Springs, Maryland.

EPILOGUE
The material contained in this portion of the Crusader story came straight from the heart of the author, a former Navy fighter pilot and Crusader pilot who, over a long flying career, survived over 2,000 flight hours, 167 combat missions and over six hundred *Essex* class carrier arrested landings in the Crusader. If any bias appears to have slipped through, it is hoped that the reader will understand the powerful bond which has developed between the F-8 Crusader and any one who ever flew that marvelous airplane! The benediction quoted in this section came from Commander "Bug" Roach, USN, Crusader pilot and LSO extraordinaire, now deceased.

APPENDICES
Appendix I
Vital statistics came from several sources and were sent to Mr. Paul Bowers, Vought Aircraft Corporation, for verification.

Appendix II
The table of deployments was drawn up by the author using data contained in Mr. Francillon's book, *Tonkin Gulf Yacht Club* as a principal source.

APPENDICES

APPENDIX I
VITAL STATISTICS

Over the years the Crusader has appeared in the sky, all over the world in fifteen different versions. In all of these modifications the basic dimensions of the airplane remained the same. They are:

Length: 54 feet, three inches.
Wingspan (folded): 22 feet, six inches.
Wingspan (spread): 35 feet, eight inches.
Height: 15 feet, nine inches.
Wing area: 375 square feet.

The features that changed over the forty-three years that it will have flown when the French Navy finally stands it down in 1998 are its weight, the wing incidence angle (in the case of the French Crusader only), the addition of ventral fins, the addition of the AAS-15 infra-red search and track pod, its ordnance capability, its power plant, its avionics, other special features and the size of the unit hydraulic tail (in the case of the F-8J and F-8E{FN}). All reconnaissance versions of the F-8 also contained an extra 1,500 pounds of fuel. Taking each of the models in the order of their appearance in the fleet, the had them following special characteristics.

XF-8A (XF8U-1): There were two pre-production models built, one of which crashed during testing killing a company pilot named Brackett. They were essentially representative of the production run of the first fleet airplanes.

F-8A (F8U-1): The first in the fleet, this first production version was powered by either a Pratt & Whitney J-57-P-4 or J-57-P-12 twin spool turbojet engine which delivered 10,000 pounds static thrust at military power and 12,000 pounds in afterburner. It was armed with the standard four Mark 12 twenty millimeter Colt cannons and also had a feature which was quickly removed forever after a disastrous series of ordnance tests. That was a fold-out rocket pack located in the belly of the airplane containing 32 2.75 inch folding fin aircraft rockets. The airplane also had the capability to carry two AIM-9 Sidewinder air-to-air missiles on the fuselage just aft of the cockpit on either side. The fire control system was an AN/APG-30 range-only radar for the gun system lead computing sight. The empty weight of the airplane was 16,150 pounds. There were 318 of these airplanes built.

RF-8A (F8U-1P): The reconnaissance version of the F-8 had the four guns removed and a camera bay in their place. The forward fuselage (the sides and bottom) were redesigned to a flat side and flat bottom shape to accommodate the camera configurations that were planned. The after fuselage fuel tank was enlarged, using some of the internal volume formerly occupied by the ammunition bays.

It was powered by the J-57-P4A engine. The radar and other fire control system components were removed. Of course the additional 1,500 pounds of fuel made the RF-8A the model of choice for Major John Glenn's cross-country speed record attempt. The empty weight of this model was 16,300 pounds. The U.S. Navy purchased 144 of these airplanes.

F-8B (F8U-1E): This model's principal improvement was the APS-67 air-intercept radar. Although it was limited in its capability (pulse radar with a 30 mile B-scale display) it was remarkably reliable. The lack of a tiltable antenna and other features of many current radars made it only amenable to enable the pilot to complete a clear air mass, rear quarter Sidewinder attack. In 1962 a feature called the approach power compensator (APC), a component of what was later called the automatic carrier landing system, (ACLS) was installed on VF-62 airplanes just prior to a Mediterranean deployment on U.S.S. *Shangri La*. The APC represented a substantial improvement in the ability of pilots to more accurately control airspeed during a carrier approach. It was also the first fleet evaluation of a component of the ACLS. The empty weight of the F-8B was 16,475 pounds. There were 130 of these airplanes built.

TF-8A (F8U-1T): Known as the "Twosader", this airplane had no armament capability and a substantial amount of fuel capacity was removed to accommodate the rear seat. It was a very popular model which provided observers a much-needed public relations aware-

Also from the publisher

TOMCAT!
THE GRUMMAN F-14 STORY
Paul T. Gillcrist, Rear Admiral (USN, Ret.)

TOMCAT! The Grumman F-14 Story is an exciting oral history of the most versatile air supremacy fighter in the world . . . much of it written from the vantage point of the cockpit!

Told in an anecdotal format, this new book is richly marbled with the salt air of fleet experience. Perhaps the F-14 program's greatest success is its overwhelming acceptance by the youngsters in the fleet. Anecdotes about the Tomcat legend abound . . . especially in the chapters devoted to its employment by the fleet in the oceans of the world.

Of particular interest are the chapters dealing with the sale of Tomcats to Iran and the subsequent integration of the airplane into Iran's armed forces. On-the-scene eye witness accounts provide a unique perspective of this fascinating aspect of the Tomcat story.

TOMCAT! contains over 150 photographs, most in color and never before published, and most of them taken by Tomcat aircrews with hand-held cameras. These provide a fascinating backdrop for the oral history contained in the book.

This is not a garden variety history about an airplane. It is a sometimes heart-stopping story of how a controversial airplane finally made it into the hearts and minds of the fleet . . . after stumbling at the starting gates of an outmoded defense acquisition system. It is the story of the best air supremacy fighter in the world!

TOMCAT!

THE GRUMMAN F-14 STORY

PAUL T. GILLCRIST, REAR ADMIRAL (USN, RET.)

THE AUTHOR

Rear Admiral Paul T. Gillcrist commanded a fleet Crusader squadron, then a carrier air wing and finally, as a flag officer, became wing commander for all Pacific Fleet fighter squadrons. During his fleet squadron command he completed three carrier deployments to the Tonkin Gulf and flew 167 combat missions in the Crusader for which he was awarded seventeen combat decorations.

The author's carrier aviation career spans an amazing twenty-seven years during which he flew virtually every airplane in the Navy's inventory from sixteen carriers. Included in his flight log book are over 2,000 flight hours and 600 carrier landings in the Crusader alone.

After retirement as Assistant Deputy Chief of Naval Operations (Air Warfare) in 1985, Rear Admiral Gillcrist formed his own aerospace consulting company and resides in Los Angeles, California. The author of *FEET WET, Reflections of a Carrier Pilot* (1990) and *TOMCAT! The Grumman F-14 Story* (1994), Admiral Gillcrist is well qualified to write the story of the Crusader!

"YEA THOUGH I FLY THRU THE VALLEY OF DEATH I SHALL FEAR NO EVIL FOR I AM THE 'MEANEST MUTHAH' IN THE VALLEY"